Lecture Notes in Artificial Inte
Edited by J. G. Carbonell and J. Siekmann

Subseries of Lecture Notes in Computer Science

Bruno Buchberger   John A. Campbell (Eds.)

Artificial Intelligence and Symbolic Computation

7th International Conference, AISC 2004
Linz, Austria, September 22-24, 2004
Proceedings

 Springer

Series Editors

Jaime G. Carbonell, Carnegie Mellon University, Pittsburgh, PA, USA
Jörg Siekmann, University of Saarland, Saarbrücken, Germany

Volume Editors

Bruno Buchberger
Johannes Kepler University
Research Institute for Symbolic Computation (RISC-Linz)
4040 Linz, Austria
E-mail: Buchberger@RISC.Uni-Linz.ac.at

John A. Campbell
University College London, Department of Computer Science
Gower Street, London WC1E 6BT, UK
E-mail: j.campbell@cs.ucl.ac.uk

Library of Congress Control Number: 2004112218

CR Subject Classification (1998): I.2.1-4, I.1, G.1-2, F.4.1

ISSN 0302-9743
ISBN 3-540-23212-5 Springer Berlin Heidelberg New York

Springer is a part of Springer Science+Business Media

springeronline.com

© Springer-Verlag Berlin Heidelberg 2004
Printed in Germany

Typesetting: Camera-ready by author, data conversion by Olgun Computergrafik
Printed on acid-free paper   SPIN: 11318781   06/3142   5 4 3 2 1 0

Preface

AISC 2004, the 7th International Conference on Artificial Intelligence and Symbolic Computation, was the latest in the series of specialized biennial conferences founded in 1992 by Jacques Calmet of the Universität Karlsruhe and John Campbell of University College London with the initial title *Artificial Intelligence and Symbolic Mathematical Computing (AISMC)*. The M disappeared from the title between the 1996 and 1998 conferences. As the editors of the AISC 1998 proceedings said, *the organizers of the current meeting decided to drop the adjective 'mathematical' and to emphasize that the conference is concerned with all aspects of symbolic computation in AI: mathematical foundations, implementations, and applications, including applications in industry and academia.*

This remains the intended profile of the series, and will figure in the call for papers for AISC 2006, which is intended to take place in China. The distribution of papers in the present volume over all the areas of AISC happens to be rather noticeably mathematical, an effect that emerged because we were concerned to select the best relevant papers that were offered to us in 2004, irrespective of their particular topics; hence the title on the cover. Nevertheless, we encourage researchers over the entire spectrum of AISC, as expressed by the 1998 quotation above, to be in touch with us about their interests and the possibility of eventual submission of papers on their work for the next conference in the series.

The papers in the present volume are evidence of the health of the field of AISC. Additionally, there are two reasons for optimism about the continuation of this situation.

The first is that almost all the items in the list of useful areas for future research that the editors of the proceedings of the first conference in 1992 suggested in a 'state of the field' paper there are represented in AISC 2004. Many have of course been present in other AISC conferences too, but never so many as in this year's conference: theorem proving, expert systems, qualitative reasoning, Gröbner bases, differential and integral expressions, computational group theory, constraint-based programming, specification (implying verification), and instances of automated learning, for example. The only major items from the 1992 list that would be needed here to make up what poker players might call a full house are knowledge representation and intelligent user interfaces for mathematical tasks and mathematical reasoning – but while a word search in this volume may not find them, ingredients of both are undoubtedly present this year. (For a hint, see the next paragraph.)

The second of our reasons for an optimistic view of AISC is the maturation of a scientific proposal or prediction that dates back to 1985. In founding the Journal of Symbolic Computation in that year, one of us proposed that SC should encompass both exact mathematical algorithmics (computer algebra) and automated reasoning. Only in recent years has an integration and interaction of these two fields started to materialize. Since 2001 in particular, this has given rise to the MKM (mathematical knowledge management) 'movement', which considers seriously the automation of the entire process of mathematical theory exploration. This is now one of the most promising areas for the application of AI methods in general (for invention or discovery of mathematical concepts, problems, theorems and algorithms) to mathematics/SC and vice versa.

We are happy to be continuing the fruitful collaboration with Springer which started with the first AISMC conference in 1992 and which permitted the publication of the proceedings in the Lecture Notes in Computer Science (LNCS 737, 958, 1138) series from 1992 to 1996 and the Lecture Notes in Artificial Intelligence (LNAI 1476, 1930, 2385) series subsequently.

We, the AISC steering committee, and the organizers of the conference, are grateful to the following bodies for their financial contributions towards its operation and success: Linzer Hochschulfonds, Upper Austrian Government, FWF (Austrian Science Foundation), Raiffeisenlandesbank Upper Austria, Siemens Austria, IBM Austria, and CoLogNET.

Our thanks are also due to the members of the program committee and several additional anonymous referees, and to those who ensured the effective running of the actual conference and its Web sites.

In this latter connection, we administered the submission and selection of papers for AISC 2004 entirely through special-purpose conference software for the first time in the history of AISC, using the START V2 conference manager described at **www.softconf.com**. This contributed substantially to the efficiency of the whole process, and allowed us to respect an unusually tight set of deadlines. We appreciate the prompt and helpful advice on using this software that we received from Rich Gerber whenever we needed it.

The effectiveness of the final stage of production of this volume was due mainly to the intensive work of Theodoros Pagtzis. We express our gratitude to him.

July 2004

Bruno Buchberger
John Campbell

Organization

AISC 2004, the 7th international conference on Artificial Intelligence and Symbolic Computation, was held at Schloss Hagenberg, near Linz, during 22–24 September 2004.

The Research Institute for Symbolic Computation (RISC) of the Johannes-Kepler Universitat Linz, and the Radon Institute for Computational and Applied Mathematics (RICAM), Austrian Academy of Sciences (Österreichische Akademie der Wissenschaften), Linz were jointly responsible for the organization and the local arrangements for the conference.

Conference Direction

Conference Chair	Bruno Buchberger
	(Johannes-Kepler-Universität Linz)
Program Chair	John Campbell
	(University College London, UK)

Local Arrangements

Local Organizers	Betina Curtis
	Ramona Pöchlinger
Websites and Publicity	Camelia Kocsis
	Koji Nakagawa
	Florina Piroi
Logic Programming Tutorial	Klaus Trümper
	(University of Texas at Dallas)

Program Committee

Luigia Carlucci Aiello	(Università 'La Sapienza', Rome)
Michael Beeson	(San José State University)
Belaid Benhamou	(Université de Provence)
Bruno Buchberger	(Johannes-Kepler-Universität Linz)
Jacques Calmet	(Universität Karlsruhe)
Bill Farmer	(McMaster University)
Jacques Fleuriot	(University of Edinburgh)
Laurent Henocque	(Université de la Méditerranée)
Tetsuo Ida	(Tsukuba University)
Michael Kohlhase	(Universität Bremen)
Steve Linton	(University of St Andrews)
Aart Middeldorp	(Universität Innsbruck)
Eric Monfroy	(Université de Nantes)

John Perram	(Syddansk Universitet)
Jochen Pfalzgraf	(Universität Salzburg)
Zbigniew Ras	(University of North Carolina)
Eugenio Roanes-Lozano	(Universidad Complutense de Madrid)
Tomas Recio	(Universidad de Cantabria)
Volker Sorge	(University of Birmingham)
John Stell	(University of Leeds)
Carolyn Talcott	(SRI International)
Dongming Wang	(Université de Paris 6)
Wolfgang Windsteiger	(Johannes-Kepler-Universität Linz)

Sponsoring Institutions

Linzer Hochschulfonds
Upper Austrian Government
FWF (Austrian Science Foundation)
Raiffeisenlandesbank Upper Austria
Siemens Austria
IBM Austria
CoLogNET

Table of Contents

Short Presentations

The Algorithmization of Physics: Math Between Science and Engineering*

Markus Rosenkranz

Johann Radon Institute for Computational and Applied Mathematics
Austrian Academy of Sciences, A-4040 Linz, Austria
Markus.Rosenkranz@oeaw.ac.at

Abstract. I give a concise description of my personal view on symbolic computation, its place within mathematics and its relation to algebra. This view is exemplified by a recent result from my own research: a new symbolic solution method for linear two-point boundary value problems. The essential features of this method are discussed with regard to a potentially novel line of research in symbolic computation.

1 Physics: The Source and Target of Math

What is the *nature of mathematics*? Over the centuries, philosophers and mathematicians have proposed various different answers to this elusive and intriguing question. Any reasonable attempt to systematically analyze these answers is a major epistemological endeavor. The goal of this presentation is more modest: I want to give you a personal (partial) answer to the question posed above, an answer that highlights some aspects of mathematics that I consider crucial from the perspective of symbolic computation. At the end of my presentation, I will substantiate my view by a recent example from my own research.

According to [4], humankind has cultivated the art of *rational problem solving* in a fundamental three-step rhythm:

1. *Observation:* The problem of the real world is specified by extracting relevant data in an abstract model.
2. *Reasoning:* The model problem is solved by suitable reasoning steps, carried out solely in the abstract model.
3. *Action:* The model solution is applied in the real world by effectuating the desired result.

In this view, *mathematics* is not limited to any particular objects like numbers or figures; it is simply "reasoning in abstract models" (item number 2 in the enumeration above). For highlighting its place in the overall picture, let us take up the example of physics – of course, one can make similar observations for other disciplines like chemistry, biology, economics or psychology.

We can see physics as a *natural science* that deals with observations about matter and energy (item number 1 in the three-step rhythm). In doing so, it

---

* This work is supported by the Austrian Science Foundation FWF in project F1322.

B. Buchberger and J.A. Campbell (Eds.): AISC 2004, LNAI 3249, pp. 1–7, 2004.

extracts various patterns from the mass of empirical data and tabulates them in natural laws. It is here that it comes in contact with mathematics, which provides a rich supply of abstract structures for clothing these laws. In this process, physicists have often stimulated deep mathematical research for establishing the concepts asked for (e.g. distributions for modeling point sources), sometimes they have also found ready-made material in an unexpected corner of "pure mathematics" (e.g. Rogers-Ramanujan identities for kinetic gas theory); most of the time, however, we see a parallel movement of mutual fertilization.

As a branch of *technical engineering*, physics is utilized for constructing the machines that we encounter in the world of technology (item number 3 in the three-step rhythm). Engineers are nowadays equipped with a huge body of powerful applied mathematics – often hidden in the special-purpose software at their disposal – for controlling some processes of nature precisely in the way desired at a certain site (e.g. the temperature profile of a chemical reactor). If we are inclined to look down to this "down-to-earth math", we should not forget that it is not only the most prominent source of our money but also the immediate reason for our present-day prosperity.

Of course, the above assignment 1 $\sim$ natural sciences (e.g. theoretical physics), 2 $\sim$ formal science (i.e. mathematics), 3 $\sim$ technical sciences (e.g. engineering physics) must be understood *cum grano salis*: Abstract "mathematical" reasoning steps are also employed in the natural and technical sciences, and a mathematician will certainly benefit from understanding the physical context of various mathematical structures. Besides this, the construction of models is also performed within mathematics when powerful concepts are invented for solving math-internal problems in the same three-step rhythm (e.g. 1 $\sim$ extension fields, 2 $\sim$ Galois groups, 3 $\sim$ solvability criterion).

2   Algebraization: The Commitment to Computing

The above view of mathematics prefers its dynamical side (problem solving) over its static one (knowledge acquisition), but actually the two sides are intimately connected (knowledge is needed for solving problems, and problems are the best filter for building up relevant knowledge bases). The *dynamic view of mathematics* is also the natural starting point for symbolic computation, as I will explicate in the next section. In fact, one can see symbolic computation as its strictest realization, deeply embedded in the overall organism of less constructive or "structural" mathematics.

Within symbolic computation, I will focus on *computer algebra* in this presentation. Strictly speaking, this means that we restrict our interest to algebraic structures (domains with functional signatures and equational axioms like rings). But this is not a dogmatic distinction, rather a point of emphasis; e.g. fields are also counted among the algebraic structures despite their non-equational axiom on reciprocals. In some sense computer algebra is the most traditional branch of symbolic computation since rewriting functions along equational chains is maybe the most natural form of "computation with symbols". What is more important, though, is the axiomatic description of the algebraic structures in use.

Judging from our present understanding, it may seem obvious that one should proceed in this way. Looking at the historical development, however, we perceive a movement of *increasing abstraction* that has not stopped at the point indicated above [9]. We can see four distinct stages in this abstraction process:

1. *Concrete Algebra* (Weber, Fricke): Virtually any ring considered was a subring of the integers or of the complex polynomials or of some algebraic number field (similar for other domains). Various "identical" results were proved separately for different instances.
2. *Abstract Algebra* (Noether, van der Waerden): Rings are described by their axioms; the classical domains mentioned above are subsumed as examples. All the proofs are now done once, within "ring theory".
3. *Universal Algebra* (Graetzer, Cohn): Classes of algebraic structures are considered collectively; rings are just one instance of an algebraic structure. Results like the homomorphism theorem can be proved for generic algebraic structures that specialize to rings, groups, and the like.
4. *Category Theory* (MacLane, Eilenberg): Categories are any collection of objects (like algebraic or non-algebraic structures) connected through arrows (like homomorphisms in algebraic structures). The objects need not have a set-theoretic texture (as the carriers of structures have).

The role of mathematics as reasoning in abstract models becomes very clear in the process of algebraization: the mathematical models are now specified precisely by way of axioms and we need no longer rely on having the same intuition about them. Let me detail this by looking at one of the most fundamental structures used in physics – the notion of the *continuum*, which provides a scale for measuring virtually all physical quantities. Its axiomatization as the complete ordered field of reals needed centuries of focused mathematical research culminating in the categoricity result. Proceeding with computer algebra, we would now strip off the topological aspects (the "complete ordered" part) from the algebraic ones (the "field" part) and then study its computable subfields (finite extensions of the rationals).

If one models physical quantities by real numbers, analyzing their mutual dependence amounts to studying real functions (real analysis), and the natural laws governing them are written as differential equations. Their application to specific situations is controlled by adjusting some data like various parameters and initial/boundary conditions. Since the latter are the most frequent data in physical problems [10], *boundary value problems* (BVPs) will serve as a fitting key example in the last section of this presentation.

3   Algorithmization: The Realization of Computing

In order to actually "compute" the solution of a problem in an abstract model, algebraization alone is not enough. We have already observed this in the above example of the continuum: The field of real numbers is regarded as an algebraic domain, but it is clearly uncomputable because of its uncountable carrier. In view of these facts, Buchberger [7, 3, 5] has defined *symbolic computation* (in particular: computer algebra) as that part of algorithmic mathematics (in particular: algorithmic algebra) which solves problems stated in non-algorithmic domains by translating them into isomorphic algorithmic representations.

Let us look at three immediate examples:

1. The traditional *definition of polynomials* introduces them as certain (infinite!) congruence classes over some term algebra in the variety of unital commutative rings [13]. The modern definition starts from a monoid ring [12], which turns out to be an isomorphic translation of the former, basically encoding the canonical representatives of the congruence classes.

2. As another example, consider the cardinal question of *ideal membership* in algebraic geometry: As infinite sets, ideals cannot directly be treated by algorithmic methods; representing them via Gröbner bases allows a finitary description and a solution of the ideal membership problem [1, 2, 6].

3. Finally, let us consider an example from a traditional domain of symbolic computation that does not belong to computer algebra, namely *automated theorem proving*. It is based on translating the non-algorithmic (semantic) concept of consequence into the algorithmic (syntactic) concept of deducibility, the isomorphism being guaranteed by Gödel's Completeness Theorem.

Returning to the example of boundary value problems introduced above, we should also note that there are actually two famous approaches for algorithmization: symbolic computation takes the path through algebraization, whereas *numerical computation* goes through approximation. Simplifying matters a bit, we could say that symbolic computation hunts down the algebraic structure of the continuum, numerical computation its topological structure[1]. But while industrial mathematics is virtually flooded with numerical solvers (mostly of finite-element type), it is strange to notice that computer algebra systems like Maple or Mathematica do not provide any command for attacking those BVPs that have a symbolic solution. My own research on symbolic functional analysis can be seen as an endeavor to change this situation.

4   Symbolic Functional Analysis: Conquering a New Territory for Computing

I will now sketch my own contribution to the exciting process of conquering more and more territory through algebraization and algorithmization. As mentioned before, it deals with certain *boundary value problems*. More precisely, we are

---

[1] The ideal algorithmic approach to problem solving would be to combine the best parts of both symbolic and numerical computation, which is the overall objective of a 10-year special research project (SFB) at Johannes Kepler University. My own research there takes place in the subproject [17] on symbolic functional analysis; for some details, see the next section.

given a function f in $C^\infty[0,1]$ say[2], and we want to find a solution u in $C^\infty[0,1]$ such that

$$Tu = f,$$
$$B_0 u = u_0, \ldots, B_{n-1} u = u_{n-1} . \tag{1}$$

Here T is a linear differential operator like $T = x^2 D^2 - 2e^x D + 1$ and $B_0, \ldots, B_{n-1}$ are boundary operators like $u \mapsto u'(0) - 2u(1)$, whose number n should coincide with the order of T. Furthermore, we require the boundary conditions to be such that the solution u exists and is unique for every choice of f; in other words, we consider only regular BVPs.

In my own understanding, BVPs are the prototype for a new kind of problem in symbolic computation: Whereas "computer algebra" focuses on algorithmically solving for numbers (typical example: Gröbner bases for triangularizing polynomial systems) and "computer analysis" does the same for functions (typical example: differential algebra for solving differential equations), the proper realm of "computer operator-theory" or *symbolic functional analyis* would be solving for operators. See [17] for more details on this three-floor conception of the algebraic part of symbolic computation.

Why are BVPs an instance of *solving for operators*? The reason is that the forcing function f in (1) is understood as a symbolic parameter: One wants to have the solution u as a term that contains f as a free variable. In other words, one needs an operator G that maps any given f to u. For making this explicit, let us rewrite the traditional formulation (1) as

$$TG = 1,$$
$$B_0 G = 0, \ldots, B_{n-1} G = 0 . \tag{2}$$

Here 1 and 0 denote the identity and zero operator, respectively. Note also that I have passed to homogeneous boundary conditions (it is always possible to reduce a fully inhomogeneous problem to such a semi-inhomogeneous one).

The *crucial idea* of my solution method for (2) is to model the above operators as noncommutative polynomials and to extract their algebraically relevant properties into a collection of identities. For details, please refer to my PhD thesis [14]; see also [16, 15]. The outcome ot this strategical plan is the noncommutative polynomial ring $\mathbb{C}\langle\{D, A, B, L, R\} \cup \{\lceil f\rceil | f \in \mathfrak{F}^\#\}\rangle$ together with a collection of 36 polynomial equalities. The indeterminates D, A, B, L, R and $\lceil f\rceil$ stand for differentiation, integral, cointegral, left boundary value, right boundary value and the multiplication operator induced by f; here f may range over any so-called analytic algebra (a natural extension of a differential algebra), e.g. the exponential polynomials. The 36 polynomial identities express properties like the product rule of differentiation, the fundamental theorem of calculus, and integration by parts.

---

[2] The smoothness conditions can be dispensed with by passing to distributions on $[0,1]$. Of course one can also choose an arbitrary finite interval $[a,b]$ instead of the unit interval. See [15] for details.

I have proved that the 36 polynomials on the left-hand side of the respective identities actually constitute a non-commutative Gröbner basis with terminating reduction (a fact that is not guaranteed in the noncommutative case, in particular not if one deals with infinitely many polynomials like the multiplication operator $\lceil f \rceil$ above). Since I have retrieved a generic formula for G, the resulting polynomial has just to be normalized with respect to this Gröbner basis; the corresponding normal form turns out to be the *classical Green's operator* with the Green's function as its kernel – a well-known picture for every physicist!

As a conclusion, let me briefly reflect on the *solution scheme* sketched above. First of all, we observe the algebraization involved in transforming the topological concepts of differentiation and integration into fitting algebraic counterparts; the creation of $\mathbb{C}\langle\{D, A, B, L, R\} \cup \{\lceil f \rceil | f \in \mathfrak{F}^{\#}\}\rangle$ can also be seen as an instance of math-internal modeling in the sense described in the first section. Second we note the crucial role of noncommutative Gröbner bases in providing the classical Green's operator G: without a confluent and noetherian system of identities the search for normal forms would not be an algorithmic process. Third we may also notice the advantage of working purely on the level of operators, as opposed to traditional solution methods on the function level that use costly determinant computations; see e.g. page 189 in [11].

Finally, let me point out that solving BVPs in this way could be the first step into a *new territory* for symbolic computation that we have baptized "symbolic functional analysis" in [17]. Its common feature would be the algorithmic study (and inversion) of some crucial operators occurring in functional analysis; its main tool would be noncommutative polynomials. Besides more general BVPs (PDEs rather than ODEs, nonlinear rather than linear, systems of equations rather than single equations, etc), some potentially interesting problems could be: single layer potentials, exterior Dirichlet problems, inverse problems like the backwards heat equation, computation of principal symbols, eigenvalue problems.

References

1. Buchberger, B.: An Algorithm for Finding a Basis for the Residual Class Ring of Zero-Dimensional Polynomial Ideal (in German). PhD Thesis, University of Innsbruck, Institute for Mathematics (1965)
2. Buchberger, B.: An Algorithmic Criterion for the Solvability of Algebraic Systems of Equations (in German). Æquationes Mathematicae, 4 (1970) 374–383, in [8] 535–545
3. Buchberger, B.: Editorial. Journal of Symbolic Computation 1/1 (1985)
4. Buchberger, B.: Logic for Computer Science. Lecture Notes, Johannes Kepler University, Linz, Austria (1991)
5. Buchberger, B.: Symbolic Computation (in German). In: Gustav Pomberger and Peter Rechenberg, Handbuch der Informatik, Birkhäuser, München (1997) 955–974
6. Buchberger, B.: Introduction to Gröbner Bases. In [8] 3–31.
7. Buchberger, B., Loos, Rüdiger: Computer Algebra – Symbolic and Algebraic Computation. In: Bruno Buchberger and George Edwin Collins and Rüdiger Loos, Algebraic Simplification, Springer, Wien (1982)

8. Buchberger, B., Winkler, F.: Gröbner Bases and Applications. In London Mathematical Society Lecture Notes 251, Cambridge University Press, Cambridge/UK (1998)

9. Corry, L.: Modern Algebra and the Rise of Mathematical Structures. Birkhäuser, Basel (1996)

10. Courant, R., Hilbert, D.: Die Methoden der mathematischen Physik, Volumes 1 & 2. Springer, Berlin (1993)

11. Kamke, E.: Differentialgleichungen und Lösungsmethoden (Volume 1). Teubner, Stuttgart, tenth edition (1983)

12. Lang, S.: Algebra. Springer, New York (2002)

13. Lausch, H., Nöbauer, W.: Algebra of Polynomials. North-Holland, Amsterdam (1973)

14. Rosenkranz, M.: The Green's Algebra – A Polynomial Approach to Boundary Value Problems. PhD Thesis, Johannes Kepler University, Linz, Austria (2003)

15. Rosenkranz, M.: A New Symbolic Method for Solving Linear Two-Point Boundary Value Problems on the Level of Operators. Journal of Symbolic Computation, submitted (2004). Also available as SFB Report 2003-41, Johannes Kepler University, Linz, Austria (2003)

16. Rosenkranz, M., Buchberger, B., Engl, Heinz W.: Solving Linear Boundary Value Problems via Non-commutative Gröbner Bases. Applicable Analysis, **82/7** (2003) 655–675

17. Rosenkranz, R., Buchberger, B., Engl, Heinz W.: Computer Algebra for Pure and Applied Functional Analysis. An FWF Proposal for F1322 Subproject of the SFB F013, Johannes Kepler University, Linz, Austria (2003)

Finite Algebras and AI: From Matrix Semantics to Stochastic Local Search

Zbigniew Stachniak

Department of Computer Science, York University
Toronto, Canada

1 Introduction

Universal algebra has underpinned the modern research in formal logic since Garrett Birkoff's pioneering work in the 1930's and 1940's. Since the early 1970's, the entanglement of logic and algebra has been successfully exploited in many areas of computer science from the theory of computation to Artificial Intelligence (AI).

The scientific outcome of the interplay between logic and universal algebra in computer science is rich and vast (cf. [2]). In this presentation I shall discuss some applications of universal algebra in AI with an emphasis on Knowledge Representation and Reasoning (KRR).

A brief survey, such as this, of possible ways in which the universal algebra theory could be employed in research on KRR systems, has to be necessarily incomplete. It is primarily for this reason that I shall concentrate almost exclusively on propositional KRR systems. But there are other reasons too. The outburst of research activities on stochastic local search for propositional satisfiability that followed the seminal paper *A New Method for Solving Hard Satisfiability Problems* by Selman, Levesque, and Mitchel (cf. [11]), provides some evidence that propositional techniques could be surprisingly effective in finding solutions to 'realistic' instances of hard problems.

2 Propositional KRR Systems

One of the main objectives of Knowledge Representation is the development of adequate and, preferably, tractable formal representational frameworks for modeling intelligent behaviors of AI agents.

In symbolic approach to knowledge representation, a KRR system consists of at least a formal knowledge representational language $\mathcal{L}$ and of an inference operation $\vdash$ on $\mathcal{L}$. Such a system may involve additional operations and relations besides $\vdash$ (such as plan generation and evaluation, belief revision, or diagnosis); for some domains, some of these additional operations can be defined or implemented in terms of 'basic' logical operations: logical inference, consistency verification, and satisfiability checking. Representing reasoning tasks as instances of logical inference, consistency, and satisfiability problems is discussed below.

B. Buchberger and J.A. Campbell (Eds.): AISC 2004, LNAI 3249, pp. 8–14, 2004.

Syntax. In the propositional case, a representational language $\mathcal{L}$, defined by a set of propositional variables Var and logical connectives $f_0, \ldots, f_n$, can be viewed as a term algebra (or *Lindenbaum's algebra of formulas*)

$$\langle Terms(Var), f_0, \ldots, f_n \rangle,$$

generated by Var, where $Terms(Var)$ denotes the set of all well-formed formulas of $\mathcal{L}$. Syntactically richer languages can be adequately modeled using, for instance, partial and many-sorted term algebras.

Inference Systems. Given a propositional language $\mathcal{L}$, a relation $\vdash$ between sets of formulas of $\mathcal{L}$ and formulas of $\mathcal{L}$ is called an *inference operationon* on $\mathcal{L}$, if for every set X of formulas:

(c1) $X \subseteq C(X)$ (*inclusion*);
(c2) $C(C(X)) \subseteq C(X)$ (*idempotence*);

where $C(X) = \{\beta : X \vdash \beta\}$. An *inference system* on $\mathcal{L}$ is a pair $\langle \mathcal{L}, \vdash \rangle$, where $\vdash$ is an inference operation on $\mathcal{L}$. Further conditions on $\vdash$ can be imposed: for every $X, Y \subseteq Terms(Var)$,

(c3) $X \subseteq Y \subseteq C(X)$ implies $C(X) = C(Y)$ (*cumulativity*);
(c4) $X \subseteq Y$ implies $C(X) \subseteq C(Y)$ (*monotonicity*);
(c5) for every endomorphism e of $\mathcal{L}$, $e(C(X)) \subseteq C(e(X))$ (*structurality*).

Every inference system satisfying (c1)–(c5) is called a *propositional logic*. Since Tarski's axiomatization of the concept of a consequence operation in formalized languages, algebraic properties of monotonic and non-monotonic inference operations have been extensively studied in the literature. (cf. [1,10,13,16]).

Matrix Semantics. The central idea behind classical matrix semantics is to view algebras similar to a language $\mathcal{L}$ as models of $\mathcal{L}$. Interpretations of formulas of $\mathcal{L}$ in an algebra $\mathcal{A}$ similar to $\mathcal{L}$ are homomorphisms of $\mathcal{L}$ into $\mathcal{A}$. When $\mathcal{A}$ is augmented with a subset d of the universe of $\mathcal{A}$, the resulting structure

$$\mathcal{M} = \langle \mathcal{A}, d \rangle,$$

called a *logical matrix* for $\mathcal{L}$, determines the inference operation $\vdash_{\mathcal{M}}$ defined in the following way: for every set $X \cup \{\alpha\}$ of formulas of $\mathcal{L}$,

$$X \vdash_{\mathcal{M}} \alpha \text{ iff for every homomorphism } h \text{ of } \mathcal{L} \text{ into } \mathcal{A}, \text{ if } h(X) \subseteq d \text{ then } h(\alpha) \in d.$$

The research on logical matrices has been strongly influenced by universal algebra and model theory. Wójcicki's monograph [16] contains a detailed account of the development of matrix semantics since its inception in the early 20th century. In AI, matrix semantics (and a closely related discipline of many-valued logics) has been successfully exploited in the areas of Automated Reasoning, KRR, and Logic Programming (cf. [3,4,5,6,9,13,15]).

Monotone Calculi. The inference opeartion $\vdash_{\mathcal{M}}$ defined by a logical matrix $\mathcal{M}$ satisfies not only (c1)–(c3) but also (c4) and (c5). Furthermore, for every propositional calculus $\langle \mathcal{L}, \vdash \rangle$ there exists a class $\mathcal{K}$ of logical matrices for $\mathcal{L}$ such that $\vdash = \bigcap \{\vdash_{\mathcal{M}} : \mathcal{M} \in \mathcal{K}\}$.

Beyond Structurality: Admissible Valuations. One way of extending matrix semantics to cover non-structural inference systems is to define the semantic entailment in terms of 'admissible interpretations', i.e., to consider *generalized matrices* of the form $\langle \mathcal{A}, d, \mathcal{H} \rangle$, where $\mathcal{A}$ and d are as above, and $\mathcal{H}$ is a subset of the set of all interpretations of $\mathcal{L}$ into $\mathcal{A}$. In this semantic framework, every inference operation that satisfies (c1)–(c4) can be defined by a class of generalized matrices. A similar approach of admitting only some interpretations to model non-structural nonmonotonic inference systems has been also developed for preferential model semantics (cf. [7]).

Beyond Monotonicity: Preferential Matrices The notion of cumulativity arose as a result of the search for desired and natural formal properties of nonmonotonic inference systems. A desired 'degree' of nonmonotonicity can be semantically modeled in terms of logical matrices of the form $\mathcal{M} = \langle \mathcal{A}, \mathcal{D}, \mathcal{H}, \prec \rangle$, where $\mathcal{A}$ and $\mathcal{H}$ are as in a generalized matrix, $\mathcal{D}$ is a family of subsets of the universe of $\mathcal{A}$, and $\prec$ is a binary (preference) relation on $\mathcal{D}$. The inference operation $\vdash_{\mathcal{M}}$ is defined as follows:

$X \vdash_{\mathcal{M}} \alpha$ *iff* for every $h \in \mathcal{H}$ and every $d \in \mathcal{D}$, if d is a minimal element of $\mathcal{D}$ (with respect to $\prec$) such that $h(X) \subseteq d$, then $h(\alpha) \in d$.

Preferential matrices have the same semantic scope as preferential model structures (cf. [8,14]).

Logical Matrices with Completion. It is not essential to interpret the underlying algebra $\mathcal{A}$ of a logical matrix $\mathcal{M}$ for a language $\mathcal{L}$ as a space of truth-values for the formulas of $\mathcal{L}$. The elements of $\mathcal{A}$ can be interpreted as propositions, events, and even infons of the Situation Theory of Barwise and Perry. If one views subsets of the universe of $\mathcal{A}$ as situations (partial or complete), then preferential matrices can be replaced by structures of the form $\mathcal{M} = \langle \mathcal{A}, \mathcal{H}, \widehat{\ } \rangle$ called *matrices with completion*, where $\mathcal{A}$, and $\mathcal{H}$ are as above and $\widehat{\ }$ is a function that maps $2^{|\mathcal{A}|}$ into $2^{|\mathcal{A}|}$ such that for every $B \subseteq |\mathcal{A}|$, $B \subseteq \widehat{B} = \widehat{\widehat{B}}$. In the language of universal algebra, $\widehat{\ }$ is a closure operator on $\mathcal{A}$. This operation can be thought of as a *completion function* that assigns an actual and complete situation $\widehat{B}$ to a (possibly partial) situation B which is a part of $\widehat{B}$. The inference operation $\vdash_{\mathcal{M}}$ associated with such a matrix is defined as follows: for every set $X \cup \{\alpha\}$ of formulas,

$X \vdash_{\mathcal{M}} \alpha$ *iff* for every $h \in \mathcal{H}, h(\alpha) \in \widehat{h(X)}$.

Matrices with completion can be used to semantically model cumulativity without any explicit reference to preference.

Beyond Matrix Semantics. The interplay between logic and universal algebra goes far beyond Matrix Semantics; a wealth of results harvested in disciplines such as type theory, term rewriting, algebraic logic, or fuzzy logic, and subjects such as bilattices, dynamic logics, or unification, have had and will continue to have a significant impact on AI research.

3   Problem Solving as Consistency Verification

Automated Reasoning deals with the development and application of computer programs to perform a variety of reasoning tasks frequently represented as instances of consistency verification problem.

Refutational Principle. Refutational theorem proving methods, such as resolution, rely on a correspondence between valid inferences and finite inconsistent sets. The *refutational principle* for an inference system $\mathcal{P} = \langle \mathcal{L}, \vdash \rangle$ states that there is an algorithm that transformes every finite set $X \cup \{\alpha\}$ of formulas into another finite set X_α of formulas in such a way that

(ref) $X \vdash \alpha$ *iff* X_α is inconsistent in $\mathcal{P}$ (i.e., for every formula $\beta, X_\alpha \vdash \beta$).

In the light of (ref), a refutational automated reasoning system answers a query $X \vdash \alpha$ by determining the consistency status of X_α.

Resolution Algebras. Let $\mathcal{L} = \langle Terms(Var), f_0, \ldots, f_n \rangle$ be a propositional language (let us assume that the disjunction, denoted by $\vee$, is among the connectives of $\mathcal{L}$). A resolution algebra for $\mathcal{L}$ is a finite algebra of the form

$$Rs = \langle \langle \{v_0, \ldots, v_k\}, \underline{f_0}, \ldots, \underline{f_n} \rangle, \mathcal{F} \rangle$$

where: $\{v_0, \ldots, v_k\}$ is a set of formulas of $\mathcal{L}$ called *verifiers*, for every $i \leq n, f_i$ and the corresponding connective f_i are of the same arity, and $\mathcal{F}$ is a subset of V. Rs defines two types of inference rules. The *resolution rule*

$$\frac{\alpha_0(p), \ldots, \alpha_k(p)}{\alpha_0(p/v_0) \vee \ldots \vee \alpha_k(p/v_k)}$$

is the case analysis on truth of a common variable p expressed using verifiers. The other inference rules are the simplification rules defined by the operations $\underline{f_0}, \ldots, \underline{f_n}$ (see [13]). A set X of formulas is refutable in Rs if and only if one of the verifiers from $\mathcal{F}$ can be derived from X using the inference rules defined by Rs.

Resolution Logics. A propositional logic $\mathcal{P} = \langle \mathcal{L}, \vdash \rangle$ is said to be a *resolution logic* if there exists a resolution algebra Rs such that for every finite set X of formulas (which do not share variables with the verifiers),

$$X \text{ is inconsistent in } \mathcal{P} \textit{ iff } X \text{ is refutable in } Rs.$$

Additional conditions to guarantee the soundness of the refutation process should also be imposed (cf. [13]). The class of resolution logics consists of those calculi which are indistinguishable on inconsistent sets from logics defined by finite matrices. Furthermore, resolution algebras for logics defined by finite logical matrices can be effectively constructed from the defining matrices (cf. [13]).

Lattices of Resolution Logics. For a logic $\mathcal{P} = \langle \mathcal{L}, \vdash \rangle$, let $\mathcal{K}_\mathcal{P}$ denote the class of all logics on $\mathcal{L}$ which have the same inconsistent sets as $\mathcal{P}$. $\mathcal{K}_\mathcal{P}$ is a bounded lattice under the ordering $\leq$ defined as follows: if $\mathcal{P}_i = \langle \mathcal{L}, \vdash_i \rangle, i = 0, 1$, then $\mathcal{P}_0 \leq \mathcal{P}_1$ iff $\mathcal{P}_1$ is inferentially at least as strong as $\mathcal{P}_0$. The lattice $\langle \mathcal{K}_\mathcal{P}, \leq \rangle$ is a convenient tool to discuss the scope of the resolution method defined in terms of resolution algebras: if $\mathcal{P}$ is a resolution logic, then so are all the logics in $\mathcal{K}_\mathcal{P}$. From the logical standpoint, the systems in $\mathcal{K}_\mathcal{P}$ can be quite different; from the refutational point of view, they can all be defined by the same resolution algebra.

Nonmonotonic Resolution Logics. Resolution algebras can also be used to implement some nonmonotonic inference systems. Let $\mathcal{P} = \langle \mathcal{L}, \vdash \rangle$ be an arbitrary cumulative inference system. The *monotone base* of $\mathcal{P}$ is the greatest logic $\mathcal{P}_B$ on $\mathcal{L}$ (with respect to $\leq$) such that $\mathcal{P}_B \leq \mathcal{P}$. The monotone bases of the so-called *supraclassical* inference systems is classical propositional logic (cf. [8]).

The *consistency preservation property* limits the inference power by which $\mathcal{P}$ and $\mathcal{P}_B$ can differ (cf. [8,13]). It states that both $\mathcal{P}$ and $\mathcal{P}_B$ have to have the same inconsistent sets of formulas. Every cumulative, structural, and proper inference system satisfies the consistency preservation property. Hence, every such system can be provided with a resolution algebra based proof system, provided that its monotone base is a resolution logic.

4   Problem Solving as Satisfiability

A reasoning task, such as a planning problem, can be solved by, first, expressing it as a satisfiability problem in some logical matrix $\mathcal{M}$ and, then, by solving it using one of the satisfiability solvers for $\mathcal{M}$. In spite of the fact that for many finite matrices $\langle \mathcal{A}, d \rangle$, the satisfiability problem:

(**SAT**$_\mathcal{M}$) for every formula α, determine whether or not there exists an interpretation h such that $h(\alpha) \in d$

is NP-complete, a number of complete and incomplete SAT$_\mathcal{M}$ solvers have been developed and their good performance in finding solutions to instances of many problems in real-world domains empirically demonstrated.

Searching for Satisfying Interpretation. Given a matrix $\mathcal{M} = \langle \mathcal{A}, d \rangle$ for a language $\mathcal{L}$, a stochastic local search algorithm for satisfiability in $\mathcal{M}$ starts by generating a random interpretation h restricted to the variables of an input formula α. Then, it locally modifies h by selecting a variable p of α, using some selection heuristic *select_var*(α, h), and changing its truth-value from $h(p)$ to some new truth-value using another selection heuristic $select\_val(\alpha, p, h)$. Such selections of variables and such changes of their truth-values are repeated until either $h(\alpha) \in d$ or the allocated time to modify h into a satisfying valuation has elapsed. The process is repeated (if needed) up to a specified number of times.

The above procedure defines informally an incomplete $SAT_\mathcal{M}$ solver (clearly, it cannot be used to determine unsatisfiability of a formula).

Polarity and $SAT_\mathcal{M}$. The classical notion of polarity of a variable p in a formula $\alpha(p)$ captures the monotonic behavior of the term operation $f_\alpha(p)$ induced by $\alpha(p)$ over p in a partially ordered algebra of truth-values. The selection heuristics $select\_var(\alpha, h)$ and $select\_val(\alpha, p, h)$ of an $SAT_\mathcal{M}$ solver can be defined in terms of polarity. This is done in the non-clausal solver polSAT for classical propositional logic as well as in its extensions to finitely-valued logics (cf. [12]).

Improving the Efficiency of Resolution with $SAT_\mathcal{M}$ Solvers. An unrestricted use of the resolution rule during the deductive process may very quickly result in combinatoric explosion of the set of deduced resolvents making the completion of a reasoning task unattainable in an acceptable amount of time. In an efficient resolution-based reasoning program the generation of resolvents that would evidently have no impact on the completion of a reasoning task must be blocked. Tautological resolvents are just that sort of formulas.

For many resolution logics the tautology problem is coNP-complete. For some of these logics, $SAT_\mathcal{M}$ solvers can be used to guide the search for refutation so that the use of tautologies during the refutation process is unlikely. At the same time the refutational completeness of the deductive process is preserved.

References

1. Blok, W.J. and Pigozzi, D. *Algebraizable Logics.* Memoirs of the American Mathematical Society, vol. 77, no 396 (1989).
2. Denecke, K. and Wismath, S.L. *Universal Algebra and Applications in Computer Science,* Chapman & Hall/CRC (2002).
3. Fitting, M. and Orłowska, E. (eds.) *Beyond Two: Theory and Applications of Multiple-Valued Logic.* Studies in Fuzziness and Soft Computing, Phisica-Verlag (2003).
4. Fitting, M. Bilattices and the semantics of logic programming. *Journal of Logic Programming* 11, pp. 91–116 (1991).
5. Ginsberg, M.L. Multivalued logics: a uniform approach to reasoning in artificial intelligence. *Comput. Intell.,* vol. 5, pp. 265–316.
6. Hähnle, R. *Automated Deduction in Multiple-Valued Logics.* Clarendon Press (1993).
7. Kraus, S., Lehmann, D. and Magidor, M. Nonmonotonic Reasoning, Preferential Models and Cumulative Logics. *Artificial Intelligence* 44, pp. 167-207 (1990).
8. Makinson, D. General Patterns in Nonmonotonic Reasoning. In *Handbook of Logic in Artificial Intelligence and Logic Programming.* Vol 3: *Nonmonotonic and Uncertain Reasoning* (Gabbay, D.M., Hogger, C.J., and Robinson J.A. eds.). Oxford University Press (1994).

9. Przymusinski, T. Three-Valued Non-Monotonic Formalisms and Logic Programming". In *Proc. of the First Int. Conf. on Principles of Knowledge Representation and Reasoning (KR'89)*, Toronto, pp. 341-348 (1989)

10. Rasiowa, H. *An Algebraic Approach to Non-Classical Logics.* North-Holland (1974).

11. Selman, B., Levesque, H., and Mitchell, D. A new method for solving hard satisfiability problems. In *Proc. AAAI-92*, pp. 440–446 (1992).

12. Stachniak, Z. Polarity-based Stochastic Local Search Algorithm for Non-clausal Satisfiability. In [3], pp. 181–192 (2002).

13. Stachniak, Z. *Resolution Proof Systems: An Algebraic Theory.* Kluwer (1996).

14. Stachniak, Z. Nonmonotonic Theories and Their Axiomatic Varieties. *J. of Logic, Language, and Computation*, vol. 4, no. 4, pp. 317–334 (1995).

15. Thistlewaite, P.B., McRobbie, M.A., and Meyer, R.K. *Automated Theorem-Proving in Non-Classical Logics.* Pitman (1988).

16. Wójcicki, R. *Theory of Logical Calculi: Basic Theory of Consequence Operations.* Kluwer (1988).

Proof Search in Minimal Logic

Helmut Schwichtenberg

Mathematisches Institut der Universität München
schwicht@mathematik.uni-muenchen.de
http://www.mathematik.uni-muenchen.de/~schwicht/

1 Introduction

We describe a rather natural proof search algorithm for a certain fragment of higher order (simply typed) minimal logic. This fragment is determined by requiring that every higher order variable Y can only occur in a context $Y\boldsymbol{x}$, where $\boldsymbol{x}$ are distinct bound variables in the scope of the operator binding Y, and of opposite polarity. Note that for first order logic this restriction does not mean anything, since there are no higher order variables. However, when designing a proof search algorithm for first order logic only, one is naturally led into this fragment of higher order logic, where the algorithm works as well.

In doing this we rely heavily on Miller's [1], who has introduced this type of restriction to higher order terms (called *patterns* by Nipkow [2]), noted its relevance for extensions of logic programming, and showed that the unification problem for patterns is solvable and admits most general unifiers. The present paper was motivated by the desire to use Miller's approach as a basis for an implementation of a simple proof search engine for (first and higher order) minimal logic. This goal prompted us into several simplifications, optimizations and extensions, in particular the following.

- Instead of arbitrarily mixed prefixes we only use those of the form $\forall\exists\forall$. Nipkow in [2] already had presented a version of Miller's pattern unification algorithm for such prefixes, and Miller in [1, Section 9.2] notes that in such a situation any two unifiers can be transformed into each other by a variable renaming substitution. Here we restrict ourselves to $\forall\exists\forall$-prefixes throughout, i.e., in the proof search algorithm as well.
- The order of events in the pattern unification algorithm is changed slightly, by postponing the raising step until it is really needed. This avoids unnecessary creation of new higher type variables. – Already Miller noted in [1, p.515] that such optimizations are possible.
- The extensions concern the (strong) existential quantifier, which has been left out in Miller's treatment, and also conjunction. The latter can be avoided in principle, but of course is a useful thing to have.

Moreover, since part of the motivation to write this paper was the necessity to have a guide for our implementation, we have paid particular attention to write at least the parts of the proofs with algorithmic content as clear and complete as possible.

B. Buchberger and J.A. Campbell (Eds.): AISC 2004, LNAI 3249, pp. 15–25, 2004.

The paper is organized as follows. Section 2 defines the pattern unification algorithm, and in section 3 its correctness and completeness is proved. Section 4 presents the proof search algorithm, and again its correctness and completeness is proved. The final section 5 contains what we have to say about extensions to $\wedge$ and $\exists$.

2   The Unification Algorithm

We work in the simply typed λ-calculus, with the usual conventions. For instance, whenever we write a term we assume that it is correctly typed. *Substitutions* are denoted by φ, ψ, ρ. The result of applying a substitution φ to a term t or a formula A is written as $t\varphi$ or $A\varphi$, with the understanding that after the substitution all terms are brought into long normal form.

Q always denotes a $\forall\exists\forall$-prefix, say $\forall\boldsymbol{x}\exists\boldsymbol{y}\forall\boldsymbol{z}$, with distinct variables. We call $\boldsymbol{x}$ the *signature variables*, $\boldsymbol{y}$ the *flexible variables* and $\boldsymbol{z}$ the *forbidden variables* of Q, and write $Q_\exists$ for the existential part $\exists\boldsymbol{y}$ of Q.

Q-terms are inductively defined by the following clauses.

- If u is a universally quantified variable in Q or a constant, and $\boldsymbol{r}$ are Q-terms, then $u\boldsymbol{r}$ is a Q-term.
- For any flexible variable y and distinct forbidden variables $\boldsymbol{z}$ from Q, $y\boldsymbol{z}$ is a Q-term.
- If r is a $Q\forall z$-term, then $\lambda z r$ is a Q-term.

Explicitly, r is a Q-term iff all its free variables are in Q, and for every subterm $y\boldsymbol{r}$ of r with y free in r and flexible in Q, the $\boldsymbol{r}$ are distinct variables either λ-bound in r (such that $y\boldsymbol{r}$ is in the scope of this λ) or else forbidden in Q.

Q-goals and *Q-clauses* are simultaneously defined by

- If $\boldsymbol{r}$ are Q-terms, then $P\boldsymbol{r}$ is a Q-goal as well as a Q-clause.
- If D is a Q-clause and G is a Q-goal, then $D \to G$ is a Q-goal.
- If G is a Q-goal and D is a Q-clause, then $G \to D$ is a Q-clause.
- If G is a $Q\forall x$-goal, then $\forall x G$ is a Q-goal.
- If $D[y := Yz]$ is a $\forall\boldsymbol{x}\exists\boldsymbol{y}, Y\forall\boldsymbol{z}$-clause, then $\forall y D$ is a $\forall\boldsymbol{x}\exists\boldsymbol{y}\forall\boldsymbol{z}$-clause.

Explicitly, a formula A is a *Q-goal* iff all its free variables are in Q, and for every subterm $y\boldsymbol{r}$ of A with y either existentially bound in A (with $y\boldsymbol{r}$ in the scope) or else free in A and flexible in Q, the $\boldsymbol{r}$ are distinct variables either λ- or universally bound in A (such that $y\boldsymbol{r}$ is in the scope) or else free in A and forbidden in Q.

A *Q-substitution* is a substitution of Q-terms.

A *unification problem* $\mathcal{U}$ consists of a $\forall\exists\forall$-prefix Q and a conjunction C of equations between Q-terms of the same type, i.e., $\bigwedge_{i=1}^{n} r_i = s_i$. We may assume that each such equation is of the form $\lambda\boldsymbol{x}r = \lambda\boldsymbol{x}s$ with the same $\boldsymbol{x}$ (which may be empty) and r, s of ground type.

A *solution* to such a unification problem $\mathcal{U}$ is a Q-substitution φ such that for every i, $r_i\varphi = s_i\varphi$ holds (i.e., $r_i\varphi$ and $s_i\varphi$ have the same normal form). We sometimes write C as $\boldsymbol{r} = \boldsymbol{s}$, and (for obvious reasons) call it a list of unification pairs.

We now define the unification algorithm. It takes a unification problem $\mathcal{U} = QC$ and returns a substitution ρ and another patter unification problem $\mathcal{U}' = Q'C'$. Note that ρ will be neither a Q-substitution nor a Q'-substitution, but will have the property that

- ρ is defined on flexible variables of Q only, and its value terms have no free occurrences of forbidden variables from Q,
- if G is a Q-goal, then $G\rho$ is a Q'-goal, and
- whenever φ' is an $\mathcal{U}'$-solution, then $(\rho \circ \varphi') \upharpoonright Q_\exists$ is an $\mathcal{U}$-solution.

To define the unification algorithm, we distinguish cases according to the form of the unification problem, and either give the transition done by the algorithm, or else state that it fails.

Case identity, i.e., $Q.r = r \wedge C$. Then

$$Q.r = r \wedge C \Longrightarrow_\varepsilon QC.$$

Case ξ, i.e., $Q.\lambda\boldsymbol{x}\,r = \lambda\boldsymbol{x}\,s \wedge C$. We may assume here that the bound variables $\boldsymbol{x}$ are the same on both sides.

$$Q.\lambda\boldsymbol{x}\,r = \lambda\boldsymbol{x}\,s \wedge C \Longrightarrow_\varepsilon Q\forall\boldsymbol{x}.r = s \wedge C.$$

Case rigid-rigid, i.e., $Q.f\boldsymbol{r} = f\boldsymbol{s} \wedge C$.

$$Q.f\boldsymbol{r} = f\boldsymbol{s} \wedge C \Longrightarrow_\varepsilon Q.\boldsymbol{r} = \boldsymbol{s} \wedge C.$$

Case flex-flex with equal heads, i.e., $Q.u\boldsymbol{y} = u\boldsymbol{z} \wedge C$.

$$Q.u\boldsymbol{y} = u\boldsymbol{z} \wedge C \Longrightarrow_\rho Q'.C\rho$$

with $\rho = [u := \lambda\boldsymbol{y}.u'\boldsymbol{w}]$, Q' is Q with $\exists u$ replaced by $\exists u'$, and $\boldsymbol{w}$ an enumeration of $\{\, y_i \mid y_i = z_i \,\}$ (note $\lambda\boldsymbol{y}.u'\boldsymbol{w} = \lambda\boldsymbol{z}.u'\boldsymbol{w}$).

Case flex-flex with different heads, i.e., $Q.u\boldsymbol{y} = v\boldsymbol{z} \wedge C$.

$$Q.u\boldsymbol{y} = v\boldsymbol{z} \wedge C \Longrightarrow_\rho Q'C\rho,$$

where ρ and Q' are defined as follows. Let $\boldsymbol{w}$ be an enumeration of the variables both in $\boldsymbol{y}$ and in $\boldsymbol{z}$. Then $\rho = [u, v := \lambda\boldsymbol{y}.u'\boldsymbol{w}, \lambda\boldsymbol{z}.u'\boldsymbol{w}]$, and Q' is Q with $\exists u, \exists v$ removed and $\exists u'$ inserted.

Case flex-rigid, i.e., $Q.u\boldsymbol{y} = t \wedge C$ with t rigid, i.e., not of the form $v\boldsymbol{z}$ with flexible v.

Subcase occurrence check: t contains (a critical subterm with head) u. Fail.

Subcase pruning: t contains a subterm $v\boldsymbol{w}_1 z\boldsymbol{w}_2$ with $\exists v$ in Q, and z free in t but not in $\boldsymbol{y}$.

$$Q.u\boldsymbol{y} = t \wedge C \Longrightarrow_\rho Q'.u\boldsymbol{y} = t\rho \wedge C\rho$$

where $\rho = [v := \lambda\boldsymbol{w}_1, z, \boldsymbol{w}_2.v'\boldsymbol{w}_1\boldsymbol{w}_2]$, Q' is Q with $\exists v$ replaced by $\exists v'$.

Subcase pruning impossible: $\lambda \boldsymbol{y} t$ (after all pruning steps are done still) has a free occurrence of a forbidden variable z. Fail.

Subcase explicit definition: otherwise.

$$Q.u\boldsymbol{y} = t \wedge C \Longrightarrow_\rho Q'C\rho$$

where $\rho = [u := \lambda \boldsymbol{y} t]$, and Q' is obtained from Q by removing $\exists u$.

This concludes the definition of the unification algorithm.

Our next task is to prove that this algorithm indeed has the three properties stated above. The first one (ρ is defined on flexible variables of Q only, and its value terms have no free occurrences of forbidden variables from Q) is obvious from the definition. We now prove the second one; the third one will be proved in the next section.

Lemma 1. *If $Q \Longrightarrow_\rho Q'$ and G is a Q-goal, then $G\rho$ is a Q'-goal.*

Proof. We distinguish cases according to the definition of the unification algorithm.

Cases identity, ξ and rigid-rigid. Then $\rho = \varepsilon$ and the claim is trivial.

Case flex-flex with equal heads. Then $\rho = [u := \lambda \boldsymbol{y}.u'\boldsymbol{w}]$ with $\boldsymbol{w}$ a sublist of $\boldsymbol{y}$, and Q' is Q with $\exists u$ replaced by $\exists u'$. Then clearly $G[u := \lambda \boldsymbol{y}.u'\boldsymbol{w}]$ is a Q'-goal (recall that after a substitution we always normalize).

Case flex-flex with different heads. Then $\rho = [u, v := \lambda \boldsymbol{y}.u'\boldsymbol{w}, \lambda \boldsymbol{z}.u'\boldsymbol{w}]$ with $\boldsymbol{w}$ an enumeration of the variables both in $\boldsymbol{y}$ and in $\boldsymbol{z}$, and Q' is Q with $\exists u, \exists v$ removed and $\exists u'$ inserted. Again clearly $G[u, v := \lambda \boldsymbol{y}.u'\boldsymbol{w}, \lambda \boldsymbol{z}.u'\boldsymbol{w}]$ is a Q'-goal.

Case flex-rigid, *Subcase* pruning: Then $\rho = [v := \lambda \boldsymbol{w}_1, z, \boldsymbol{w}_2.v'\boldsymbol{w}_1\boldsymbol{w}_2]$, and Q' is Q with $\exists v$ replaced by $\exists v'$. Suppose G is a Q-goal. Then clearly $G[v := \lambda \boldsymbol{w}_1, z, \boldsymbol{w}_2.v'\boldsymbol{w}_1\boldsymbol{w}_2]$ is a Q'-goal.

Case flex-rigid, *Subcase* explicit definition: Then $\rho = [u := \lambda \boldsymbol{y} t]$ with a Q-term $\lambda \boldsymbol{y} t$ without free occurrences of forbidden variables, and Q' is obtained from Q by removing $\exists u$. Suppose G is a Q-goal. Then clearly $G[u := \lambda \boldsymbol{y} t]$ form) is a Q'-goal.

Let $Q \longrightarrow_\rho Q'$ mean that for some C, C' we have $QC \Longrightarrow_\rho Q'C'$. Write $Q \longrightarrow_\rho^* Q'$ if there are $\rho_1, \ldots, \rho_n$ and $Q_1, \ldots, Q_{n-1}$ such that

$$Q \longrightarrow_{\rho_1} Q_1 \longrightarrow_{\rho_2} \cdots \longrightarrow_{\rho_{n-1}} Q_{n-1} \longrightarrow_{\rho_n} Q',$$

and $\rho = \rho_1 \circ \cdots \circ \rho_n$.

Corollary 1. *If $Q \longrightarrow_\rho^* Q'$ and G is a Q-goal, then $G\rho$ is a Q'-goal.*

3   Correctness and Completeness of the Unification Algorithm

Lemma 2. *Let a unification problem $\mathcal{U}$ consisting of a $\forall \exists \forall$-prefix Q and a list $\boldsymbol{r} = \boldsymbol{s}$ of unification pairs be given. Then either*

– *the unification algorithm makes a transition $\mathcal{U} \Longrightarrow_\rho \mathcal{U}'$, and*

$$\Phi' : \mathcal{U}'\text{-solutions} \to \mathcal{U}\text{-solutions}$$
$$\varphi' \mapsto (\rho \circ \varphi') {\restriction} Q_\exists$$

is well-defined and we have $\Phi: \mathcal{U}\text{-solutions} \to \mathcal{U}'\text{-solutions}$ such that Φ' is inverse to Φ, i.e. $\Phi'(\Phi\varphi) = \varphi$, or else
– *the unification algorithm fails, and there is no $\mathcal{U}$-solution.*

Proof. Case identity, i.e., $Q.r = r \wedge C \Longrightarrow_\varepsilon QC$. Let Φ be the identity.

Case ξ, i.e., $Q.\lambda x\, r = \lambda x\, s \wedge C \Longrightarrow_\varepsilon Q\forall x.r = s \wedge C$. Let again Φ be the identity.

Case rigid-rigid, i.e., $Q.f\boldsymbol{r} = f\boldsymbol{s} \wedge C \Longrightarrow_\varepsilon Q.\boldsymbol{r} = \boldsymbol{s} \wedge C$. Let again Φ be the identity.

Case flex-flex with equal heads, i.e., $Q.u\boldsymbol{y} = u\boldsymbol{z} \wedge C \Longrightarrow_\rho Q'.C\rho$ with $\rho = [u := \lambda \boldsymbol{y}.u'\boldsymbol{w}]$, Q' is Q with $\exists u$ replaced by $\exists u'$, and $\boldsymbol{w}$ an enumeration of those y_i which are identical to z_i (i.e., the variable at the same position in $\boldsymbol{z}$). Notice that $\lambda \boldsymbol{y}.u'\boldsymbol{w} = \lambda \boldsymbol{z}.u'\boldsymbol{w}$.

1. Φ' is well-defined: Let φ' be a $\mathcal{U}'$-solution, i.e., assume that $C\rho\varphi'$ holds. We must show that $\varphi := (\rho \circ \varphi') {\restriction} Q_\exists$ is a $\mathcal{U}$-solution.

For $u\boldsymbol{y} = u\boldsymbol{z}$: We must show $(u\varphi)\boldsymbol{y} = (u\varphi)\boldsymbol{z}$. But $u\varphi = u\rho\varphi' = (\lambda \boldsymbol{y}.u'\boldsymbol{w})\varphi'$. Hence $(u\varphi)\boldsymbol{y} = (u\varphi)\boldsymbol{z}$ by the construction of $\boldsymbol{w}$.

For $(r = s) \in C$: We need to show $(r = s)\varphi$. But by assumption $(r = s)\rho\varphi'$ holds, and $r = s$ has all its flexible variables from $Q_\exists$.

2. Definition of $\Phi: \mathcal{U}\text{-solutions} \to \mathcal{U}'\text{-solutions}$. Let a Q-substitution φ be given such that $(u\boldsymbol{y} = u\boldsymbol{z})\varphi$ and $C\varphi$. Define $u'(\Phi\varphi) := \lambda \boldsymbol{w}.(u\varphi)\boldsymbol{w}0$ (w.l.o.g), and $v(\Phi\varphi) := v$ for every other variable v in $Q_\exists$.

$\Phi\varphi =: \varphi'$ is a $\mathcal{U}'$-solution: Let $(r = s) \in C$. Then $(r = s)\varphi$ by assumption, for φ is a Q-substitution such that $C\varphi$ holds. We must show

$$(r = s)\rho\varphi'.$$

Notice that our assumption $(u\varphi)\boldsymbol{y} = (u\varphi)\boldsymbol{z}$ implies that the normal form of both sides can only contain the variables in $\boldsymbol{w}$. Therefore

$$\begin{aligned}
u\rho\varphi' &= (\lambda \boldsymbol{y}.u'\boldsymbol{w})\varphi' \\
&= \lambda \boldsymbol{y}.(\lambda \boldsymbol{w}.(u\varphi)\boldsymbol{w}0)\boldsymbol{w} \\
&= \lambda \boldsymbol{y}.(u\varphi)\boldsymbol{w}0 \\
&= \lambda \boldsymbol{y}.(u\varphi)\boldsymbol{y} \\
&= u\varphi
\end{aligned}$$

and hence $(r = s)\rho\varphi'$.

3. $\Phi'(\Phi\varphi) = \varphi$: So let φ be an $\mathcal{U}$-solution, and $\varphi' := \Phi\varphi$. Then

$$\begin{aligned}
u(\Phi'\varphi') &= u\big((\rho \circ \varphi') {\restriction} Q_\exists\big) \\
&= u\rho\varphi' \\
&= u\varphi, \qquad\qquad \text{as proved in 2.}
\end{aligned}$$

For every other variable v in $Q_\exists$ we obtain

$$v(\Phi'\varphi') = v((\rho \circ \varphi') \restriction Q_\exists)$$
$$= v\rho\varphi'$$
$$= v\varphi'$$
$$= v\varphi.$$

Case flex-flex with different heads, i.e., $\mathcal{U}$ is $Q.u\boldsymbol{y} = v\boldsymbol{z} \wedge C$. Let $\boldsymbol{w}$ be an enumeration of the variables both in $\boldsymbol{y}$ and in $\boldsymbol{z}$. Then $\rho = [u, v := \lambda\boldsymbol{y}.u'\boldsymbol{w}, \lambda\boldsymbol{z}.u'\boldsymbol{w}]$, Q' is Q with $\exists u, \exists v$ removed and $\exists u'$ inserted, and $\mathcal{U}' = Q'C\rho$.

1. Φ' is well-defined: Let φ' be a $\mathcal{U}'$-solution, i.e., assume that $C\rho\varphi'$ holds. We must show that $\varphi := (\rho \circ \varphi') \restriction Q_\exists$ is a $\mathcal{U}$-solution.

For $u\boldsymbol{y} = v\boldsymbol{z}$: We need to show $(u\varphi)\boldsymbol{y} = (v\varphi)\boldsymbol{z}$. But $(u\varphi)\boldsymbol{y} = (u\rho\varphi')\boldsymbol{y} = (\lambda\boldsymbol{y}.(u'\varphi')\boldsymbol{w})\boldsymbol{y} = (u'\varphi')\boldsymbol{w}$, and similarly $(v\varphi)\boldsymbol{z} = (u'\varphi')\boldsymbol{w}$.

For $(r = s) \in C$: We need to show $(r = s)\varphi$. But since u' is a new variable, φ and $\rho \circ \varphi'$ coincide on all variables free in $r = s$, and we have $(r = s)\rho\varphi'$ by assumption.

2. Definition of Φ: $\mathcal{U}$-solutions $\to$ $\mathcal{U}'$-solutions. Let a Q-substitution φ be given such that $(u\boldsymbol{y} = v\boldsymbol{z})\varphi$ and $C\varphi$. Define

$$u'(\Phi\varphi) := \lambda\boldsymbol{w}.(u\varphi)\boldsymbol{w}\boldsymbol{0} \quad \text{w.l.o.g.; } \boldsymbol{0} \text{ arbitrary}$$
$$v'(\Phi\varphi) := \lambda\boldsymbol{w}.(v\varphi)\boldsymbol{0}\boldsymbol{w}$$
$$w(\Phi\varphi) := w\varphi \qquad \text{otherwise, i.e., } w \neq u', v', u \text{ flexible.}$$

Since by assumption $(u\varphi)\boldsymbol{y} = (v\varphi)\boldsymbol{z}$, the normal forms of both $(u\varphi)\boldsymbol{y}$ and $(v\varphi)\boldsymbol{z}$ can only contain the common variables $\boldsymbol{w}$ from $\boldsymbol{y}, \boldsymbol{z}$ free. Hence, for $\varphi' := \Phi\varphi$, $u\rho\varphi' = u\varphi$ by the argument in the previous case, and similarly $v\rho\varphi' = v\varphi$. Since $r\varphi = s\varphi$ ($(r = s) \in C$ arbitrary) by assumption, and ρ only affects u and v, we obtain $r\rho\varphi' = s\rho\varphi'$, as required. $\Phi'(\Phi\varphi) = \varphi$ can now be proved as in the previous case.

Case flex-rigid, $\mathcal{U}$ is $Q.u\boldsymbol{y} = t \wedge C$.

Subcase occurrence check: t contains (a critical subterm with head) u. Then clearly there is no Q-substitution φ such that $(u\varphi)\boldsymbol{y} = t\varphi$.

Subcase pruning: Here t contains a subterm $v\boldsymbol{w}_1 z\boldsymbol{w}_2$ with $\exists v$ in Q, and z free in t. Then $\rho = [v := \lambda\boldsymbol{w}_1, z, \boldsymbol{w}_2.v'\boldsymbol{w}_1\boldsymbol{w}_2]$, Q' is Q with $\exists v$ replaced by $\exists v'$, and $\mathcal{U}' = Q'.u\boldsymbol{y} = t\rho \wedge C\rho$.

1. Φ' is well-defined: Let φ' be a $\mathcal{U}'$-solution, i.e., $(u\varphi')\boldsymbol{y} = t\rho\varphi'$, and $r\rho\varphi' = s\rho\varphi'$ for $(r = s) \in C$. We must show that $\varphi := (\rho \circ \varphi') \restriction Q_\exists$ is a $\mathcal{U}$-solution.

For $u\boldsymbol{y} = t$: We need to show $(u\varphi)\boldsymbol{y} = t\rho\varphi'$. But

$$(u\varphi)\boldsymbol{y} = (u\rho\varphi')\boldsymbol{y}$$
$$= (u\varphi')\boldsymbol{y} \quad \text{since } \rho \text{ does not touch } u$$
$$= t\rho\varphi' \quad \text{by assumption.}$$

For $(r = s) \in C$: We need to show $(r = s)\varphi$. But since v' is a new variable, $\varphi = (\rho \circ \varphi') \restriction Q_\exists$ and $\rho \circ \varphi'$ coincide on all variables free in $r = s$, and the claim follows from $(r = s)\rho\varphi'$.

2. Definition of Φ: $\mathcal{U}$-solutions $\to \mathcal{U}'$-solutions. For a $\mathcal{U}$-solution φ define

$$v'(\Phi\varphi) := \lambda \boldsymbol{w}_1, \boldsymbol{w}_2.(v\varphi)\boldsymbol{w}_1 0 \boldsymbol{w}_2$$
$$w(\Phi\varphi) := w\varphi \qquad\qquad \text{otherwise, i.e., } w \neq v', v \text{ flexible.}$$

Since by assumption $(u\varphi)\boldsymbol{y} = t\varphi$, the normal form of $t\varphi$ cannot contain z free. Therefore, for $\varphi' := \Phi\varphi$,

$$
\begin{aligned}
v\rho\varphi' &= (\lambda \boldsymbol{w}_1, z, \boldsymbol{w}_2.v'\boldsymbol{w}_1\boldsymbol{w}_2)\varphi' \\
&= \lambda \boldsymbol{w}_1, z, \boldsymbol{w}_2.(\lambda \boldsymbol{w}_1, \boldsymbol{w}_2.(v\varphi)\boldsymbol{w}_1 0 \boldsymbol{w}_2)\boldsymbol{w}_1\boldsymbol{w}_2 \\
&= \lambda \boldsymbol{w}_1, z, \boldsymbol{w}_2.(v\varphi)\boldsymbol{w}_1 0 \boldsymbol{w}_2 \\
&= \lambda \boldsymbol{w}_1, z, \boldsymbol{w}_2.(v\varphi)\boldsymbol{w}_1 z \boldsymbol{w}_2 \\
&= v\varphi.
\end{aligned}
$$

Hence $\varphi' = \Phi\varphi$ satisfies $(u\varphi')\boldsymbol{y} = t\rho\varphi'$. For $r = s$ this follows by the same argument. $\Phi'(\Phi\varphi) = \varphi$ can again be proved as in the previous case.

Subcase pruning impossible: Then $\lambda \boldsymbol{y} t$ has an occurrence of a universally quantified (i.e., forbidden) variable z. Therefore clearly there is no Q-substitution φ such that $(u\varphi)\boldsymbol{y} = t\varphi$.

Subcase explicit definition. Then $\rho = [u := \lambda \boldsymbol{y} t]$, Q' is obtained from Q by removing $\exists u$, and $\mathcal{U}' = Q'C\rho$. Note that ρ is a Q'-substitution, for we have performed the pruning steps.

1. Φ' is well-defined: Let φ' be a $\mathcal{U}'$-solution, i.e., $r\rho\varphi' = s\rho\varphi'$ for $(r = s) \in C$. We must show that $\varphi := (\rho \circ \varphi') \restriction Q_\exists$ is an $\mathcal{U}$-solution.

For $u\boldsymbol{y} = t$: We need to show $(u\rho\varphi')\boldsymbol{y} = t\rho\varphi'$. But

$$
\begin{aligned}
(u\rho\varphi')\boldsymbol{y} &= ((\lambda \boldsymbol{y} t)\varphi')\boldsymbol{y} \\
&= t\varphi' \\
&= t\rho\varphi' \qquad\qquad \text{since } u \text{ does not appear in } t.
\end{aligned}
$$

For $(r = s) \in C$: We need to show $(r = s)\varphi$. But this clearly follows from $(r = s)\rho\varphi'$.

2. Definition of Φ: $\mathcal{U}$-solutions $\to \mathcal{U}'$-solutions, and proof of $\Phi'(\Phi\varphi) = \varphi$. For a $\mathcal{U}$-solution φ define $\Phi\varphi = \varphi\restriction Q_\exists$. Then

$$u\rho\varphi' = \lambda \boldsymbol{y} t\varphi' = \lambda \boldsymbol{y} t\varphi = u\varphi,$$

and clearly $v\rho\varphi' = v\varphi$ for all other flexible φ. For $(r = s) \in C$, from $r\varphi = s\varphi$ we easily obtain $r\varphi' = s\varphi'$.

It is not hard to see that the unification algorithm terminates, by defining a measure that decreases with each transition.

Corollary 2. *Given a unification problem $\mathcal{U} = QC$, the unification algorithm either returns $\#f$, and there is no $\mathcal{U}$-solution, or else returns a pair (Q', ρ) with a "transition" substitution ρ and a prefix Q' (i.e., a unification problem $\mathcal{U}'$*

with no unification pairs) such that for any Q'-substitution φ', $(\rho \circ \varphi') \restriction Q_\exists$ is an $\mathcal{U}$-solution, and every $\mathcal{U}$-solution can be obtained in this way. Since the empty substitution is a Q'-substitution, $\rho \restriction Q_\exists$ is an $\mathcal{U}$-solution, which is most general in the sense stated.

$\mathsf{unif}(Q, \boldsymbol{r} = \boldsymbol{s})$ denotes the result of the unification algorithm at $\mathcal{U} = Q\boldsymbol{r} = \boldsymbol{s}$.

4   Proof Search

A *Q-sequent* has the form $\mathcal{P} \Rightarrow G$, where $\mathcal{P}$ is a list of Q-clauses and G is a Q-goal.

We write $M[\mathcal{P}]$ to indicate that all assumption variables in the derivation M concern clauses in $\mathcal{P}$.

Write $\vdash^n S$ for a set S of sequents if there are derivations $M_i^{G_i}[\mathcal{P}_i]$ in long normal form for all $(\mathcal{P}_i \Rightarrow G_i) \in S$ such that $\sum \#M_i \le n$. Let $\vdash^{<n} S$ mean $\exists m{<}n \vdash^m S$.

We now prove correctness and completeness of the proof search procedure: correctness is the if-part of the two lemmata to follow, and completeness the only-if-part.

Lemma 3. *Let Q be a $\forall\exists\forall$-prefix, $\{\mathcal{P} \Rightarrow \forall\boldsymbol{x}.\boldsymbol{D} \to A\} \cup S$ Q-sequents with $\boldsymbol{x}, \boldsymbol{D}$ not both empty. Then we have for every substitution φ:*

$$\varphi \text{ is a } Q\text{-substitution such that } \vdash^n \left(\{\mathcal{P} \Rightarrow \forall\boldsymbol{x}.\boldsymbol{D} \to A\} \cup S\right)\varphi$$

if and only if

$$\varphi \text{ is a } Q\forall\boldsymbol{x}\text{-substitution such that } \vdash^{<n} \left(\{\mathcal{P} \cup \boldsymbol{D} \Rightarrow A\} \cup S\right)\varphi.$$

Proof. "*If*". Let φ be a $Q\forall\boldsymbol{x}$-substitution and $\vdash^{<n} \left(\{\mathcal{P} \cup \boldsymbol{D} \Rightarrow A\} \cup S\right)\varphi$. So we have

$$N^{A\varphi}[\boldsymbol{D}\varphi \cup \mathcal{P}\varphi].$$

Since φ is a $Q\forall\boldsymbol{x}$-substitution, no variable in $\boldsymbol{x}$ can be free in $\mathcal{P}\varphi$, or free in $y\varphi$ for some $y \in \mathsf{dom}(\varphi)$. Hence

$$M^{(\forall\boldsymbol{x}.\boldsymbol{D}\to A)\varphi}[\mathcal{P}\varphi] := \lambda\boldsymbol{x}\lambda\boldsymbol{u}^{\boldsymbol{D}\varphi} N$$

is a correct derivation.

"*Only if*". Let φ be a Q-substitution and $\vdash^n \left(\{\mathcal{P} \Rightarrow \forall\boldsymbol{x}.\boldsymbol{D} \to A\} \cup S\right)\varphi$. This means we have a derivation (in long normal form)

$$M^{(\forall\boldsymbol{x}.\boldsymbol{D}\to A)\varphi}[\mathcal{P}\varphi] = \lambda\boldsymbol{x}\lambda\boldsymbol{u}^{\boldsymbol{D}\varphi}.N^{A\varphi}[\boldsymbol{D}\varphi \cup \mathcal{P}\varphi].$$

Now $\#N < \#M$, hence $\vdash^{<n} \left(\{\mathcal{P} \cup \boldsymbol{D} \Rightarrow A\} \cup S\right)\varphi$, and φ clearly is a $Q\forall\boldsymbol{x}$-substitution.

Lemma 4. *Let Q be a $\forall\exists\forall$-prefix, $\{\mathcal{P} \Rightarrow Pr\} \cup S$ Q-sequents and φ a substitution. Then*

$$\varphi \text{ is a } Q\text{-substitution such that } \vdash^n \left(\{\mathcal{P} \Rightarrow Pr\} \cup S\right)\varphi$$

if and only if there is a clause $\forall\boldsymbol{x}.\boldsymbol{G} \to Ps$ in $\mathcal{P}$ such that the following holds. Let $\boldsymbol{z}$ be the final universal variables in Q, $\boldsymbol{x}$ be new ("raised") variables such that $X_i\boldsymbol{z}$ has the same type as x_i, let Q^* be Q with the existential variables extended by $\boldsymbol{x}$, and let $*$ indicate the substitution $[x_1, \ldots, x_n := X_1\boldsymbol{z}, \ldots, X_n\boldsymbol{z}]$. Then $\mathsf{unif}(Q^*, \boldsymbol{r} = \boldsymbol{s}^*) = (Q', \rho)$ and there is a Q'-substitution φ' such that $\vdash^{<n} \left(\{\mathcal{P} \Rightarrow \boldsymbol{G}^*\} \cup S\right)\rho\varphi'$, and $\varphi = (\rho \circ \varphi') \!\restriction\! Q_\exists$.

Proof. "If". Let $\mathsf{unif}(Q^*, \boldsymbol{r} = \boldsymbol{s}^*) = (Q', \rho)$, and assume that φ' is a Q'-substitution such that $N_i \vdash \left(\mathcal{P} \Rightarrow \boldsymbol{G}^*\right)\rho\varphi'$. Let $\varphi := (\rho \circ \varphi') \!\restriction\! Q_\exists$. From $\mathsf{unif}(Q^*, \boldsymbol{r} = \boldsymbol{s}^*) = (Q', \rho)$ we know $\boldsymbol{r}\rho = \boldsymbol{s}^*\rho$, hence $\boldsymbol{r}\varphi = \boldsymbol{s}^*\rho\varphi'$. Then

$$u^{(\forall\boldsymbol{x}.\boldsymbol{G}\to Ps)\varphi}\left((\boldsymbol{x}\rho\varphi')\boldsymbol{z}\right)\boldsymbol{N}^{\boldsymbol{G}^*\rho\varphi'}$$

derives $Ps^*\rho\varphi'$ (i.e., $Pr\varphi$) from $\mathcal{P}\varphi$.

"Only if". Assume φ is a Q-substitution such that $\vdash (\mathcal{P} \Rightarrow Pr)\varphi$, say by $u^{(\forall\boldsymbol{x}.\boldsymbol{G}\to Ps)\varphi}\boldsymbol{t}\boldsymbol{N}^{(\boldsymbol{G}\varphi)[\boldsymbol{x}:=\boldsymbol{t}]}$, with $\forall\boldsymbol{x}.\boldsymbol{G} \to Ps$ a clause in $\mathcal{P}$, and with additional assumptions from $\mathcal{P}\varphi$ in $\boldsymbol{N}$. Then $\boldsymbol{r}\varphi = (\boldsymbol{s}\varphi)[\boldsymbol{x} := \boldsymbol{t}]$. Since we can assume that the variables $\boldsymbol{x}$ are new and in particular not range variables of φ, with

$$\vartheta := \varphi \cup [\boldsymbol{x} := \boldsymbol{t}]$$

we have $\boldsymbol{r}\varphi = \boldsymbol{s}\vartheta$. Let $\boldsymbol{z}$ be the final universal variables in Q, $\boldsymbol{x}$ be new ("raised") variables such that $X_i\boldsymbol{z}$ has the same type as x_i, let Q^* be Q with the existential variables extended by $\boldsymbol{x}$, and for terms and formulas let $*$ indicate the substitution $[x_1, \ldots, x_n := X_1\boldsymbol{z}, \ldots, X_n\boldsymbol{z}]$. Moreover, let

$$\vartheta^* := \varphi \cup [X_1, \ldots, X_n := \lambda\boldsymbol{z}.t_1, \ldots, \lambda\boldsymbol{z}.t_n].$$

Then $\boldsymbol{r}\vartheta^* = \boldsymbol{r}\varphi = \boldsymbol{s}\vartheta = \boldsymbol{s}^*\vartheta^*$, i.e., ϑ^* is a solution to the unification problem given by Q^* and $\boldsymbol{r} = \boldsymbol{s}$. Hence by Lemma 2 $\mathsf{unif}(Q^*, \boldsymbol{r} = \boldsymbol{s}^*) = (Q', \rho)$ and there is a Q'-substitution φ' such that $\vartheta^* = (\rho \circ \varphi') \!\restriction\! Q^*_\exists$, hence $\varphi = (\rho \circ \varphi') \!\restriction\! Q_\exists$. Also, $(\boldsymbol{G}\varphi)[\boldsymbol{x} := \boldsymbol{t}] = \boldsymbol{G}\vartheta = \boldsymbol{G}^*\vartheta^* = \boldsymbol{G}^*\rho\varphi'$.

A *state* is a pair (Q, S) with Q a prefix and S a finite set of Q-sequents. By the two lemmas just proved we have *state transitions*

$$(Q, \{\mathcal{P} \Rightarrow \forall\boldsymbol{x}.\boldsymbol{D} \to A\} \cup S) \mapsto^\varepsilon (Q\forall\boldsymbol{x}, \{\mathcal{P} \cup \boldsymbol{D} \Rightarrow A\} \cup S)$$
$$(Q, \{\mathcal{P} \Rightarrow Pr\} \cup S) \mapsto^\rho (Q', (\{\mathcal{P} \Rightarrow \boldsymbol{G}^*\} \cup S)\rho),$$

where in the latter case there is a clause $\forall\boldsymbol{x}.\boldsymbol{G} \to Ps$ in $\mathcal{P}$ such that the following holds. Let $\boldsymbol{z}$ be the final universal variables in Q, $\boldsymbol{x}$ be new ("raised") variables such that $X_i\boldsymbol{z}$ has the same type as x_i, let Q^* be Q with the existential variables extended by $\boldsymbol{x}$, and let $*$ indicate the substitution $[x_1, \ldots, x_n := X_1\boldsymbol{z}, \ldots, X_n\boldsymbol{z}]$, and $\mathsf{unif}(Q^*, \boldsymbol{r} = \boldsymbol{s}^*) = (Q', \rho)$.

Notice that by Lemma 1, if $\mathcal{P} \Rightarrow Pr$ is a Q-*sequent* (which means that $\bigwedge\mathcal{P} \to Pr$ is a Q-goal), then $(\mathcal{P} \Rightarrow \boldsymbol{G}^*)\rho$ is a Q'-sequent.

Theorem 1. *Let Q be a prefix, and S be a set of Q-sequents. For every substitution φ we have: φ is a Q-substitution satisfying $\vdash S\varphi$ iff there is a prefix Q', a substitution ρ and a Q'-substitution φ' such that*

$$(Q, S) \mapsto^{\rho*} (Q', \emptyset),$$
$$\varphi = (\rho \circ \varphi') {\restriction} Q_\exists.$$

Examples. 1. The sequent $\forall y.\forall z Ryz \rightarrow Q, \forall y_1, y_2 Ry_1 y_2 \Rightarrow Q$ leads first to $\forall y_1, y_2 Ry_1 y_2 \Rightarrow Ryz$ under $\exists y \forall z$, then to $y_1 = y \wedge y_2 = z$ under $\exists y \forall z \exists y_1, y_2$, and finally to $Y_1 z = y \wedge Y_2 z = z$ under $\exists y, Y_1, Y_2 \forall z$, which has the solution $Y_1 = \lambda zy$, $Y_2 = \lambda zz$.

2. $\forall y.\forall z Ryz \rightarrow Q, \forall y_1 Ry_1 y_1 \Rightarrow Q$ leads first to $\forall y_1 1 Ry_1 y_1 \Rightarrow Ryz$ under $\exists y \forall z$, then to $y_1 = y \wedge y_1 = z$ under $\exists y \forall z \exists y_1$, and finally to $Y_1 z = y \wedge Y_1 z = z$ under $\exists y, Y_1 \forall z$, which has no solution.

3. Here is a more complex example (derived from proofs of the Orevkov-formulas), for which we only give the derivation tree.

$$
\cfrac{
  \cfrac{
    \cfrac{\forall y.(\forall z Ryz \rightarrow \bot) \rightarrow \bot}{(\forall z R0z \rightarrow \bot) \rightarrow \bot}
    \qquad
    \cfrac{
      \cfrac{
        \cfrac{\forall y.(\forall z_1 Ryz_1 \rightarrow \bot) \rightarrow \bot}{(\forall z_1 Rzz_1 \rightarrow \bot) \rightarrow \bot}
        \quad
        \cfrac{
          \cfrac{\cfrac{\forall z S0z \rightarrow \bot}{S0z_1 \rightarrow \bot} \quad (*) \quad \cfrac{R0z \quad Rzz_1}{S0z_1}}{\bot}
        }{\cfrac{Rzz_1 \rightarrow \bot}{\forall z_1 Rzz_1 \rightarrow \bot}}
      }{R0z \rightarrow \bot}
    }{\forall z R0z \rightarrow \bot}
  }{\bot}
}{}
$$

where $(*)$ is a derivation from $\mathrm{Hyp}_1 : \forall z, z_1 . R0z \rightarrow Rzz_1 \rightarrow S0z_1$.

5   Extension by $\wedge$ and $\exists$

The extension by conjunction is rather easy; it is even superfluous in principle, since conjunctions can always be avoided at the expense of having lists of formulas instead of single formulas.

However, having conjunctions available is clearly useful at times, so let's add it. This requires the notion of an *elaboration path* for a formula (cf. [1]). The reason is that the property of a formula to have a unique atom as its *head* is lost when conjunctions are present. An elaboration path is meant to give the directions (left or right) to go when we encounter a conjunction as a strictly positive subformula. For example, the elaboration paths of $\forall x A \wedge (B \wedge C \rightarrow D \wedge \forall y E)$ are (left), (right, left) and (right, right). Clearly, a formula is equivalent to the conjunction (over all elaboration paths) of all formulas obtained from it by following an elaboration path (i.e., always throwing away the other part of the conjunction). In our example,

$$\forall x A \wedge (B \wedge C \rightarrow D \wedge \forall y E) \leftrightarrow \forall x A \wedge (B \wedge C \rightarrow D) \wedge (B \wedge C \rightarrow \forall y E).$$

In this way we regain the property of a formula to have a unique head, and our previous search procedure continues to work.

For the existential quantifier $\exists$ the problem is of a different nature. We chose to introduce $\exists$ by means of axiom schemata. Then the problem is which of such schemes to use in proof search, given a goal G and a set $\mathcal{P}$ of clauses. We might proceed as follows.

List all prime, positive and negative existential subformulas of $\mathcal{P} \Rightarrow G$, and remove any formula from those lists which is of the form of another one[1]. For every positive existential formula – say $\exists x B$ – add (the generalization of) the existence introduction scheme

$$\exists^+_{x,B} \colon \forall x.B \to \exists x B$$

to $\mathcal{P}$. Moreover, for every negative existential formula – say $\exists x A$ – and every (prime or existential) formula C in any of those two lists, exept the formula $\exists x A$ itself, add (the generalization of) the existence elimination scheme

$$\exists^-_{x,A,C} \colon \exists x A \to (\forall x.A \to C) \to C$$

to $\mathcal{P}$. Then start the search algorithm as described in section 4. The normal form theorem for the natural deduction system of minimal logic with $\exists$ then guarantees completeness.

However, experience has shown that this complete search procedure tends to be trapped in too large a search space. Therefore in our actual implementation we decided to only take instances of the existence elimination scheme with *existential* conclusions.

Acknowledgements

I have benefitted from a presentation of Miller's [1] given by Ulrich Berger, in a logic seminar in München in June 1991.

References

1. Dale Miller. A logic programming language with lambda–abstraction, function variables and simple unification. *Journal of Logic and Computation*, 2(4):497–536, 1991.
2. Tobias Nipkow. Higher-order critical pairs. In R. Vemuri, editor, *Proceedings of the Sixth Annual IEEE Symposium on Logic in Computer Science*, pages 342–349, Los Alamitos, 1991. IEEE Computer Society Press.

---

[1] To do this, for patterns the dual of the theory of "most general unifiers", i.e., a theory of "most special generalizations", needs to be developed.

Planning and Patching Proof[*]

Alan Bundy

School of Informatics, University of Edinburgh
3.09 Appleton Tower, 11 Crichton Street, Edinburgh, EH8 9LE, UK
A.Bundy@ed.ac.uk

Abstract. We describe proof planning: a technique for both describing the hierarchical structure of proofs and then using this structure to guide proof attempts. When such a proof attempt fails, these failures can be analyzed and a patch formulated and applied. We also describe rippling: a powerful proof method used in proof planning. We pose and answer a number of common questions about proof planning and rippling.

1 Introduction

The Program Committee Chair of AISC-04 suggested that the published version of my talk might be:

> " ... a short paper telling our audience what are the highlights of the publication landscape for the subject of your presentation, what they should read if they want to become better informed by systematic reading, and why they should read the cited material (i.e. why it represents the highlights), could have both immediate use and longer-term educational use for people who don't attend the conference but buy or read the proceedings later."

Below I have attempted to fulfill this brief. I have organized the paper as a 'Frequently Asked Questions' about proof planning, in general, and rippling, in particular.

2 Proof Planning

2.1 Introduction

What Is Proof Planning? Proof planning is a technique for guiding the search for a proof in automated theorem proving. A proof plan is an outline or plan of a proof. To prove a conjecture, proof planning constructs a proof plan for a proof and uses it to guide the construction of the proof itself. Proof planning reduces the amount of search and curbs the combinatorial explosion. It also helps pinpoint the cause of any proof attempt failure, suggesting a patch to facilitate a renewed attempt.

---

[*] The research reported in this paper was supported by EPSRC grant GR/S01771.

Common patterns in proofs are identified and represented in computational form as general-purpose tactics, i.e. programs for directing the proof search process. These tactics are then formally specified with methods using a meta-language. Standard patterns of proof failure and appropriate patches to the failed proofs attempts are represented as critics. To form a proof plan for a conjecture the proof planner reasons with these methods and critics. The proof plan consists of a customized tactic for the conjecture, whose primitive actions are the general-purpose tactics. This customized tactic directs the search of a tactic-based theorem prover.

For a general, informal introduction to proof planning see [Bundy, 1991]. Proof planning was first introduced in [Bundy, 1988]. An earlier piece of work that led to the development of proof planning was the use of meta-level inference to guide equation solving, implemented in the Press system (see [Sterling et al, 1989]).

Has Proof Planning Been Implemented? Yes, in the *Oyster/Clam* system [Bundy et al, 1990] and λ*Clam* system at Edinburgh and the Omega system at Saarbrücken [Benzmüller et al, 1997]. *Clam* and λ*Clam* are the proof planners. They constructs a customized tactic for a conjecture and then a proof checker, such as *Oyster*, executes the tactic.

In principle, *Clam* could be interfaced to any tactic-based theorem prover. To test this assertion, we interfaced *Clam* to the Cambridge HOL theorem prover [Boulton et al, 1998]. We are currently building a proof planner, called IsaPlanner, in Isabelle [Dixon & Fleuriot, 2003].

How Has Proof Planning Been Evaluated? One of the main domains of application has been in inductive reasoning [Bundy, 2001], with applications to software and hardware verification, synthesis and transformation, but it has also been applied to co-induction [Dennis et al, 2000], limit theorems, diagonalization arguments, transfinite ordinals, summing series, equational reasoning, meta-logical reasoning, algebra, *etc.* A survey of such applications can be found in chapter 5 of [Bundy et al, 2005].

Can Proof Planning Be Applied to Non-mathematical Domains? Yes. We have had some success applying proof planning to game playing (Bridge [Frank et al, 1992,Frank & Basin, 1998] and Go [Willmott et al, 2001]) and to configuration problems [Lowe et al, 1998]. It is potentially applicable wherever there are common patterns of reasoning. Proof planning can be used to match the problem to the reasoning method in a process of meta-level reasoning. Proof planning gives a clean separation between the factual and search control information, which facilitates their independent modification.

What Is the Relation Between Proof Planning and Rippling? Rippling is a key method in our proof plans for induction. It is also useful in non-inductive domains. However, you can certainly envisage a proof planning system which did not contain a rippling method (the Saarbrücken Omega system, for instance) and you can envisage using rippling, e.g. as a tactic, in a non-proof planning system. So there is no necessary connection. For more on rippling see §3.

What Are the Scope and Limitations of Proof Planning? A critical evaluation of proof planning can be found in [Bundy, 2002].

2.2   Discovery and Learning

Is It Possible to Automate the Learning of Proof Plans? Proof plans can be learnt from example proofs. In the case of equation-solving methods, this was demonstrated in [Silver, 1985]; in the case of inductive proof methods it was demonstrated in [Desimone, 1989]. Both projects used forms of explanation-based generalization. We also have a current project on the use of data-mining techniques (both probabilistic reasoning and genetic programming) to construct tactics from large corpora of proofs, [Duncan *et al*, 2004].

The hardest aspect of learning proof plans is coming up with the key meta-level concepts to describe the preconditions of the methods. An example of such a meta-level concept is that of 'wave-front' idea used in rippling. We have not made much progress on automating the learning of these.

How Can Humans Discover Proof Plans? This is an art similar to the skill used by a good mathematics teacher when analyzing a student's proof or explaining a new method of proof to a class. The key is identifying the appropriate meta-level concepts to generalize from particular examples. Armed with the right concepts, standard inductive learning techniques can form the right generalization (see §2.2).

2.3   Drawbacks and Limitations

What Happens if the Attempt to Find a Proof Plan Fails? In certain circumstances proof critics can suggest an appropriate patch to a partial proof plan. Suppose the preconditions of a method succeed, but this method is unpacked into a series of sub-methods one of which fails, i.e. the preconditions of the sub-method fail. Critics are associated with some of these patterns of failure. For instance, one critic may fire if the first two preconditions of a method succeed, but the last one fails. It will then suggest an appropriate patch for this kind of failure, e.g. suggest the form of a missing lemma, or suggest generalizing the conjecture. The patch is instituted and proof planning continues.

The original critics paper is [Ireland, 1992]. A more recent paper is [Ireland & Bundy, 1996]. Two important application of critics are: discovering loop invariants in the verification of imperative programs [Stark & Ireland, 1998]; and the correction of false conjectures [Monroy *et al*, 1994].

In other circumstances, a subgoal may be reached to which no method or critic is applicable. It may be possible to back-up to a choice point in the search, i.e. a place where two or more methods or critics were applicable. However, the search space defined by the methods and critics is typically much smaller than the search space defined by the object-level rules and axioms; that is both the strength and the weakness of proof planning. The total search space is cropped to the portion where the proof is most likely to occur – reducing the combinatorial explosion, but losing completeness. It is always possible to regain completeness by supplementing the methods with a default, general-purpose exhaustive search method, but some would regard this as a violation of the spirit of proof planning. For more discussion of these points see [Bundy, 2002].

Is It Possible to Discover New Kinds of Proof in a Proof Planning System? Since general-purpose proof plans represent common patterns in proofs, then, by definition, they cannot discover new kinds of proof. This limitation could be overcome in several ways. One would be to include a default method which invoked some general search technique. This might find a new kind of proof by accident. Another might be to have meta-methods which constructed new methods. For instance, a method for one domain might be applied to another by generalizing its preconditions[1]. Or a method might be learnt from an example proof (see §2.2). Proof plans might, for instance, be learnt from proofs constructed by general search. For more discussion of these points see [Bundy, 2002].

Isn't Totally Automated Theorem Proving Infeasible? For the foreseeable future theorem provers will require human interaction to guide the search for non-trivial proofs. Fortunately, proof planning is also useful in interactive theorem provers. Proof plans facilitate the hierarchical organization of a partial proof, assisting the user to navigate around it and understand its structure. They also provide a language for chunking the proof and for describing the interrelation between the chunks. Interaction with a semi-automated theorem prover can be based on this language. For instance, the user can: ask why a proof method failed to apply; demand that a heuristic precondition is overridden; use the analysis from proof critics to patch a proof; etc.

The XBarnacle system is an semi-automated theorem prover based on proof planning [Lowe & Duncan, 1997]. There is also a version of XBarnacle with interaction critics, where the user assists the prover to find lemmas and generalizations [Jackson & Lowe, 2000].

Doesn't Proof Planning Promote Cheating by Permitting Ad Hoc Adjustments to Enable a Prover to 'Discover' Particular Proofs? Not if the recommended methodology is adopted. [Bundy, 1991] specifies a set of

---

[1] Often new departures come in mathematics when mathematicians switch from one area to another, bringing their proof methods with them.

criteria for assessing proof plans. These include generality and parsimony, which discourage the creation of ad hoc methods designed to guide particular theorems. Rather, they encourage the design of a few, general-purpose methods which guide a wide range of theorems. The expectancy criterion promotes the association of a method with an explanation of why it works. This discourages the design of methods which often succeed empirically, but for poorly understood reasons. Of course, these criteria are a matter of degree, so poor judgement may produce methods which other researchers regard as ad hoc. The criteria of proof planning then provide a basis for other researchers to criticize such poor judgement.

2.4   Is All This of Relevance to Me?

Would a Proof Planning Approach Be Appropriate for My Application? The properties of a problem that indicate that proof planning might be a good solution are: 1. A search space which causes a combinatorial explosion when searched without heuristic guidance; 2. The existence of heuristic tactics which enable expert problem solvers to search a much smaller search space defined by these tactics; 3. The existence of specifications for each tactic to determine when it is appropriate to apply it and what effect it will have if it succeeds.

How Would I Go About Developing Proof Plans for My Domain? The key problem is to identify the tactics and their specifications. This is usually done by studying successful human problem solving and extracting the tactics. Sometimes there are texts describing the tactics, e.g. in bridge and similar games. Sometimes knowledge acquisition techniques, like those used in expert systems, are needed, e.g. analysis of problem solving protocols, exploratory interviews with human experts.

3   Rippling

3.1   Introduction

What Is Rippling? A technique for controlling term rewriting using annotations to restrict application and to ensure termination, see Fig. 1. A goal expression is rippled with respect to one or more given expressions. Each given expression embeds in the goal expression. Annotations in the goal mark those subexpressions which correspond to bits of the given (the skeleton) and those which do not (the wave-fronts). The goal is rewritten so that the embeddings are preserved, i.e. rewriting can only move the wave-fronts around within the skeleton - wave-fronts can change but the skeleton cannot. Furthermore, wave-fronts are given a direction (outwards or inwards) and movement can only be in that direction. Outward wave-fronts can mutate to inward, but not vice versa. This ensures termination. Rippling can be implemented by putting wave annotation into the rewrite rules to turn them into wave-rules, see Fig. 2. The successful application of wave-rule requires that any wave-front in the wave-rule must match

$$t <> (Y <> Z) = (t <> Y) <> Z$$

$$\vdash \boxed{h :: t}^{\uparrow} <> (y <> z) = (\boxed{h :: t}^{\uparrow} <> y) <> z$$

$$\vdash \boxed{h :: t <> (y <> z)}^{\uparrow} = \boxed{h :: t <> y}^{\uparrow} <> z$$

$$\vdash \boxed{h :: t <> (y <> z)}^{\uparrow} = \boxed{h :: (t <> y) <> z}^{\uparrow}$$

$$\vdash \boxed{h = h \wedge t <> (y <> z) = (t <> y) <> z}^{\uparrow}$$

Fig. 1. This example is taken from the step case of the induction proof of the associativity of append, where $<>$ is infix list append and $::$ is infix list cons. The hollow grey boxes in the induction conclusion represent wave-fronts. Orange boxes are used when colour is available. Notice how these grow in size until a copy of the induction hypothesis appears inside them.

$$\boxed{H :: T}^{\uparrow} <> L \Rightarrow \boxed{H :: T <> L}^{\uparrow}$$

$$rev(\boxed{H :: T}^{\uparrow}) \Rightarrow \boxed{rev(T) <> (H :: nil)}^{\uparrow}$$

$$\boxed{X_1 :: X_2}^{\uparrow} = \boxed{Y_1 :: Y_2}^{\uparrow} \Rightarrow \boxed{X_1 = Y_1 \wedge X_2 = Y_2}^{\uparrow}$$

$$X <> \boxed{(Y <> Z)}^{\uparrow} \Rightarrow \boxed{(X <> Y) <> Z}^{\uparrow}$$

Fig. 2. Rewrite rules from the recursive definition of $<>$ and rev, the replacement rule for equality (backwards) and the associativity of $<>$ are annotated as wave rules. The bits not in wave-fronts are called the skeleton. Note that the skeleton is the same on each side of the wave rule, but that more of it is surrounded by the wave-front on the right hand side compared to the left hand side.

a wave-front in the goal. Rippling was originally developed for guiding the step cases of inductive proofs, in which the givens are the induction hypotheses and the goal is the induction conclusion.

For an informal introduction to rippling with a large selection of examples see [Bundy *et al*, 1993]. For a more formal account see: [Basin & Walsh, 1996]. A thorough account will shortly be available in [Bundy *et al*, 2005].

Why Is It Called Rippling? Raymond Aubin coined the term 'rippling-out', in his 1976 Edinburgh PhD thesis, to describe the pattern of movement of what we now call wave-fronts, during conventional rewriting with constructor-style recursive definitions. In [Bundy, 1988], we turned this on its head by taking such a movement of wave-fronts as the definition of rippling rather than the effect of rewriting. This enabled the idea to be considerably generalized. Later we invented 'rippling-sideways', 'rippling-in', etc and so generalized the combined technique to 'rippling'.

3.2   Relation to Standard Rewriting Techniques

Are Wave-Rules Just the Step Cases of Recursive Definitions? No. Many lemmas and other axioms can also be annotated as wave-rules. Examples include: associative laws; distributive laws; replacement axioms for equality; many logical axioms; etc. Equations that cannot be expressed as wave-rules include commutative laws and (usually) the step cases from mutually recursive definitions. The latter can be expressed as wave-rules in an abstraction in which the mutually defined functions are regarded as indistinguishable. Lots of example wave-rules can be found in [Bundy *et al*, 2005].

How Does Rippling Differ from the Standard Application of Rewrite Rules? Rippling differs from standard rewriting in two ways. Firstly, the wave annotation may prevent the application of a wave-rule which, viewed only as a rewrite rule, would otherwise apply. This will happen if the left-hand side of the wave-rule contains a wave-front which does not match a wave-front in the expression being rewritten. Secondly, equations can usually be oriented as wave-rules in both directions, but without loss of termination. The wave annotations prevent looping. An empirical comparison of rippling and rewriting can be found in [Bundy & Green, 1996].

Since Rippling Is Terminating, Is It Restricted to Terminating Sets of Rewrite Rules? No. If all the rewrite rules in a non-terminating set can be annotated as wave-rules then the additional conditions of wave annotation matching, imposed by rippling, will ensure that rippling still terminates. Examples are provided by the many rewrite rules that can be annotated as wave-rules in both directions, and may even both be used in the same proof, without loss of termination.

Couldn't We Simply Perform Rippling Using a Suitable Order, e.g. Recursive Path Ordering, Without the Need for Annotations? No, each skeleton gives (in essence) a different termination ordering which guides the proof towards fertilization with that skeleton. Different annotations on the same term can result in completely different rewritings.

Is Rippling Restricted to First-Order, Equational Rewriting? No, there are at least two approaches to higher-order rippling. One is based on viewing wave annotation as representing an embedding of the given in the goal [Smaill & Green, 1996]. The other is based on a general theory of colouring λcalculus terms in different ways [Hutter & Kohlhase, 1997].

Rippling can also be extended to support reasoning about logic programs – and other situations where values are passed between conjoined relations via shared existential variables, as opposed to being passed between nested functions. Relational rippling adapts rippling to this environment [Bundy & Lombart, 1995].

3.3 Wave Annotations

Is the Concept of Wave-Rule a Formal or Informal One? 'Wave-rule' can be defined formally. It is a rewrite rule containing wave annotation in which the skeletons are preserved and the wave-front measure of the right-hand side is less than that of the left-hand side. Informally, the skeleton consists of those bits of the expression outside of wave-fronts or inside the wave-holes. The measure records the position of the wave-fronts in the skeleton. It decreases when outwards directed wave-fronts move out or downwards wave-fronts move in. A formal definition of skeleton and of the wave-front measure can be found in [Basin & Walsh, 1996] and in chapter 4 of [Bundy *et al*, 2005].

Where Do the Wave Annotations in Wave-Rules and in Induction Rules Come from? Wave annotation can be inserted in expressions by a family of difference unification algorithms invented by Basin and Walsh (see [Basin & Walsh, 1993]). These algorithms are like unification but with the additional ability to hide non-matching structure in wave-fronts. Ground difference matching can be used to insert wave annotation into induction rules and ground difference unification for wave-rules. 'Ground' means that no instantiation of variables occurs. 'Matching' means that wave-fronts are inserted only into the induction conclusion and not the induction hypothesis. 'Unification' means that wave-fronts are inserted into both sides of wave-rules. Note that the process of inserting wave annotations can be entirely automated.

3.4 Performance

Has Rippling Been Used to Prove Any Hard Theorems? Yes. Rippling has been used successfully in the verification of the Gordon microprocessor and the synthesis of a decision procedure and of the rippling tactic itself. It has also been used outwith inductive proofs for the summing of series and the Lim+ theorem. A survey of some of these successes can be found in chapter 5 of [Bundy *et al*, 2005].

Can Rippling Fail? Yes, if there is no wave-rule available to move a wave-front. In this case we apply critics to try to patch the partial proof. For inductive proofs, for instance, these may: generalize the induction formula; revise the induction rule; introduce a case split; or introduce and prove an intermediate lemma, according to the precise circumstances of the breakdown. The fact that a failed ripple provides so much information to focus the attempt to patch the proof is one of the major advantages of rippling. More details about critics, including some hard examples which have been proved with their use, can be found in [Ireland & Bundy, 1996] and in chapter 3 of [Bundy *et al*, 2005].

3.5 Miscellaneous

How Is Rippling Used to Choose Induction Rules? There is a one-level look-ahead into the rippling process to see what induction rules would permit rippling to take place. In particular, which wave-fronts placed around which induction variables would match with corresponding wave-fronts in induction rules. We call this ripple analysis. It is similar to the use of recursive definitions to suggest induction rules, as pioneered by Boyer and Moore, but differs in that all available wave-rules are used in ripple analysis, and not just recursive definitions. More detail of ripple analysis can be found in [Bundy *et al*, 1989], although that paper is rather old now and is not a completely accurate description of current rippling implementations. In particular, the term 'ripple analysis' was not in use when that paper was written and it is misleadingly called 'recursion analysis' there. A recent alternative approach is to postpone the choice of induction rule by using meta-variables as place holders for the induction term and then instantiating these meta-variables during rippling: thus tailoring the choice of induction rule to fit the needs of rippling [Kraan *et al*, 1996,Gow, 2004].

3.6 And Finally ...

Why Do You Use Orange Boxes to Represent Wave-Fronts? [2]
The boxes form hollow squares, which help to display the outwards or inwards movement of wave-fronts. In the days of hand-written transparencies, orange was used because it is one of the few transparent overhead pen colours, allowing the expression in the wave-front to show through[3].

For more information about the research work outlined above, electronic versions of some of the papers and information about down loading software, see the web site of my research group at http://dream.dai.ed.ac.uk/.

References

[Basin & Walsh, 1993] Basin, David A. and Walsh, Toby. (1993). Difference unification. In Bajcsy, R., (ed.), *Proc. 13th Intern. Joint Conference on Artificial Intelligence (IJCAI '93)*, volume 1, pages 116–122, San Mateo, CA. Morgan Kaufmann. Also available as Technical Report MPI-I-92-247, Max-Planck-Institut für Informatik.

[Basin & Walsh, 1996] Basin, David and Walsh, Toby. (1996). A calculus for and termination of rippling. *Journal of Automated Reasoning*, 16(1–2):147–180.

---

[2] The boxes in Figs 1 and 2 are grey rather than orange because colour representation is not possible in this book.

[3] Famously, Pete Madden coloured his wave-fronts red for a conference talk. The underlying expressions were made invisible, ruining his presentation.

[Benzmüller *et al*, 1997] Benzmüller, C., Cheikhrouhou, L., Fehrer, D., Fiedler, A., Huang, X., Kerber, M., Kohlhase, K., Meier, A, Melis, E., Schaarschmidt, W., Siekmann, J. and Sorge, V. (1997). *Ω*mega: Towards a mathematical assistant. In McCune, W., (ed.), *14th International Conference on Automated Deduction*, pages 252–255. Springer-Verlag.

[Boulton *et al*, 1998] Boulton, R., Slind, K., Bundy, A. and Gordon, M. (September/October 1998). An interface between CLAM and HOL. In Grundy, J. and Newey, M., (eds.), *Proceedings of the 11th International Conference on Theorem Proving in Higher Order Logics (TPHOLs'98)*, volume 1479 of *Lecture Notes in Computer Science*, pages 87–104, Canberra, Australia. Springer.

[Bundy & Green, 1996] Bundy, Alan and Green, Ian. (December 1996). An experimental comparison of rippling and exhaustive rewriting. Research paper 836, Dept. of Artificial Intelligence, University of Edinburgh.

[Bundy & Lombart, 1995] Bundy, A. and Lombart, V. (1995). Relational rippling: a general approach. In Mellish, C., (ed.), *Proceedings of IJCAI-95*, pages 175–181. IJCAI.

[Bundy, 1988] Bundy, A. (1988). The use of explicit plans to guide inductive proofs. In Lusk, R. and Overbeek, R., (eds.), *9th International Conference on Automated Deduction*, pages 111–120. Springer-Verlag. Longer version available from Edinburgh as DAI Research Paper No. 349.

[Bundy, 1991] Bundy, Alan. (1991). A science of reasoning. In Lassez, J.-L. and Plotkin, G., (eds.), *Computational Logic: Essays in Honor of Alan Robinson*, pages 178–198. MIT Press. Also available from Edinburgh as DAI Research Paper 445.

[Bundy, 2001] Bundy, Alan. (2001). The automation of proof by mathematical induction. In Robinson, A. and Voronkov, A., (eds.), *Handbook of Automated Reasoning, Volume 1*. Elsevier.

[Bundy, 2002] Bundy, A. (2002). *A Critique of Proof Planning*, pages 160–177. Springer.

[Bundy *et al*, 1989] Bundy, A., van Harmelen, F., Hesketh, J., Smaill, A. and Stevens, A. (1989). A rational reconstruction and extension of recursion analysis. In Sridharan, N. S., (ed.), *Proceedings of the Eleventh International Joint Conference on Artificial Intelligence*, pages 359–365. Morgan Kaufmann. Also available from Edinburgh as DAI Research Paper 419.

[Bundy *et al*, 1990] Bundy, A., van Harmelen, F., Horn, C. and Smaill, A. (1990). The Oyster-Clam system. In Stickel, M. E., (ed.), *10th International Conference on Automated Deduction*, pages 647–648. Springer-Verlag. Lecture Notes in Artificial Intelligence No. 449. Also available from Edinburgh as DAI Research Paper 507.

[Bundy *et al*, 1993] Bundy, A., Stevens, A., van Harmelen, F., Ireland, A. and Smaill, A. (1993). Rippling: A heuristic for guiding inductive proofs. *Artificial Intelligence*, 62:185–253. Also available from Edinburgh as DAI Research Paper No. 567.

[Bundy *et al*, 2005] Bundy, A., Basin, D., Hutter, D. and Ireland, A. (2005). *Rippling: Meta-level Guidance for Mathematical Reasoning*. Cambridge University Press.

[Dennis *et al*, 2000] Dennis, L., Bundy, A. and Green, I. (2000). Making a productive use of failure to generate witness for coinduction from divergent proof attempts. *Annals of Mathematics and Artificial Intelligence*, 29:99–138. Also available as paper No. RR0004 in the Informatics Report Series.

[Desimone, 1989] Desimone, R. V. (1989). Explanation-Based Learning of Proof Plans. In Kodratoff, Y. and Hutchinson, A., (eds.), *Machine and Human Learning*. Kogan Page. Also available as DAI Research Paper 304. Previous version in proceedings of EWSL-86.

[Dixon & Fleuriot, 2003] Dixon, L. and Fleuriot, J. D. (2003). IsaPlanner: A prototype proof planner in Isabelle. In *Proceedings of CADE'03*, Lecture Notes in Computer Science.

[Duncan et al, 2004] Duncan, H., Bundy, A., Levine, J., Storkey, A. and Pollet, M. (2004). The use of data-mining for the automatic formation of tactics. In *Workshop on Computer-Supported Mathematical Theory Development*. IJCAR-04.

[Frank & Basin, 1998] Frank, I. and Basin, D. (1998). Search in games with incomplete information: A case study using bridge card play. *Artificial Intelligence*, 100(1–2):87–123.

[Frank et al, 1992] Frank, I., Basin, D. and Bundy, A. (1992). An adaptation of proof-planning to declarer play in bridge. In *Proceedings of ECAI-92*, pages 72–76, Vienna, Austria. Longer Version available from Edinburgh as DAI Research Paper No. 575.

[Gow, 2004] Gow, Jeremy. (2004). *The Dynamic Creation of Induction Rules Using Proof Planning*. Unpublished Ph.D. thesis, Division of Informatics, University of Edinburgh.

[Hutter & Kohlhase, 1997] Hutter, D. and Kohlhase, M. (1997). A colored version of the λ-Calculus. In McCune, W., (ed.), *14th International Conference on Automated Deduction*, pages 291–305. Springer-Verlag. Also available as SEKI-Report SR-95-05.

[Ireland & Bundy, 1996] Ireland, A. and Bundy, A. (1996). Productive use of failure in inductive proof. *Journal of Automated Reasoning*, 16(1–2):79–111. Also available from Edinburgh as DAI Research Paper No 716.

[Ireland, 1992] Ireland, A. (1992). The Use of Planning Critics in Mechanizing Inductive Proofs. In Voronkov, A., (ed.), *International Conference on Logic Programming and Automated Reasoning – LPAR 92, St. Petersburg*, Lecture Notes in Artificial Intelligence No. 624, pages 178–189. Springer-Verlag. Also available from Edinburgh as DAI Research Paper 592.

[Jackson & Lowe, 2000] Jackson, M. and Lowe, H. (June 2000). XBarnacle: Making theorem provers more accessible. In McAllester, D., (ed.), *CADE17*, number 1831 in Lecture Notes in Computer Science, Pittsburg. Springer.

[Kraan et al, 1996] Kraan, I., Basin, D. and Bundy, A. (1996). Middle-out reasoning for synthesis and induction. *Journal of Automated Reasoning*, 16(1–2):113–145. Also available from Edinburgh as DAI Research Paper 729.

[Lowe & Duncan, 1997] Lowe, H. and Duncan, D. (1997). XBarnacle: Making theorem provers more accessible. In McCune, William, (ed.), *14th International Conference on Automated Deduction*, pages 404–408. Springer-Verlag.

[Lowe et al, 1998] Lowe, H., Pechoucek, M. and Bundy, A. (1998). Proof planning for maintainable configuration systems. *Artificial Intelligence in Engineering Design, Analysis and Manufacturing*, 12:345–356. Special issue on configuration.

[Monroy et al, 1994] Monroy, R., Bundy, A. and Ireland, A. (1994). Proof Plans for the Correction of False Conjectures. In Pfenning, F., (ed.), *5th International Conference on Logic Programming and Automated Reasoning, LPAR'94*, Lecture Notes in Artificial Intelligence, v. 822, pages 54–68, Kiev, Ukraine. Springer-Verlag. Also available from Edinburgh as DAI Research Paper 681.

[Silver, 1985] Silver, B. (1985). *Meta-level inference: Representing and Learning Control Information in Artificial Intelligence*. North Holland, Revised version of the author's PhD thesis, Department of Artificial Intelligence, U. of Edinburgh, 1984.

[Smaill & Green, 1996] Smaill, Alan and Green, Ian. (1996). Higher-order annotated terms for proof search. In von Wright, Joakim, Grundy, Jim and Harrison, John, (eds.), *Theorem Proving in Higher Order Logics: 9th International Conference, TPHOLs'96*, volume 1275 of *Lecture Notes in Computer Science*, pages 399–414, Turku, Finland. Springer-Verlag. Also available as DAI Research Paper 799.

[Stark & Ireland, 1998] Stark, J. and Ireland, A. (1998). Invariant discovery via failed proof attempts. In Flener, P., (ed.), *Logic-based Program Synthesis and Transformation*, number 1559 in LNCS, pages 271–288. Springer-Verlag.

[Sterling *et al*, 1989] Sterling, L., Bundy, Alan, Byrd, L., O'Keefe, R. and Silver, B. (1989). Solving symbolic equations with PRESS. *J. Symbolic Computation*, 7:71–84. Also available from Edinburgh as DAI Research Paper 171.

[Willmott *et al*, 2001] Willmott, S., Richardson, J. D. C., Bundy, A. and Levine, J. M. (2001). Applying adversarial planning techniques to Go. *Journal of Theoretical Computer Science*, 252(1-2):45–82. Special issue on algorithms, Automata, complexity and Games.

A Paraconsistent Higher Order Logic[*]

Jørgen Villadsen

Computer Science, Roskilde University
Building 42.1, DK-4000 Roskilde, Denmark
jv@ruc.dk

Abstract. Classical logic predicts that everything (thus nothing useful at all) follows from inconsistency. A paraconsistent logic is a logic where an inconsistency does not lead to such an explosion, and since in practice consistency is difficult to achieve there are many potential applications of paraconsistent logics in knowledge-based systems, logical semantics of natural language, etc.
Higher order logics have the advantages of being expressive and with several automated theorem provers available. Also the type system can be helpful.
We present a concise description of a paraconsistent higher order logic with countably infinite indeterminacy, where each basic formula can get its own indeterminate truth value. The meaning of the logical operators is new and rather different from traditional many-valued logics as well as from logics based on bilattices. Thus we try to build a bridge between the communities of higher order logic and many-valued logic.
A case study is studied and a sequent calculus is proposed based on recent work by Muskens.

Many non-classical logics are, at the propositional level, funny toys which work quite good, but when one wants to extend them to higher levels to get a real logic that would enable one to do mathematics or other more sophisticated reasonings, sometimes dramatic troubles appear.
J.-Y. Béziau: *The Future of Paraconsistent Logic*
Logical Studies Online Journal 2 (1999) p. 7

A preliminary version appeared in the informal proceedings of the workshop on *Paraconsistent Computational Logic* PCL 2002 (editors Hendrik Decker, Jørgen Villadsen, Toshiharu Waragai) http://www.ruc.dk/~jv/pcl.pdf

1 Introduction

Classical logic predicts that everything (thus nothing useful at all) follows from inconsistency. A paraconsistent logic is a logic where an inconsistency does not lead to such an explosion.

In a paraconsistent logic the meaning of some of the logical operators must be different from classical logic in order to block the explosion, and since there

---

[*] This research was partly sponsored by the IT University of Copenhagen.

B. Buchberger and J.A. Campbell (Eds.): AISC 2004, LNAI 3249, pp. 38–51, 2004.
© Springer-Verlag Berlin Heidelberg 2004

are many ways to change the meaning of these operators there are many different paraconsistent logics. We present a paraconsistent higher order logic ∇ based on the (simply) typed λ-calculus [10, 4]. Although it is a generalization of Łukasiewicz's three-valued logic the meaning of the logical operators is new. The results extend the three-valued and four-valued logics first presented in [19, 20] to infinite-valued logics.

One advantage of a higher order logic is that the logic is very expressive in the sense that most mathematical structures, functions and relations are available (for instance arithmetic). Another advantage is that there are several automated theorem provers for classical higher order logic, e.g. TPS [5], LEO [9], HOL, PVS, IMPS, and Isabelle (see [23] for further references). It should be possible to modify some of these to our paraconsistent logic; in particular the generic theorem prover Isabelle [18] already implements several object logics.

We are inspired by the notion of indeterminacy as discussed by Evans [11]. Even though the higher order logic ∇ is paraconsistent some of its extensions, like ∇_Δ, are classical. We reuse the symbols ∇ and Δ later for related purposes.

We also propose a sequent calculus for the paraconsistent higher order logic ∇ based on the seminal work by Muskens [17]. In the sequent $\Theta \vdash \Gamma$ we understand Θ as a conjunction of a set of formulas and Γ as a disjunction of a set of formulas. We use $\Theta \Vdash \Gamma$ as a shorthand for $\Theta, \omega \vdash \Gamma$, where ω is an axiom which provides countably infinite indeterminacy such that each basic formula can get its own indeterminate truth value (or as we prefer: truth code).

As mentioned above higher order logic includes much, if not all, of ordinary mathematics, and even though ∇ is paraconsistent we can use it for classical mathematics by keeping the truth values determinate. Hence we shall not consider here paraconsistent mathematics. Using the standard foundation of mathematics (axiomatic set theory) it is possible to show that ∇ is consistent.

The essential point is that the higher-order issues and many-valued issues complement each other in the present framework:

- On the one hand we can view ∇ as a paraconsistent many-valued extension of classical higher order logic.

- On the other hand we can view ∇ as a paraconsistent many-valued propositional logic with features from classical higher order logic.

First we introduce a case study and motivate our definitions of the logical operators. Then we describe the syntax and semantics of the typed λ-calculus and introduce the primitives of the paraconsistent higher order logic ∇, in particular the modality and implications available. Finally we present a sequent calculus for ∇ and the extensions ∇_ω, ∇_Δ, $\nabla_\dagger$ and $\nabla_\ddagger$.

2   A Case Study

Higher order logic is not really needed for the case study – propositional logic is enough – but the purpose of the case study is mainly to illustrate the working of the paraconsistency. The features of first order logic and higher order logic, including mathematical concepts, seem necessary in general.

Imagine a system that collects information on the internet – say, from news media – about the following scenario: Agent X thinks that his supervisor hides a secret that the public ought to know about. For simplicity we ignore all temporal aspects and take all reports to be in present tense. We also do not take into account a more elaborate treatment of conditionals / counterfactuals.

Assume that the following pieces of information are available:

#123   If X leaks the secret then he keeps his integrity.

#456   If X does not leak the secret then he keeps his job.

#789   If X does not keep his job then he does not keep his integrity.

#1000   X does not keep his job.

The numbers indicate that the information is collected over time and perhaps from various sources.

Classically the information available to the system is inconsistent. This is not entirely obvious, especially not when numerous other pieces of information are also available to the system and when the system is operating under time constraints. Note that the information consists of both simple facts (#1000) as well as rules (#123, #456, #789). A straightforward formalization is as follows:

$$L \to I \qquad \theta_1$$
$$\neg L \to J \qquad \theta_2$$
$$\neg J \to \neg I \qquad \theta_3$$
$$\neg J \qquad \theta_0$$

Here the propositional symbol L means that X leaks the secret, I that X keeps his integrity, and J that X keeps his job. As usual $\to$ is implication, $\wedge$ is conjunction, and $\neg$ is negation.

If we use classical logic on the formulas $\theta_0, \theta_1, \theta_2, \theta_3$ the system can conclude $L, \neg L, I, \neg I, J, \neg J$, and whatever other fact or rule considered. Of course we might try to revise the information $\theta_0, \theta_1, \theta_2, \theta_3$, but it is not immediately clear what would be appropriate. In the present paper we propose to use a paraconsistent logic ∇ such that the system can conclude only $\neg J$ or θ_0, which is reasonable (the logic ∇ is monotonic and any formula φ entails itself).

The paraconsistent logic ∇ is an extension of classical logic in the sense that classical reasoning is easily possible, just add the special formulas ΔL, ΔI, and ΔJ to the formulas $\theta_0, \theta_1, \theta_2, \theta_3$ and L, I, and J behave classically.

We now turn to the motivation of the logical operators, which are to be defined using so-called key equalities. We return to the case study in section 6.

3   Overall Motivation

Classical logic has two truth values, namely $\bullet$ and $\circ$ (truth and falsehood), and the designated truth value $\bullet$ yields the logical truths. We use the symbol $\top$ for the truth value $\bullet$ and $\bot$ for $\circ$ (later these symbols are seen as abbreviations for specific formulas).

But classical logic cannot handle inconsistency since an explosion occurs. In order to handle inconsistency we allow additional truth values and the first question is:

1. How many additional values do we need?

It seems reasonable to consider countably infinitely many additional truth values – one for each proper constant we might introduce in the theory for the knowledge base. Each proper constant (a proposition, a property or a relation) can be inconsistent "independently" of other proper constants. We are inspired by the notion of indeterminacy as discussed by Evans [11]. Hence in addition to the determinate truth values $\Delta = \{\bullet, \circ\}$ we also consider the indeterminate truth values $\nabla = \{I, II, III, \ldots\}$ to be used in case of inconsistencies. We refer to the determinate and indeterminate truth values $\Delta \cup \nabla$ as the truth codes. We can then use, say, $(\Delta \cup \nabla) \setminus \{\bullet\}$ as substitutes for the natural numbers $\omega = \{0, 1, 2, 3, \ldots\}$.

The second question is:

2. How are we going to define the connectives?

One way to proceed is as follows. First we want De Morgan laws to holds; hence $\varphi \lor \psi \equiv \neg(\neg\varphi \land \neg\psi)$. For implication we have the classically acceptable $\varphi \to \psi \equiv \varphi \leftrightarrow \varphi \land \psi$. For negation we propose to map $\bullet$ to $\circ$ and vice versa, leaving the other values unchanged (after all, we want the double negation law $\varphi \leftrightarrow \neg\neg\varphi$ to hold for all formulas φ). For conjunction we want the idempotent law to hold and $\bullet$ should to be neutral, and $\circ$ is the default result. For biimplication we want reflexivity and $\bullet$ should to be neutral, $\circ$ should be negation, and again $\circ$ is the default result. The universal quantification is defined using the same principles as a kind of generalized conjunction and the existential quantification follows from a generalized De Morgan law.

We do not consider a separate notion of entailment – we simply say that φ entails ψ iff $\varphi \to \psi$ holds. While it is true that $\varphi \land \neg\varphi$ does not entail arbitrary ψ we do have that $\neg\varphi$ entails $\varphi \to \psi$, hence we do not have a relevant logic [1] in general (but only for so-called first degree entailment). Our logic validates clear "fallacies of relevance" like the one just noted, or like the inference from φ to $\psi \to \psi$, but these do not seem problematic for the applications discussed above.

Our logic is a generalization of Lukasiewicz's three-valued logic (originally proposed 1920–30), with the intermediate value duplicated many times and ordered such that none of the copies of this value imply other ones, but it differs from Lukasiewicz's many-valued logics as well as from logics based on bilattices [8, 12, 6, 13].

4   Conjunction, Disjunction, and Negation

The motivation for our logical operators is to be found in the key equalities shown to the right of the following semantic clauses (the basic semantic clause and the clause $[\![\top]\!] = \bullet$ are omitted; further clauses are discussed later). Also $\varphi \Leftrightarrow \neg\neg\varphi$ is considered to be a key equality as well.

$$[\![\neg\varphi]\!] = \begin{cases} \bullet & \text{if } [\![\varphi]\!] = \circ \\ \circ & \text{if } [\![\varphi]\!] = \bullet \\ [\![\varphi]\!] & \text{otherwise} \end{cases} \qquad \begin{array}{l} \top \Leftrightarrow \neg\bot \\ \bot \Leftrightarrow \neg\top \end{array}$$

$$[\![\varphi \wedge \psi]\!] = \begin{cases} [\![\varphi]\!] & \text{if } [\![\varphi]\!] = [\![\psi]\!] \\ [\![\psi]\!] & \text{if } [\![\varphi]\!] = \bullet \\ [\![\varphi]\!] & \text{if } [\![\psi]\!] = \bullet \\ \circ & \text{otherwise} \end{cases} \qquad \begin{array}{l} \varphi \Leftrightarrow \varphi \wedge \varphi \\ \psi \Leftrightarrow \top \wedge \psi \\ \varphi \Leftrightarrow \varphi \wedge \top \end{array}$$

In the semantic clauses several cases may apply if and only if they agree on the result. The semantic clauses work for classical logic and also for our logic.

We have the following standard abbreviations:

$$\bot \equiv \neg\top \qquad \varphi \vee \psi \equiv \neg(\neg\varphi \wedge \neg\psi) \qquad \exists v.\varphi \equiv \neg\forall v.\neg\varphi$$

The universal quantification $\forall v.\varphi$ will be introduced later (as a kind of generalized conjunction). A suitable abbreviation for $\top$ is also provided later.

As explained we have an infinite number of truth values (truth codes) in general, but the special cases of three-valued and four-valued logics are interesting too. In order to investigate finite truth tables we first add just $[\![\dagger]\!] = \mathsf{I}$ as an indeterminacy. We do not have $\varphi \vee \neg\varphi$. Unfortunately we do have that $\varphi \wedge \neg\varphi$ entails $\psi \vee \neg\psi$ (try with $\bullet$, $\circ$ and I using the truth tables and use the fact that any φ entails itself). The reason for this problem is that in a sense there is not only a single indeterminacy, but a unique one for each basic formula.

However, in many situations only two indeterminacies are ever needed, corresponding to the left and right hand side of the implication. Hence we add $[\![\ddagger]\!] = \mathsf{II}$ as the alternative indeterminacy.

$\wedge$	$\bullet$	$\circ$	I	II		$\vee$	$\bullet$	$\circ$	I	II		$\neg$	
$\bullet$	$\bullet$	$\circ$	I	II		$\bullet$	$\bullet$	$\bullet$	$\bullet$	$\bullet$		$\bullet$	$\circ$
$\circ$	$\circ$	$\circ$	$\circ$	$\circ$		$\circ$	$\bullet$	$\circ$	I	II		$\circ$	$\bullet$
I	I	$\circ$	I	$\circ$		I	$\bullet$	I	I	$\bullet$		I	I
II	II	$\circ$	$\circ$	II		II	$\bullet$	II	$\bullet$	II		II	II

Keep in mind that truth tables are never taken as basic – they are simply calculated from the semantic clauses taking into account also the abbreviations.

5 Implication, Biimplication, and Modality

As for conjunction and negation the motivation for the biimplication operator $\leftrightarrow$ (and the implication operator $\rightarrow$ as defined later) is based on the few key equalities shown to the right of the following semantic clause.

$$\llbracket \varphi \leftrightarrow \psi \rrbracket \;=\; \begin{cases} \bullet & \text{if } \llbracket \varphi \rrbracket = \llbracket \psi \rrbracket \\ \llbracket \psi \rrbracket & \text{if } \llbracket \varphi \rrbracket = \bullet \\ \llbracket \varphi \rrbracket & \text{if } \llbracket \psi \rrbracket = \bullet \\ \llbracket \neg\psi \rrbracket & \text{if } \llbracket \varphi \rrbracket = \circ \\ \llbracket \neg\varphi \rrbracket & \text{if } \llbracket \psi \rrbracket = \circ \\ \circ & \text{otherwise} \end{cases} \qquad \begin{aligned} \top &\Leftrightarrow \varphi \leftrightarrow \varphi \\ \psi &\Leftrightarrow \top \leftrightarrow \psi \\ \varphi &\Leftrightarrow \varphi \leftrightarrow \top \\ \neg\psi &\Leftrightarrow \bot \leftrightarrow \psi \\ \neg\varphi &\Leftrightarrow \varphi \leftrightarrow \bot \end{aligned}$$

As before, several cases may apply if and only if they agree on the result and the semantic clause works for classical logic too.

The semantic clause is an extension of the clause for equality $=$:

$$\llbracket \varphi = \psi \rrbracket \;=\; \begin{cases} \bullet \text{ if } \llbracket \varphi \rrbracket = \llbracket \psi \rrbracket \\ \circ \text{ otherwise} \end{cases}$$

We have the following abbreviations:

$$\varphi \Leftrightarrow \psi \;\equiv\; \varphi = \psi \qquad\qquad \varphi \Rightarrow \psi \;\equiv\; \varphi \Leftrightarrow \varphi \wedge \psi \qquad\qquad \Box\varphi \;\equiv\; \varphi = \top$$

$$\varphi \to \psi \;\equiv\; \varphi \leftrightarrow \varphi \wedge \psi \qquad\qquad \sim\!\varphi \;\equiv\; \neg\Box\varphi$$

The logical necessity operator $\Box$ is a so-called S5 modality.

We could also have used $(\varphi \Rightarrow \psi) \wedge (\psi \Rightarrow \varphi)$ for $\varphi \Leftrightarrow \psi$ (using $=$ instead of $\Leftrightarrow$ in the definition of $\Rightarrow$). Besides, $\Leftrightarrow$ binds very loosely, even more loosely than $\leftrightarrow$ does, and $=$ binds tightly. The binding priority is the only difference between $=$ and $\Leftrightarrow$ for formulas, but $\Leftrightarrow$ and $\leftrightarrow$ are quite different as can be seen from the truth tables.

$\Leftrightarrow$	$\bullet$	$\circ$	I	II
$\bullet$	$\bullet$	$\circ$	$\circ$	$\circ$
$\circ$	$\circ$	$\bullet$	$\circ$	$\circ$
I	$\circ$	$\circ$	$\bullet$	$\circ$
II	$\circ$	$\circ$	$\circ$	$\bullet$

$\Rightarrow$	$\bullet$	$\circ$	I	II
$\bullet$	$\bullet$	$\circ$	$\circ$	$\circ$
$\circ$	$\bullet$	$\bullet$	$\bullet$	$\bullet$
I	$\bullet$	$\circ$	$\bullet$	$\circ$
II	$\bullet$	$\circ$	$\circ$	$\bullet$

$\Box$	
$\bullet$	$\bullet$
$\circ$	$\circ$
I	$\circ$
II	$\circ$

$\leftrightarrow$	$\bullet$	$\circ$	I	II
$\bullet$	$\bullet$	$\circ$	I	II
$\circ$	$\circ$	$\bullet$	I	II
I	I	I	$\bullet$	$\circ$
II	II	II	$\circ$	$\bullet$

$\to$	$\bullet$	$\circ$	I	II
$\bullet$	$\bullet$	$\circ$	I	II
$\circ$	$\bullet$	$\bullet$	$\bullet$	$\bullet$
I	$\bullet$	I	$\bullet$	I
II	$\bullet$	II	II	$\bullet$

$\sim$	
$\bullet$	$\circ$
$\circ$	$\bullet$
I	$\bullet$
II	$\bullet$

We could try the following standard abbreviations:

$$\varphi \overset{?}{\to} \psi \;\equiv\; \neg\varphi \vee \psi \qquad\qquad \varphi \overset{?}{\leftrightarrow} \psi \;\equiv\; (\varphi \overset{?}{\to} \psi) \wedge (\psi \overset{?}{\to} \varphi)$$

But here we have neither $\varphi \overset{?}{\to} \varphi$ nor $\varphi \overset{?}{\leftrightarrow} \varphi$, since the diagonals differ from $\bullet$ at indeterminacies.

$$
\begin{array}{c|cccc}
\overset{?}{\leftrightarrow} & \bullet & \circ & | & || \\
\hline
\bullet & \bullet & \circ & | & || \\
\circ & \circ & \bullet & | & || \\
| & | & | & | & \bullet \\
|| & || & || & \bullet & ||
\end{array}
\qquad
\begin{array}{c|cccc}
\overset{?}{\to} & \bullet & \circ & | & || \\
\hline
\bullet & \bullet & \circ & | & || \\
\circ & \bullet & \bullet & \bullet & \bullet \\
| & \bullet & | & | & \bullet \\
|| & \bullet & || & \bullet & ||
\end{array}
\qquad
\begin{array}{c|cccc}
\leftrightsquigarrow & \bullet & \circ & | & || \\
\hline
\bullet & \bullet & \circ & | & || \\
\circ & \circ & \bullet & \bullet & \bullet \\
| & | & \bullet & \bullet & \bullet \\
|| & || & \bullet & \bullet & \bullet
\end{array}
\qquad
\begin{array}{c|cccc}
\rightsquigarrow & \bullet & \circ & | & || \\
\hline
\bullet & \bullet & \circ & | & || \\
\circ & \bullet & \bullet & \bullet & \bullet \\
| & \bullet & \bullet & \bullet & \bullet \\
|| & \bullet & \bullet & \bullet & \bullet
\end{array}
$$

We instead use the following abbreviations:

$$\varphi \rightsquigarrow \psi \equiv {\sim}\varphi \vee \psi \qquad\qquad \varphi \leftrightsquigarrow \psi \equiv (\varphi \rightsquigarrow \psi) \wedge (\psi \rightsquigarrow \varphi)$$

Although $\varphi \rightsquigarrow \psi$ does not entail $\neg\psi \rightsquigarrow \neg\varphi$, we do have $\varphi \leftrightsquigarrow \varphi$ and $\varphi \rightsquigarrow \varphi$, and this implication is very useful as we shall see in a moment.

We also use the predicate Δ for determinacy and ∇ for indeterminacy with the abbreviations (note that Δ and ∇ are used for predicates and for sets of truth codes):

$$\Delta\varphi \equiv \Box(\varphi \vee \neg\varphi) \qquad\qquad \nabla\varphi \equiv \neg\Delta\varphi$$

$$
\begin{array}{cc}
\nabla & \quad\quad \Delta \\
\end{array}
$$

$$
\begin{array}{cc|cc}
\bullet & \circ & \bullet & \bullet \\
\circ & \circ & \circ & \bullet \\
| & \bullet & | & \circ \\
|| & \bullet & || & \circ
\end{array}
$$

We now come to the central abbreviations based directly on the semantic clause above:

$$
\begin{aligned}
\varphi \leftrightarrow \psi \equiv\ & (\varphi = \psi \rightsquigarrow \top) \wedge \\
& (\varphi \rightsquigarrow \psi) \wedge \\
& (\psi \rightsquigarrow \varphi) \wedge \\
& (\neg\varphi \rightsquigarrow \neg\psi) \wedge \\
& (\neg\psi \rightsquigarrow \neg\varphi) \wedge \\
& (\neg(\varphi = \psi) \wedge \nabla\varphi \wedge \nabla\psi \rightsquigarrow \bot)
\end{aligned}
$$

We could also use $(\varphi \leftrightsquigarrow \psi) \wedge (\neg\varphi \leftrightsquigarrow \neg\psi) \wedge (\varphi = \psi \vee \Delta\varphi \vee \Delta\psi)$ for $\varphi \leftrightarrow \psi$.

6   A Case Study – Continued

Recall that classical logic explodes in the presence of the formulas $\theta_0, \theta_1, \theta_2, \theta_3$ since $\theta_0, \theta_1, \theta_2, \theta_3$ entails any formula φ. In ∇ we have several counter-examples as follows.

The reason why we do not have $(\theta_0 \wedge \theta_1 \wedge \theta_2 \wedge \theta_3) \to J$ is that $[\![L]\!] = \bullet$, $[\![I]\!] = |$, $[\![J]\!] = \circ$ is a counter-example. This can be seen from the truth tables – the result is $|$ which is not designated. The same counter-example also shows that the system cannot conclude $\neg L$ since $[\![\neg L]\!] = \circ = [\![J]\!]$ and it cannot conclude J as just explained.

The system cannot conclude L (take $[\![L]\!] = \circ$, $[\![I]\!] = \circ$, $[\![J]\!] = \mathsf{I}$ as a counter-example) and neither I nor $\neg I$ (take $[\![L]\!] = \mathsf{I}$, $[\![I]\!] = \mathsf{II}$, $[\![J]\!] = \mathsf{I}$ in both cases).

There is quite some flexibility with respect to the formalization – as an example we consider changing θ_0 to $\Box\neg J$ where the logical necessity operator $\Box$ expresses that the fact is "really" true. Now the previous counter-example to L is no good and since it can be shown that there is no other counter-example, the system concludes L (it also concludes $\neg I$, and of course still $\neg J$). Recall that classical logic is useless in this case study. The previous counter-example to $\neg L$ and J is ok, and there is a counter-example $[\![L]\!] = \mathsf{I}$, $[\![I]\!] = \circ$, $[\![J]\!] = \circ$ to I.

We think that the case study shows that the logic ∇ is inconsistency-tolerant in an interesting way, but of course further investigations are needed to clarify the potential.

7   A Sequent Calculus

We base the paraconsistent higher order logic ∇ on the (simply) typed λ-calculus [10] (see also [7], especially for the untyped λ-calculus and for the notion of combinators which we use later).

Classical higher-order logic is often built from a very few primitives, say equality $=$ and the selection operator $\imath$ as in Q_0 [4], but it does not seem like we can avoid taking, say, negation, conjunction and universal quantification as primitives for ∇. Also we prefer to extend the selection operator $\imath$ to the (global) choice operator ε described later.

We use the following well-known abbreviations in order to replace negation and conjunction by joint denial (also known as Sheffer's stroke):

$$\neg\varphi \equiv \varphi|\varphi \qquad \varphi \wedge \psi \equiv \neg(\varphi|\psi)$$

We also have a so-called indeterminacy generation operator ∂ as a primitive. We use the following abbreviations:

$$\overline{\varphi} \equiv \neg\varphi \qquad \dot{\varphi} \equiv \partial\varphi \qquad \ddot{\varphi} \equiv \partial\dot{\varphi} \qquad \ldots$$

The indeterminacy generation operator is injective and we can use it for the natural numbers. We say much more about it later.

The truth tables in case of four truth values are the following.

$\mid$	$\bullet$	$\circ$	I	II
$\bullet$	$\circ$	$\bullet$	I	II
$\circ$	$\bullet$	$\bullet$	$\bullet$	$\bullet$
I	I	$\bullet$	I	$\bullet$
II	II	$\bullet$	$\bullet$	II

$=$	$\bullet$	$\circ$	I	II
$\bullet$	$\bullet$	$\circ$	$\circ$	$\circ$
$\circ$	$\circ$	$\bullet$	$\circ$	$\circ$
I	$\circ$	$\circ$	$\bullet$	$\circ$
II	$\circ$	$\circ$	$\circ$	$\bullet$

∂	
$\bullet$	$\bullet$
$\circ$	I
I	II
II	$\circ$

The truth table only displays equality $=$ between formulas (the biimplication operator $\Leftrightarrow$), but it is applicable to any type. We have the abbreviation:

$$\top \equiv (\lambda x.x) = (\lambda x.x)$$

Here $\lambda x.x$ is the identity function in the λ-calculus (any type for x will do).

7.1 Syntax

We define the following sets of types and terms – the latter for each type τ and in the case of the λ abstraction we additionally require that τ is of the form $\alpha\beta$.

$$\mathcal{T} = o \mid \mathcal{TT} \mid \mathcal{S} \qquad \mathcal{L}_\tau = \mathcal{L}_{\gamma\tau}\mathcal{L}_\gamma \mid \lambda\mathcal{V}_\alpha.\mathcal{L}_\beta \mid \mathcal{C}_\tau \mid \mathcal{V}_\tau$$

Here $\mathcal{S}$ is the set of sorts (empty in the propositional case, where the only basic type is o for formulas), $\mathcal{C}_\tau$ and $\mathcal{V}_\tau$ are the sets of term constants and variables of type τ (the set of variables must be countably infinite), and $\alpha, \beta, \gamma, \tau \in \mathcal{T}$.

We often write $\tau_1 \ldots \tau_m\gamma$ instead of the type $\tau_1(\ldots(\tau_m\gamma))$ and $\varphi\psi_1 \ldots \psi_n$ instead of the term $((\varphi\psi_1)\ldots)\psi_n$. Note that the relational types are $\tau_1 \ldots \tau_n o$ (also called predicates).

If we add a sort of individuals ι to the propositional higher order logic ∇ we obtain the higher order logic ∇^ι (further sorts can be added, but for our purposes they are not needed).

7.2 Semantics

As usual Y^X is the set of functions from X to Y.

A universe U is an indexed set of type universes $U_\tau \neq \emptyset$ such that $U_{\alpha\beta} \subseteq U_\beta^{U_\alpha}$. The universe is full if $\subseteq$ is replaced by $=$.

A basic interpretation I on a universe U is a function $I\colon \bigcup\mathcal{C}_\tau \to \bigcup U_\tau$ such that $I\kappa_\tau \in U_\tau$ for $\kappa_\tau \in \mathcal{C}_\tau$. Analogously, an assignment A on a universe U is a function $A\colon \bigcup\mathcal{V}_\tau \to \bigcup U_\tau$ such that $Av_\tau \in U_\tau$ for $v_\tau \in \mathcal{V}_\tau$.

A model $M \equiv \langle U, I \rangle$ consists of a basic interpretation I on a universe U such that for all assignments A on the universe U the interpretation $\llbracket \cdot \rrbracket^{M,A}\colon \bigcup\mathcal{L}_\tau \to \bigcup U_\tau$ has $\llbracket \varphi_\tau \rrbracket^{M,A} \in U_\tau$ for all terms $\varphi_\tau \in \mathcal{L}_\tau$, where (we use the λ-calculus in the meta-language as well):

$$\llbracket \kappa \rrbracket = I\kappa$$
$$\llbracket v \rrbracket = Av$$
$$\llbracket \lambda v_\alpha.\,\varphi_\beta \rrbracket = \lambda u.\,\llbracket \varphi \rrbracket^{A[v \mapsto u]}$$
$$\llbracket \varphi_{\gamma\tau}\psi_\gamma \rrbracket = \llbracket \varphi \rrbracket \llbracket \psi \rrbracket$$

For clarity we omit some types and parameters.

What we call just a model is also known as a general model, and a full model is then a standard model. An arbitrary basic interpretation on a universe is sometimes considered a very general model.

7.3 Primitives

We use five primitive combinators of the following types ($\tau \in \mathcal{T}$):

D	ooo	Joint denial – Sheffer's stroke
Q	$\tau\tau o$	Equality
A	$(\tau o)o$	Universal quantification
C	$(\tau o)\tau$	Global choice
V	oo	Indeterminacy generation

We have the following abbreviations (we omit the types):

$$\varphi|\psi \equiv \mathsf{D}\varphi\psi \qquad \varphi = \psi \equiv \mathsf{Q}\varphi\psi \qquad \check{\varphi} \equiv \mathsf{Q}\varphi$$

$$\forall v.\varphi \equiv \mathsf{A}\,\lambda v.\varphi \qquad \tilde{\varphi} \equiv \mathsf{C}\varphi \qquad \varepsilon v.\varphi \equiv \mathsf{C}\,\lambda v.\varphi \qquad \partial\varphi \equiv \mathsf{V}\varphi$$

Only a few of these abbreviations need explanation. The (global) choice operator ε chooses *some* value v for which φ is satisfied (v can be free in φ); if no such value exists then an arbitrary value is chosen (of the right type). The choice is global in the sense that all choices are the same for equivalent φ's, hence for instance we have $(\varepsilon x.\bot) = (\varepsilon x.\bot)$.

The notation $\check{\varphi}$ turns φ into a singleton set with itself as the sole member and $\tilde{\varphi}$ is its inverse, since $\tilde{\check{\varphi}} = \varphi$, which is called the selection property, cf. the selection operator $\imath$ in Q_0 [4]. But $\tilde{\varphi}$ is of course also defined for non-singleton sets, namely as the (global) choice operator just described. We say a little more about these matters when we come to the choice rules.

We can even eliminate the λ-notation if we use two additional primitive combinators, the so-called S and K combinators of suitable types. For example, the identity function $\lambda x.x$ is available as the abbreviation $\mathsf{I} \equiv \mathsf{SKK}$, cf. [7].

7.4 Structural Rules

In the sequent $\Theta \vdash \Gamma$ we understand Θ as a conjunction of a set of formulas and Γ as a disjunction of a set of formulas, and we have the usual rules for a monotonic sequent calculus:

$$\frac{\Theta,\varphi \vdash \Gamma \quad \Theta \vdash \varphi,\Gamma}{\Theta \vdash \Gamma}\ \text{Cut} \qquad \frac{\Theta \vdash \Gamma}{\Theta,\varphi \vdash \Gamma} \qquad \frac{\Theta \vdash \Gamma}{\Theta \vdash \varphi,\Gamma}$$

Notice that $\varphi \vdash \varphi$ follows from these rules and the rules for equality below.

7.5 Fundamental Rules

We use the abbreviation:

$$\varphi \overset{\forall}{=} \psi \equiv (\lambda pq.\, \forall x.\, px = qx)\varphi\psi$$

We have the usual conversion and extensionality axioms of the λ-calculus:

$$(\lambda v.\varphi)\,\psi = \varphi[\psi/v] \qquad \varphi \overset{\forall}{=} \psi \vdash \varphi = \psi$$

Here $\varphi[\psi/v]$ means the substitution of ψ for the variable v in φ (the notation presupposes that ψ is substitutable for v in φ). For later use we note that if the notation for an arbitrary so-called eigen-variable π is used in place of ψ then it must not occur free in other formulas in the given axiom/rule. Also $\varphi[\psi]$ means $\varphi[\psi/v]$ for an arbitrary variable v with respect to the given axioms/rule.

We have the usual reflexivity and substitution axioms for equality:

$$\varphi = \varphi \qquad \varphi = \psi,\ \theta[\varphi] \vdash \theta[\psi]$$

7.6 Logical Rules

Let $\overline{\Theta} = \{\overline{\theta} \mid \theta \in \Theta\}$. Negation is different from classical logic. We follow [17] and add only the following rules:

$$\frac{\overline{\Gamma} \vdash \Theta}{\overline{\Theta} \vdash \Gamma} \qquad \frac{\Gamma \vdash \overline{\Theta}}{\Theta \vdash \overline{\Gamma}}$$

Conjunction and universal quantification are straightforward:

$$\varphi, \psi \vdash \varphi \wedge \psi \qquad \frac{\Theta, \varphi, \psi \vdash \Gamma}{\Theta, \varphi \wedge \psi \vdash \Gamma}$$

$$\forall v.\varphi \vdash \varphi[\psi/v] \qquad \frac{\Theta \vdash \varphi[\pi/v], \Gamma}{\Theta \vdash \forall v.\varphi, \Gamma}$$

Remember that the eigen-variable condition is built into the notation.

We also have to provide axioms for the negation and conjunction in case of indeterminacy:

$$\nabla x \rightsquigarrow \neg x = x \qquad x \neq y \wedge \nabla x \wedge \nabla y \rightsquigarrow x | y$$

7.7   Choice Rules

We have the following choice axioms [10] corresponding to the Axiom of Choice in axiomatic set theory:

$$pv \rightsquigarrow p\tilde{p}$$

Notice that due to the use of $\rightsquigarrow$ we can only make a choice if $\exists v.\,\square(pv)$. If we used a different implication the choice might not be possible at all.

7.8   Generation Rules

We use the following abbreviations:

$$\infty \equiv \top \qquad 0 \equiv \bot \qquad 1 \equiv \dot{0} \qquad 2 \equiv \dot{1} \qquad 3 \equiv \dot{2} \qquad \ldots$$

$$\mathbb{N} \equiv \lambda x.\, x \neq \infty \qquad \mathbb{T} \equiv \lambda x.\top \qquad \emptyset \equiv \lambda x.\bot$$

We have the following important axioms:

$$\dot{x} = \dot{y} \rightsquigarrow x = y \qquad \dot{\infty} = \infty$$

$$p\infty \wedge p0 \wedge (\forall x.\, px \rightsquigarrow p\dot{x}) \rightsquigarrow py$$

The first axiom ensures the injective property and the second axiom makes the third axiom, the induction principle, work as expected.

Hence $2 + 2 = 4$ can be stated in ∇ (seeing $+$ as a suitable abbreviation). It can also be proved, but many other theorems of ordinary mathematics can not be proved, of course (it does not contain arithmetic in general).

Since $\varphi \vdash \varphi$ we have among others $1 \vdash 1$, but this is just a curiosity.

7.9   Countably Infinite Indeterminacy

Let ω be the axiom:

$$(\nabla x \rightsquigarrow \nabla \dot{x}) \wedge \exists y. \nabla y$$

No ambiguity is possible with respect to the use of ω for the set of natural numbers, and the motivation is that the axiom ω introduces a countably infinite type in ∇. The first part says that once indeterminate always indeterminate. The second part of the axiom ω says that indeterminacy exists. In other words, we can say that ω yields ∇-*confinement* and ∇-*existence*.

With the axiom ω we extend ∇ to the indeterminacy theory ∇_ω (propositional higher order logic with countably infinite indeterminacy) such that all theorems of ordinary mathematics can be proved (the axioms can be shown consistent in axiomatic set theory [16], which is the standard foundation of mathematics and stronger than ∇_ω, cf. the second Gödel incompleteness theorem).

Although the propositional higher order logic ∇ is our starting point, the indeterminacy theory ∇_ω is going to be our most important formal system and we use IT $= \{\varphi \mid \omega \vdash \varphi\}$ as a shorthand for its theorems and $\Theta \Vdash \Gamma$ instead of $\Theta, \omega \vdash \Gamma$. In particular we previously could have used $\theta_0, \theta_1, \theta_2, \theta_3 \Vdash \varphi$ in the case study.

We allow a few more abbreviations:

$$\hat{\varphi} \equiv \varphi \wedge \neg \varphi \qquad \check{\varphi} \equiv \varphi \vee \neg \varphi$$

We can now state the interesting property of ∇_ω (coming from ∇) succinctly:

$$\hat{\varphi} \not\Vdash \check{\varphi}$$

7.10   Classical Logic

Let Δ be the axiom:

$$\Delta x$$

The Δ axiom is equivalent to ∇-*non-existence*, namely $\not\exists x. \nabla x$, and with the axiom Δ we extend ∇ to the classical propositional higher order logic ∇_Δ which was thoroughly investigated in [14,2].

Finally we can combine the extensions ∇_Δ and ∇^ι into the classical higher order logic ∇_Δ^ι, also known as Q_0 based on the typed λ-calculus, and often seen as a restriction of the transfinite type theory Q [3,21] by removing the transfinite types. Q_0 is implemented in several automated theorem provers with many active users [18,5,9,23]. Classical second order logic, first order logic, elementary logic (first order logic without functions and equality) and propositional logic can be seen as restrictions of Q_0.

In contrast to the paraconsistent ∇_ω the classical ∇_Δ^ι is not a foundation of mathematics, but we obtain the type theory Q_0^σ by replacing the sort ι with the sort σ and adding the relevant Peano postulates $\dot{x} \neq 0 \wedge (x \neq y \rightsquigarrow \dot{x} \neq \dot{y})$ in our notation, cf. [4, pp. 209/217] for the details.

7.11   Other Logics

In order to investigate finite truth tables, namely the three-valued and four-valued logics discussed in previous sections, we have the following abbreviations:

$$\dagger \equiv \dot{\perp} \qquad \ddagger \equiv \ddot{\perp}$$

We get the four-valued logic $\nabla_\ddagger$ by adding the following axiom to ∇:

$$\Delta x \ \vee \ x = \dagger \ \vee \ x = \ddagger$$

Likewise we get the three-valued logic $\nabla_\dagger$ by adding the following axiom to ∇:

$$\Delta x \ \vee \ x = \dagger$$

But here $\ddagger = \perp$ due to the injection property of the indeterminacy generation.

8   Conclusions and Future Work

We have proposed a paraconsistent higher order logic ∇_ω with countably infinite indeterminacy and described a case study (see [20, 22] for further applications).

We have presented a sequent calculus for the paraconsistent logic ∇ and the simple axiom ω turning ∇ into the ∇_ω that can serve as a foundation of mathematics. Another axiom Δ turns ∇ into the classical logic ∇_Δ. We would like to emphasize that it is not at all obvious how to get from ∇_Δ to ∇ when the usual axiomatics and semantics of ∇_Δ do not deal with the axiom Δ separately as we do here. Corresponding to the proof-theoretical $\vdash$ we have the model-theoretical $\vDash$ based on the type universes, and soundness and completeness results are to be investigated (the latter with respect to general models of ∇ only).

We also intend to compare the paraconsistent higher order logic ∇_ω with work on multi-valued higher order resolution [15].

References

1. A. R. Anderson and N. D. Belnap Jr. *Entailment: The Logic of Relevance and Necessity*. Princeton University Press, 1975.
2. P. B. Andrews. A reduction of the axioms for the theory of propositional types. *Fundamenta Mathematicae*, 52:345–350, 1963.
3. P. B. Andrews. *A Transfinite Type Theory with Type Variables*. North-Holland, 1965.
4. P. B. Andrews. *An Introduction to Mathematical Logic and Type Theory: To Truth through Proof*. Academic Press, 1986.
5. P. B. Andrews, M. Bishop, S. Issar, D. Nesmith, F. Pfenning, and H. Xi. TPS: A theorem proving system for classical type theory. *Journal of Automated Reasoning*, 16:321–353, 1996.
6. O. Arieli and A. Avron. Bilattices and paraconsistency. In D. Batens, C. Mortensen, G. Priest, and J. Van-Bengedem, editors, *Frontiers in Paraconsistent Logic*, pages 11–27. Research Studies Press, 2000.

7. H. P. Barendregt. *The Lambda Calculus, Its Syntax and Semantics*. North-Holland, revised edition, 1984.

8. N. D. Belnap Jr. A useful four-valued logic. In J. M. Dunn and G. Epstein, editors, *Modern Uses of Multiple-Valued Logic*, pages 8–37. D. Reidel, 1977.

9. C. Benzmüller and M. Kohlhase. LEO: A higher-order theorem prover. In *15th International Conference on Automated Deduction (CADE-15)*, pages 139–143. Springer-Verlag, 1998. LNCS 1421.

10. A. Church. A formulation of the simple theory of types. *Journal of Symbolic Logic*, 5:56–68, 1940.

11. G. Evans. Can there be vague objects? *Analysis*, 38(4):208, 1978.

12. M. Ginsberg. Multivalued logics: A uniform approach to inference in artificial intelligence. *Computer Intelligence*, 4:265–316, 1988.

13. S. Gottwald. *A Treatise on Many-Valued Logics*. Research Studies Press, 2001.

14. L. Henkin. A theory of propositional types. *Fundamenta Mathematicae*, 52:323–344, 1963.

15. M. Kohlhase and Ortwin Scheja. Higher-order multi-valued resolution. *Journal of Applied Non-Classical Logic*, 9(4), 1999.

16. E. Mendelson. *Introduction to Mathematical Logic*. Chapman & Hall, 4th edition, 1997.

17. R. Muskens. *Meaning and Partiality*. CSLI Publications, Stanford, California, 1995.

18. L. C. Paulson. *Isabelle – A Generic Theorem Prover*. Springer-Verlag, 1994. LNCS 828.

19. J. Villadsen. Combinators for paraconsistent attitudes. In P. de Groote, G. Morrill, and C. Retoré, editors, *Logical Aspects of Computational Linguistics*, pages 261–278. Springer-Verlag, 2001. LNCS 2099.

20. J. Villadsen. Paraconsistent query answering systems. In T. Andreasen, A. Motro, H. Christiansen, and H. L. Larsen, editors, *International Conference on Flexible Query Answering Systems*, pages 370–384. Springer-Verlag, 2002. LNCS 2522.

21. J. Villadsen. Supra-logic: Using transfinite type theory with type variables for paraconsistency. In *III World Congress on Paraconsistency*, 2003. Toulouse, France.

22. J. Villadsen. Paraconsistent assertions. In *Second German Conference on Multi-agent System Technologies*. Springer-Verlag, 2004. To appear in LNCS.

23. F. Wiedijk. Comparing mathematical provers. In *Mathematical Knowledge Management (MKM 2003)*, pages 188–202. Springer-Verlag, 2003. LNCS 2594.

Abstraction Within Partial Deduction for Linear Logic

Peep Küngas

Norwegian University of Science and Technology
Department of Computer and Information Science
peep@idi.ntnu.no

Abstract. Abstraction has been used extensively in Artificial Intelligence (AI) planning, human problem solving and theorem proving. In this article we show how to apply abstraction within Partial Deduction (PD) formalism for Linear Logic (LL). The proposal is accompanied with formal results identifying limitations and advantages of the approach.
We adapt a technique from AI planning for constructing abstraction hierarchies, which are then exploited during PD. Although the complexity of PD for propositional LL is generally decidable, by applying abstraction the complexity is reduced to polynomial in certain cases.

1 Introduction

Partial Deduction (PD) (or partial evaluation of logic programs, which was first introduced by Komorowski [8]) is known as one of optimisation techniques in logic programming. Given a logic program, PD derives a more specific program while preserving the meaning of the original program. Since the program is more specialised, it is usually more efficient than the original program.

A formalism of PD for !-Horn fragment [5] of Linear Logic [3] (LL) is given in [11]. Also soundness and completeness of the formalism are proved there. However, the formalism is still limited with the computational complexity arising from the underlying logic. Since propositional !-Horn fragment of LL (HLL) can be encoded as a Petri net, the complexity of HLL is equivalent to the complexity of Petri net reachability checking and thus decidable [5]. Therefore, we consider it very important to identify methodologies, which would help to decrease the effects inherited from the complexity.

Abstraction techniques, constituting a subset of divide-and-conquer approaches, are widely viewed as methods for making intractable problems tractable. Using abstraction techniques we may cut solution search space from b^d to $kb^{d/k}$ [9, 15], where b and d are respectively the branching factor and the depth of the search tree and k is the ratio of the abstraction space to the base space in an abstraction hierarchy.

Korf [9] showed that when *optimal* abstraction hierarchies are used, it is possible to reduce the expected search time from $O(n)$ to $O(log\ n)$. This improvement makes combinatorial problems tractable. For instance, if n is a function

B. Buchberger and J.A. Campbell (Eds.): AISC 2004, LNAI 3249, pp. 52–65, 2004.
© Springer-Verlag Berlin Heidelberg 2004

exponential in the problem size, then *log n* is just linear, according to Korf. The essential reason why abstraction reduces complexity is that the total complexity is the sum of the complexities of the multiple searches, not their product.

In the following we shall show how to generate automatically abstraction hierarchies for LL applications. Then, when applying PD, first an initial (partial) proof is constructed in the most abstract space in that hierarchy and then gradually the proof is extended while moving from higher levels of abstraction towards the base level given by the initial problem. A (partial) proof found at a certain abstraction level may be viewed as a proof including "gaps", which have to be filled at lower levels of abstractions.

The rest of the paper is organised as follows. Section 2 introduces preliminaries and basic definitions. Section 3 introduces a method for constructing abstraction hierarchies. Section 4 focuses on PD with abstraction hierarchies. Section 5 demonstrates the application of abstraction to PD. Section 6 reviews the related work and concludes the paper.

2 Preliminaries and Definitions

2.1 Horn Linear Logic

In the following we are considering !-Horn fragment [5] of LL (HLL) consisting of multiplicative conjunction ($\otimes$), linear implication ($\multimap$) and "of course" operator (!). In terms of resource acquisition the logical expression $A \otimes B \vdash C \otimes D$ means that resources C and D are obtainable only if both A and B are obtainable. After the sequent has been applied, A and B are consumed and C and D are produced.

While implication $A \multimap B$ as a computability statement clause in HLL could be applied only once, $!(A \multimap B)$ may be used an unbounded number of times. Therefore the latter formula could be represented with an extralogical LL axiom $\vdash A \multimap B$. When $A \multimap B$ is applied, then literal A becomes deleted from and B inserted to the current set of literals. If there is no literal A available, then the clause cannot be applied. In HLL ! cannot be applied to other formulae than linear implications.

Whenever a compact representation is needed we shall write the set of extralogical axioms $\{\vdash A \otimes B \multimap C, \vdash D \otimes E \multimap F, \ldots\}$ and the sequent $X \vdash Y$ to be proved as a single sequent $X \otimes !(A \otimes B \multimap C) \otimes !(D \otimes E \multimap F) \otimes \ldots \vdash Y$. To allow shorter representation of formulae, we are using in the following sometime abbreviation $a^n = \underbrace{a \otimes \ldots \otimes a}_{n}$, for $n > 0$.

2.2 Partial Deduction for HLL

In this section we present the definitions of the basic concepts of PD for HLL. We adopt the PD framework as it was formalised in [11].

Definition 1. *Computation Specification Clause (CSC) is a LL sequent*

$$\vdash I \multimap_f O,$$

where I and O are multiplicative conjunctions of literals and f is a function, which implements the computation step.

The succedent and the antecedent of CSC C are denoted with $succ(C)$ and $ant(C)$ respectively.

Definition 2. *Computation Specification (CS) is a finite set of CSCs.*

Definition 3. *Computation Specification Application (CSA) is defined as*

$$\Gamma; S \vdash G,$$

where Γ is a CS, S is the initial state and G is the goal state of computation. Both S and G are represented with multiplicative conjunctions of literals.

We are going to use letters S, G and Γ throughout the paper in this context. PD back- and forward chaining steps, respectively $\mathcal{R}_b(L_i)$ and $\mathcal{R}_f(L_i)$, are defined [10] with the following rules:

$$\frac{S \vdash B \otimes C}{S \vdash A \otimes C} \; \mathcal{R}_b(L_i) \qquad \frac{A \otimes C \vdash G}{B \otimes C \vdash G} \; \mathcal{R}_f(L_i)$$

L_i in the inference figures is a labelling of a particular LL axiom representing a CSC in the form $\vdash B \multimap_{L_i} A$. $\mathcal{R}_f(L_i)$ and $\mathcal{R}_b(L_i)$ apply clause L_i to move the initial state towards the goal state or the other way around. A, B and C are multiplicative conjunctions. This brings us to the essence of PD, which is program manipulation, in our case basically modification of initial and goal states. As a side-effect of PD a modified program is created.

2.3   Abstraction

Definition 4. *Size of a CSA A is the number of different literals A involves and is denoted with $\mathcal{S}(A)$.*

Definition 5. *Abstraction $\mathcal{A}$ is a set of literals, which are allowed in a CSA, if particular abstraction is applied.*

Definition 6. *Abstraction level is a position in an abstraction hierarchy. The lowest abstraction level is denoted with 0 and represents the original problem space.*

Definition 7. *Abstraction hierarchy $\mathcal{H}$ for a CSA A is a total order of abstractions such that $\forall \mathcal{A}_i, \mathcal{A}_j \in \mathcal{H} \wedge i \neq j \wedge (\mathcal{S}(\mathcal{A}_i(A)) < \mathcal{S}(\mathcal{A}_j(A))) \Rightarrow \mathcal{A}_j(A) \prec \mathcal{A}_i(A)$, whereas i and j are abstraction levels.*

Due to the way we construct abstraction hierarchies, it is not possible that $\mathcal{S}(\mathcal{A}_i(A)) \equiv \mathcal{S}(\mathcal{A}_j(A))$, unless $i \equiv j$.

Definition 8. *If l is a literal, then $Level(l)$ is the highest abstraction level where l can appear.*

To explain the notion of level let us consider an abstraction hierarchy in Table 1. There $Level(X) \equiv 2$, meaning that literal X occurs in all abstracted versions of a net starting from level 2. Similarly $Level(F) \equiv 1$ and $Level(M) \equiv 0$ in the same table.

Definition 9. *State S at abstraction level i is denoted with $\mathcal{A}_i(S) \equiv \bigotimes\limits_{j=1}^{n} \mathcal{A}_i(S)(l_j)$, where l_j is the j-th literal in S, S has n literals, and*

$$\mathcal{A}_i(S)(l) = \begin{cases} l \text{ , if } Level(l) \geq i \\ 1 \text{ , otherwise} \end{cases}$$

In the preceding 1 is a constant of HLL. In the following we may omit some instances of 1 in formulae to keep the representation simpler. Anyway, our formalism is still consistent since there are rules in HLL facilitating the removal of 1 from formulae and we assume that these rules are applied implicitly.

Definition 10. *Abstracted CSC $\vdash I \multimap O$ at abstraction level i is defined as $\mathcal{A}_i(\vdash I \multimap O) \equiv \vdash \mathcal{A}_i(I) \multimap \mathcal{A}_i(O)$.*

Definition 11. *Abstracted CSA $A = \Gamma; S \vdash G$ at abstraction level i is defined as $\mathcal{A}_i(A) = \Gamma'; \mathcal{A}_i(S) \vdash \mathcal{A}_i(G)$, where $\Gamma' = \bigcup\limits_{\forall c \in \Gamma \wedge \mathcal{A}_i(c) \neq \vdash 1 \multimap 1} \mathcal{A}_i(c)$.*

At abstraction level 0 an original CSA is presented – $\mathcal{A}_0(A) \equiv A$. We write Γ_i and L_i to denote respectively a set of CSCs and literals of a CSA $\mathcal{A}_i(A)$.

Definition 12. *Serialised proof fragment (SPF) is a sequence $\langle S_0, C_1, S_1, \ldots, C_n, S_n \rangle$ starting with a state S_0 and ending with a state S_n. Between every two states there is a CSC $C_i, i = 1 \ldots n$ such that S_{i-1} is the state where C_i was applied with a PD rule and the state S_i is the result achieved by applying C_i. Whenever a compact representation of a SPF is required, we shall write $\langle C_1, \ldots, C_n \rangle$.*

Definition 13. *Partial proof is a pair (H, T), where both H and T are SPFs with the first element of H being the initial state and the last element of T respectively the goal state. Initially the partial proof has only one element in both H and T – the initial state in H and the goal state in T.*

The proof is extended in the following way. PD forward step can be applied only to the last element of H. Symmetrically, PD backward step can be applied only to the first element of T. In the former case the applied CSC and a new state are inserted to the end of H. In the latter case the new state and the applied CSC are inserted to the beginning of T.

Definition 14. *Complete proof is a partial proof with the last element of H and the first element of T being equal.*

If there is no need to distinguish partial and complete proofs, we write just proof.

Definition 15. *New CSCs at abstraction level n are defined with $\Gamma_n^{new} = C_i \mid$ $(C_i \in \Gamma) \wedge (\mathcal{A}_{n+1}(C_i) \equiv\vdash 1 \multimap 1) \wedge (\mathcal{A}_n(C_i) \neq \mathcal{A}_{n+1}(C_i)).$*

Definition 16. *SPF $s = \langle C_1, \ldots, C_n \rangle, C_j \in \Gamma, j = 1 \ldots n$ at abstraction level i is defined as $\mathcal{A}_i(s) = \langle C_j \mid 0 < j \leq |s|, C_j \in \Gamma_i \rangle.$*

Basically it means that at abstraction level i in a partial proof s only these CSCs are allowed, which exist in the abstracted CSA at abstraction level i. Other CSCs are just discarded. Opposite operation to abstraction is refinement.

Definition 17. *Refinement $\mathcal{R}_k^l, k > l$ of a SPF $s = \langle C_1, \ldots, C_n \rangle, C_j \in \Gamma, j = 1 \ldots n$ from abstraction level k to abstraction level l is defined as a sequence $\mathcal{R}_k^l(s) = \langle \alpha_0, C_1, \alpha_1, \ldots \alpha_{n-1}, C_n, \alpha_n \rangle$, where $\alpha_i, i = 0 \ldots n$ is a sequence of CSCs from $A \in \bigcup\limits_{i=k-1}^{l} \Gamma_i.$*

This means that during refinement only new CSCs at particular abstraction level may be inserted to partial proofs. In the following we write $\mathcal{R}^j$ instead of $\mathcal{R}_{j+1}^j$.

3   Generating Abstraction Hierarchies

In this section we describe how to construct abstraction hierarchies for CSAs. These hierarchies are later used to gradually refine an abstract solution during PD. The abstraction method, we propose here, has been inspired from an abstraction method [7] from the field of AI planning.

Given a problem space, which consists of a CSA, our algorithm reformulates the original problem into more abstract ones. The main effect of abstraction is the elimination of inessential program clauses at every abstraction level and thus the division of the original search space into smaller, sequentially searchable ones. The original problem represents the lowest abstraction level.

Ordered monotonicity property is used as the basis for generating abstraction hierarchies. This property captures the idea that if an abstract solution is refined, the structure of the abstract solution should be maintained. Hence elements in the proof fragments of a proof, would not be reordered while extending this sequence at abstraction level $i - 1$. The process of refining an abstract solution requires the application of additional CSCs to achieve the literals ignored at more abstract levels.

Definition 18. *Ordered monotonic refinement $\mathcal{R}$ is a refinement of an abstract solution s so that $\mathcal{A}_i(\mathcal{R}_i^j(s)) = s, j \leq i$, where s is a proof fragment, i denotes the abstraction level, where s was constructed and j is the target abstraction level.*

Definition 19. *Ordered monotonic hierarchy is an abstraction hierarchy with the property that for every solvable problem there exists an abstract solution that has a sequence of ordered monotonic refinements into the base space.*

Definition 20. *Let A and B be arbitrary vertices in a directed graph. Then we say that they are strongly connected, if there exists a cycle with A as its initial and final vertex such that this cycle includes Venice B.*

An ordered monotonic abstraction hierarchy is constructed by dividing literals a CSA between abstraction levels such that the literals at level i do not interact with literals at level $i + 1$. We say that places A and B do not interact with each other, if they are not strongly connected in the constraint graph of the particular CSA. Therefore, the ordered monotonicity property guarantees that the goals and subgoals arising during the process of refining an abstract solution will not interact with the conditions already achieved at more abstract levels. This sort of abstraction is considered as *Theorem Increasing* in [4], stating that if a theorem is not provable in an abstract space, it neither is in the base space.

```
Algorithm DetermineConstraints(graph, Γ, G)

inputs: a set of CSCs Γ and a goal state G
output: constraints, which guarantee ordered monotonicity

begin
for ∀literal ∈ G
    if not(ConstraintsDetermined(literal, graph)) then
        ConstraintsDetermined(literal, graph) ← true
        for ∀csc ∈ Γ
            if literal ∈ succ(csc) then
                for ∀l ∈ succ(csc)
                    AddDirectedEdge(literal, l, graph)
                end for
                for ∀l ∈ ant(csc)
                    AddDirectedEdge(literal, l, graph)
                end for
                DetermineConstraints(graph, Γ, ant(csc))
            end if
        end for
    end if
end for
return graph
end DetermineConstraints
```

Fig. 1. Building a constraint graph.

Our algorithm first generates a graph representing dependencies (see Fig. 1) between literals in a CSA, and then, by using that graph, finally generates an ordered monotonic abstraction hierarchy (see Fig. 2).

Antecedents of CSCs if form $K \multimap 1$ (where K is a multiplicative conjunction of literals), are inserted as they occur in the goal state because they, although possibly needed for achieving the goal, contain possibly literals not included in the constraint graph. The latter is due to the fact that the constraint graph is

Algorithm *CreateHierarchy*(Γ, G)

inputs: a set of CSCs Γ and a goal state G
output: an ordered monotonic abstraction hierarchy

begin
graph ← *DetermineConstraints*({}, Γ, G)
components ← *FindStronglyConnectedComponents*(*graph*)
partialOrder ← *ConstructReducedGraph*(*graph, components*)
absHierarchy ← *TopologicalSort*(*partialOrder*)
return *absHierarchy*
end *CreateHierarchy*

Fig. 2. Creating an abstraction hierarchy.

extended by observing the succedents of CSCs and in the case it consists only of 1, the CSC would not be considered. Anyway these CSCs may be needed during reasoning.

3.1 The Role of the Initial State

While building a constraint graph for abstraction, dependencies between literals are detected. If it should happen that at least one literal $l \in S$ is not included in the constraint graph and it does not occur in the goal state G either, then there is no proof for the particular CSA. This applies iff there are no CSCs, which could consume literal l.

Theorem 1. *Given a CSA and a set of edges D_e of the constraint graph $\mathcal{D}$, which was constructed for the CSA, and if $\exists l.(l \in S \wedge l \notin G \wedge l \notin D_e \wedge \forall c \in \Gamma.(succ(c) \neq 1)$, then there is no proof for the CSA.*

Proof. While finding dependencies between literals through constraint graph construction, roughly a way for literal propagation is estimated for reaching the goal G and literals on the way are inserted to the graph. Therefore, if not all literals $l \in S$ are included in the constraint graph, then there is no way to find a proof for $\Gamma; S \vdash G$.

Anyway, some literals in the initial state S may be not engaged during PD and thus they exist in both states S and G. In that case the missing literal from a constraint graph does not indicate that the state G is not reachable. Similarly, CSCs C with $succ(C) \equiv 1$ have to be considered, since they only consume literals and therefore are rejected, when generating a constraint graph.

This case is illustrated in Fig. 3(b), where a constraint graph is generated for CSA $\Gamma; A \otimes B \otimes C \vdash F$. As it can be seen in Fig. 3(b) the literal B, although being in the state S, is not included in the constraint graph. The same applies for literal A. Therefore the CSA has no proof.

$$\Gamma = \begin{array}{l} \vdash A \multimap E \\ \vdash B \otimes C \multimap D \\ \vdash C \multimap E \otimes F \end{array}$$

(a)

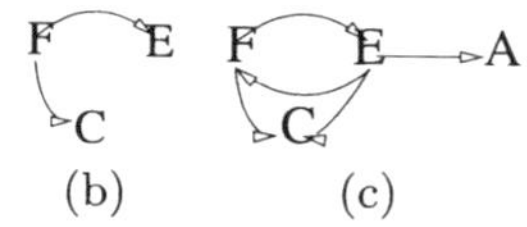

(b)

(c)

Fig. 3. Γ (a), constraint graph for CSA $\Gamma; A \otimes B \otimes C \vdash F$ (b), and constraint graph for CSA $\Gamma; A \otimes C \vdash F \otimes E^2$ (c).

3.2 Removing Redundant CSCs

After the possibly required literals have been determined, we can throw away all CSCs, which include at least one literal which is not included in the constraint graph. In that way the search space would be pruned and search made more efficient. Provability of a particular CSA would not be affected by removing these CSCs.

Theorem 2. *Given a CSA and a set of edges D_e of the constraint graph $\mathcal{D}$, which was constructed for the CSA, we can discard all CSCs $c \in \Gamma$, which satisfy condition $\exists l.((l \in succ(c) \vee l \in ant(c)) \wedge l \notin D_e \wedge succ(c) \neq \emptyset)$ without affecting the provability of the CSA.*

Proof. If there is a CSC $c \in \Gamma$ of a CSA such that $\exists l.((l \in ant(c) \vee l \in succ(c)) \wedge l \notin D_e)$, then it means that c was not considered during construction of constraint graph $\mathcal{D}$. Therefore c is not considered relevant for finding a proof for the particular CSA and can be discarded.

CSC reduction is illustrated in Fig. 3(c), where a constraint graph is generated for CSA $\Gamma; A \otimes C \vdash F \otimes E^2$. Since literals B and D are not present in the constraint graph, they are considered irrelevant for PD. Therefore all CSCs c such that $B \in ant(c)$ or $D \in ant(c)$ or $B \in succ(c)$ or $D \in succ(c)$ can be removed without affecting the provability result of the original problem. Hence $\vdash B \otimes C \multimap D$ can be removed from Γ.

3.3 The Computational Complexity of Constructing Abstraction Hierarchies

According to [7] the complexity of building the constraint graph is $O(n * o * l)$, where n is the number of different literals in a CSC, o is the maximum number of CSCs relevant for achieving any given literal, and l is the total number of different literals in succedents of relevant CSCs. Building a hierarchy is also $O(n * o * l)$ since the number of edges in the graph is bounded by $n * o * l$ and the complexity of the graph algorithms used is linear.

4 PD with Abstraction Hierarchies

To prove hierarchically, sequents are first mapped to the highest level of abstraction. This is done by substituting literals, not essential at that abstraction level, in the initial theorem to be proved by the unit of $\otimes$, which is 1. In this way corresponding abstracted initial and final states are formed. Analogously we substitute literals in CSCs to form a set of abstracted CSCs.

Due to ordered monotonicity at every level of abstraction only *new* CSCs can be inserted into a proof, because the old ones have already been placed at their respective positions in the proof, if they were needed. By permitting only newly introduced CSCs at each level the branching factor of proof search space is decreased.

Every proof found at an higher abstraction level may generate a sequence of subgoals for a lower level abstraction, if literals have been restored to CSCs used in these proofs. Thus, through using abstraction, also the distance between subgoal states is reduced. While extending a proof with new CSCs at particular abstraction level, gaps in a proof are filled. The gaps were introduced by moving to the lower abstraction level. The high-level algorithm for theorem proving with abstraction is presented in Fig. 4.

```
Algorithm AbstractProver(S, G, H, Γ)
inputs: the initial and the goal state, an abstraction hierarchy, Γ
output: P //a set of valid proofs
begin
level ← highestLevel(hierarchy)
proof ← {}
absInit ← A_level(S)
absGoal ← A_level(G)
P ← Solve(absInit, proof, absGoal, Γ_level^new)
for l ← level − 1 to 0
    P_2 ← {}
    for ∀p ∈ P
        absInit ← A_l(S)
        absGoal ← A_l(G)
        P_2 ← P_2 ∪ ExtendProof(0, absInit, absGoal, p, Γ_l^new, l)
    end for
    P ← P_2
end for
return P
end AbstractProver
```

Fig. 4. A pseudocode for theorem proving with abstraction.

AbstractProver goes incrementally through all abstraction levels starting from the highest and ending at the base abstraction level. At every abstraction level it computes a set of abstract proofs, which are refined at lower levels until a proof has been computed or it is determined that there is no solution for a particular CSA.

The main operation of algorithm *ExtendProof* is to detect and fill gaps in proofs when refining them. The complexity of the algorithm *AbstractProver* together with the *ExtendProof* depends on the number of abstraction levels l_h, on the serialised proof length l_p (only CSCs $C_i, i = 1 \ldots n$ used there are counted) and on the number of CSCs in Γ, which is denoted with l_a.

An abstract proof, found at an abstraction level higher than 0 may be viewed as a sequence including "gaps", which have to be filled at a lower level of abstraction. It has to be emphasised that at one abstraction level several solutions may be found and not all of these, if any, lead to a solution at less abstract levels. Thus several abstract solutions may have to be refined before a solution for a less abstract problem is found.

Ordered monotonicity determines that while extending a proof s at a lower abstraction level, we can insert only *new* CSCs, whereas the CSCs which are already in s, determine new subgoals, which have to be solved at the lower abstraction level. Thus in that way we reduce one monolithic PD problem into several smaller PD problems and thereby reduce the distance between subgoals. By dividing CSCs between different abstraction levels the branching factor of the search space is decreased. Following the former idea we define optimal abstraction hierarchy as an abstraction hierarchy, where at each level exactly one new CSC is introduced and a CSA at the highest abstraction level has exactly one CSC.

Definition 21. *Optimal abstraction hierarchy $\mathcal{H}_o$ of a CSA is an abstraction hierarchy with $n = |\Gamma|$ abstraction levels starting from level 0. Therefore, in $\mathcal{H}_o$, $|\Gamma_i \setminus (\Gamma_i \cap \Gamma_{i+1})| = 1, i = 0 \ldots n - 2$ and $|\Gamma_{n-1}| = 1$.*

Theorem 3. *Given that an optimal abstraction hierarchy $\mathcal{H}_o$ is used, computational complexity of PD problem with our algorithm is $O(|\Gamma| * |s|)$, where $|\Gamma|$ is the number of CSC in a CSA and $|s|$ is the expected length of the complete proof s.*

Proof. We define the exponential complexity of PD for HLL as $l_t^{l_s}$, since it could be modeled through Petri net reachability checking. l_t in the preceding is the number of CSC in Γ and $l_s = |s|$ is the length of a complete proof s. Since at the lowest abstraction level of $\mathcal{H}_o$ we have l_s CSCs in the sequence s, there are at every abstraction level maximally l_s gaps, which have to be filled. By assuming that there are l_h abstraction levels in $\mathcal{H}_o$, the resulting complexity is $O(l_h * l_s * (l_t/l_h)^{l_s})$. Since we assumed the usage of an *optimal* abstraction hierarchy ($l_t \equiv l_h$), the exponential complexity of PD is reduced to $O(l_h * l_s * 1^{l_s}) = O(l_h * l_s)$, which is polynomial.

Some restrictions to CSAs, to achieve an optimal abstraction hierarchy, are summarised with the following proposition and theorem.

Proposition 1. *There are no strongly connected literals in a constraint graph of a CSA, if (1) $\forall l \in L.(| \{c \mid c \in \Gamma \land l \in succ(c)\} | \leq 1) \land (| \{c \mid c \in \Gamma \land l \in ant(c)\} | \leq 1)$ and (2) $\forall c \in \Gamma.(\mathcal{S}(succ(c)) \leq 1) \land (\mathcal{S}(ant(c)) \leq 1)$.*

Proof. One can easily see that if the preconditions (1) and (2) are satisfied, a tree with branching factor 1 is constructed during dependency graph construction. Therefore there are no strongly connected components.

Theorem 4. *If the preconditions of Proposition 1 are satisfied, then our algorithm in Fig. 2 gives us an optimal abstraction hierarchy.*

Proof. The proof follows from Proposition 1 and Definition 21.

The last theorem identifies that in order to take advantage of optimal abstraction hierarchies, we have to constrain heavily the CSCs in a CSA. However, even if an optimal abstraction hierarchy is not constructed for a CSA, the non-optimal one may still help to reduce the complexity of PD.

5   An Abstraction Example

In order to demonstrate PD with abstraction, let us consider the following CSA:

$$CSA_0 ::= E \otimes M \otimes X^3 \otimes !(E \otimes N \multimap H) \otimes !(H \multimap E \otimes I \otimes M) \otimes !(M \multimap N) \otimes$$

$$\otimes !(I^7 \multimap F) \otimes !(F \otimes X \multimap Y^2) \otimes !(Y^2 \otimes X^2 \multimap X^2) \vdash E \otimes M \otimes X^2$$

The corresponding constraint graph, achieved by using the algorithm in Fig. 1, is presented in Fig. 5(a). A directed edge from node A to node B in the graph indicates that A cannot occur lower in the abstraction hierarchy than B.

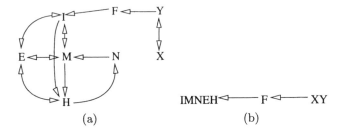

Fig. 5. The constraint graph (a) and the abstraction hierarchy derived from it (b).

One corresponding abstraction hierarchy derived from the graph presented in Fig. 5(a) is depicted in Fig. 5(b). Using this abstraction hierarchy, we introduce 2 new abstracted CSAs CSA_1 and CSA_2 for abstraction levels 1 and 2 respectively:

$$CSA_1 ::= X^3 \otimes !(Y^2 \otimes X^2 \multimap X^2) \otimes !(F \otimes X \multimap Y^2) \otimes !(1 \multimap F) \vdash X^2$$

$$CSA_2 ::= X^3 \otimes !(Y^2 \otimes X^2 \multimap X^2) \otimes !(X \multimap Y^2) \vdash X^2$$

The literals and CSCs available at different abstraction levels are represented in Table 1. The value "—" there represents that a CSC was abstracted to $\vdash 1 \multimap 1$ and was thus then discarded, because this axiom is already included in HLL.

In Table 1 the CSCs at abstraction level 2 are obtained by substituting all literals except X and Y with constant 1. For instance, $\vdash H \multimap E \otimes I \otimes M$ is first abstracted to $\vdash 1 \multimap 1 \otimes 1 \otimes 1$, because literals H, E, I and M are not allowed at abstraction level 2. Since the abstracted CSC is basically equivalent to $\vdash 1 \multimap 1$, it is discarded at this abstraction level. Analogously in the CSCs at abstraction level 1 only literals F, X and Y are permitted. Finally at abstraction level 0 all literals are allowed, because it is the lowest level of abstraction in the particular abstraction hierarchy.

Table 1. An abstraction hierarchy.

	Level 0	Level 1	Level 2
Literals	I, E, M, H, N, F, X, Y	F, X, Y	X, Y
CSC_1	$\vdash E \otimes N \multimap H$	—	—
CSC_2	$\vdash H \multimap E \otimes I \otimes M$	—	—
CSC_3	$\vdash M \multimap N$	—	—
CSC_4	$\vdash I^7 \multimap F$	$\vdash 1 \multimap F$	—
CSC_5	$\vdash F \otimes X \multimap Y^2$	$\vdash F \otimes X \multimap Y^2$	$\vdash X \multimap Y^2$
CSC_6	$\vdash Y^2 \otimes X^2 \multimap X^2$	$\vdash Y^2 \otimes X^2 \multimap X^2$	$\vdash Y^2 \otimes X^2 \multimap X^2$
$S \vdash G$	$E \otimes M \otimes X^3 \vdash E \otimes M \otimes X^2$	$X^3 \vdash X^2$	$X^3 \vdash X^2$

We start PD from level 2 (the highest abstraction level) by finding a proof for $\mathcal{CSA}_2$. The resulting complete proof (head H and tail T are not distinguished) is represented as:

$$\langle X^3, CSC_5, X^2 \otimes Y^2, CSC_6, X^2 \rangle.$$

Now, when moving from abstraction level 2 to level 1, although the CSA to be proved would remain the same, the representation of CSC_5 has changed. Thus this part of the proof has to be extended. Therefore the resulting proof for $X^3 \vdash X^2$ at abstraction level 1 is in serialised form:

$$\langle X^3, CSC_4, X^3 \otimes F, CSC_5, X^2 \otimes Y^2, CSC_6, X^2 \rangle.$$

At abstraction level 0, in the original problem space, only CSCs CSC_1, CSC_2 and CSC_3 are new and may be used to extend the partial proof obtained at abstraction level 1 to prove the initial problem $E \otimes M \otimes X^3 \vdash E \otimes M \otimes X^2$. Thus the complete proof at abstraction level 0 would be:

$$\langle E \otimes M \otimes X^3, \mathcal{P}_f, E \otimes M \otimes X^3 \otimes I^7, CSC_4, E \otimes M \otimes X^3 \otimes F, CSC_5,$$

$$, E \otimes M \otimes X^2 \otimes Y^2, CSC_6, E \otimes M \otimes X^2 \rangle,$$

where

$$\mathcal{P}_f = \{CSC_3, E \otimes N \otimes X^3 \otimes I^i, CSC_1, H \otimes X^3 \otimes I^i, CSC_2, E \otimes M \otimes X^3 \otimes I^{i+1}\}^7,$$

where in turn $i = 0 \ldots 6$ denotes the iteration index.

6   Related Work and Conclusions

The first explicit use of abstraction in automated deduction was in the planning version of GPS [13]. ABSTRIPS [16] was the first system that attempted to automate the formation of abstraction spaces, but only partially automated the process. Besides, ABSTRIPS produced only *relaxed models* [6], by abstracting away literals in preconditions, thus the problem space was not abstracted at all. However, the algorithm, we apply here, produces *reduced models*, where literals are abstracted away at particular abstraction level from all formulae. Other approaches to automatic construction of abstraction spaces include [1, 2, 12].

In this paper we presented a method to apply abstractions to PD problems by using information about dependencies between literals. Then through hierarchical PD the complexity of overall theorem proving would be decreased.

If an abstract space is formed by dropping conditions from the original problem space as we did, information is lost and axioms in the abstract space can be applied in situations in which they cannot be applied in the original space. Thus, using an abstraction space formed by dropping information, it is difficult to guarantee that if there exists an abstract solution, there exists also a solution in the base space. This problem is called the false proof problem [14, 4].

Acknowledgements

This work is partially supported by the Norwegian Research Foundation in the framework of Information and Communication Technology (IKT-2010) program – the ADIS project. I would like to thank the anonymous referees for their constructive comments and suggestions.

References

1. J. S. Anderson and A. M. Farley. Plan abstraction based on operator generalization. In *Proceedings of the Seventh National Conference on Artificial Intelligence, Saint Paul, MN, 1988*, pages 100–104, 1988.
2. J. Christensen. *Automatic Abstraction in Planning*. PhD thesis, Department of Computer Science, Stanford University, 1991.
3. J.-Y. Girard. Linear logic. *Theoretical Computer Science*, 50:1–102, 1987.
4. F. Giunchiglia and T. Walsh. A theory of abstraction. *Artificial Intelligence*, 57:323–389, 1992.
5. M. I. Kanovich. Linear logic as a logic of computations. *Annals of Pure and Applied Logic*, 67:183–212, 1994.
6. C. A. Knoblock. An analysis of ABSTRIPS. In J. Hendler, editor, *Proceedings of the First International Conference on Artificial Intelligence Planning Systems (AIPS'92), College Park, Maryland, June 15–17, 1992*, pages 126–135, 1992.
7. C. A. Knoblock. Automatically generating abstractions for planning. *Artificial Intelligence*, 68:243–302, 1994.
8. J. Komorowski. *A Specification of An Abstract Prolog Machine and Its Application to Partial Evaluation*. PhD thesis, Department of Computer and Information Science, Linkoping University, Linkoping, Sweden, 1981.
9. R. E. Korf. Planning as search: A quantitative approach. *Artificial Intelligence*, 33:65–88, 1987.

10. P. Küngas and M. Matskin. Linear logic, partial deduction and cooperative problem solving. In *Proceedings of the First International Workshop on Declarative Agent Languages and Technologies (in conjunction with AAMAS 2003), DALT'2003, Melbourne, Australia, July 15, 2003*, volume 2990 of *Lecture Notes in Artificial Intelligence*. Springer-Verlag, 2004.

11. P. Küngas and M. Matskin. Partial deduction for linear logic – the symbolic negotiation perspective. In *Proceedings of the Second International Workshop on Declarative Agent Languages and Technologies (in conjunction with AAMAS 2004), DALT'2004, New York, USA, July 19, 2004*. Springer-Verlag, 2004. To appear.

12. A. Y. Levy. Creating abstractions using relevance reasoning. In *Proceedings of the Twelfth National Conference on Artificial Intelligence (AAAI'94)*, pages 588–594, 1994.

13. A. Newell and H. A. Simon. *Human Problem Solving*. Prentice-Hall, 1972.

14. D. A. Plaisted. Theorem proving with abstraction. *Artificial Intelligence*, 16:47–108, 1981.

15. D. Ruby and D. Kibler. Learning subgoal sequences for planning. In *Proceedings of the Eleventh International Joint Conference on Artificial Intelligence (IJCAI'89), Detroit, Michigan, USA, 20–25 August, 1989*, volume 1, pages 609–614, 1989.

16. E. D. Sacerdoti. Planning in a hierarchy of abstraction spaces. *Artificial Intelligence*, 5:115–135, 1974.

A Decision Procedure for Equality Logic
with Uninterpreted Functions

Olga Tveretina

Department of Computer Science, TU Eindhoven, P.O. Box 513
5600 MB Eindhoven, The Netherlands
o.tveretina@tue.nl

Abstract. The equality logic with uninterpreted functions (EUF) has
been proposed for processor verification. A procedure for proving satisfiability of formulas in this logic is introduced. Since it is based on the
DPLL method, the procedure can adopt its heuristics. Therefore the procedure can be used as a basis for efficient implementations of satisfiability
checkers for EUF. A part of the introduced method is a technique for
reducing the size of formulas, which can also be used as a preprocessing
step in other approaches for checking satisfiability of EUF formulas.

Keywords: equality logic with uninterpreted functions, satisfiability,
DPLL procedure.

1 Introduction

The equality logic with uninterpreted functions (EUF) has been proposed for
verifying hardware [4]. This type of logic is mainly used for proving the equivalence between systems. When verifying the equivalence between two formulas it
is often possible to eliminate functions replacing them with uninterpreted functions. The abstraction process does not preserve validity and may transform a
valid formula into the invalid formula. However, in some application domains
the process of abstraction is justified.

An EUF formula is a Boolean formula over atoms that are equalities between
terms. In this logic, formulas have truth values while terms have values from some
domain. For example, the formula:

$$f(x) \not\approx f(z) \land x \approx y \land y \approx z$$

is unsatisfiable.

Here we write '$\approx$' for equality rather than '$=$' to avoid confusion with other
applications of the symbol '$=$', and we use the notation $s \not\approx t$ as an abbreviation
of $\neg(s \approx t)$.

In the past years, various procedures for checking the satisfiability of such
formulas have been suggested. Barrett et al. [2] proposed a decision procedure
based on computing congruence closure in combination with case splitting.

In [1] Ackermann showed that the problem of deciding the validity of an
EUF formula can be reduced to checking the satisfiability of the equality formula. Many current approaches [6,3] use a transformation of EUF formulas into

B. Buchberger and J.A. Campbell (Eds.): AISC 2004, LNAI 3249, pp. 66–79, 2004.

function-free formulas of the equality logic. Then the equality logic formula can be transformed into a propositional one and a standard satisfiability checker can be applied. Goel et al. [6] and Bryant et al. [3] reduced an equality formula to a propositional one by adding transitivity constraints. In this approach it is analyzed which transitivity properties may be relevant.

A different approach is called range allocation [9,10]. In this approach a formula structure is analyzed to define a small domain for each variable. Then a standard BDD-based tool is used to check satisfiability of the formula under the domain.

Another approach is given in [7]. This approach is based on BDD computation, with some extra rules for dealing with transitivity. Unfortunately, the unicity of the reduced BDDs is lost.

In this paper an approach based on the Davis-Putnam-Logemann-Loveland procedure (DPLL) [5] is introduced.

The DPLL procedure was introduced in the early 60s as a proof procedure for first-order logic. Nowadays, only its propositional logic core component is widely used in efficient provers. The success was a motivation for developing a DPLL-based procedure for the equality logic with uninterpreted functions. The main idea of the DPLL method is to choose an atom from the formula and proceed with two recursive calls: one obtained by assuming this atom and another by assuming the negation of the atom. The procedure terminates with the answer "unsatisfiable" if the empty clause is derived for all branches, and otherwise it returns "satisfiable".

The first technique based on the DPLL method for EUF is introduced in [8]. The proposed DPLL procedure calls the congruence closure module for positive equations.

In this paper the different procedure based on the DPLL method is proposed. The main problem dealt with in the paper is: given an EUF formula, decide whether it is satisfiable or not. Similar to the propositional logic every EUF formula can be transformed into an EUF formula in conjunctive normal form (CNF) such that the original formula is satisfiable if and only if the CNF is satisfiable. Hence we may, and shall, concentrate on satisfiability of a formula in conjunctive normal form. The idea of UIF-DPLL is to split a literal that occurs in purely positive clauses of length more than one, and to apply the reduction rules.

This paper is organized as follows. In Section 2 basic definitions are given. In Section 3 the DPLL method is described. The UIF-DPLL calculus is presented in Section 4. In Section 5 the UIF-DPLL procedure is introduced and a proof of its soundness and completeness is given. The technique for reducing the size of a formula and an optimized procedure are presented in Section 6. Some concluding remarks are in Section 7.

2 Basic Definitions and Preliminaries

Each EUF formula can be straightforwardly converted into an equivalent CNF in the same manner as in propositional logic. The well-known Tseitin transformation [12] transforms an arbitrary propositional formula into a CNF in such a way that the original formula is satisfiable if and only if the CNF is satisfiable. Both the size of the resulting CNF and the complexity of the transformation procedure is linear in the size of the original formula. In this transformation new propositional variables are introduced. So applying it directly to EUF formulas will yield a CNF in which the atoms are both equalities and propositional variables. However, if we have n propositional variables $p_1, \ldots, p_n$ we can introduce $n + 1$ fresh variables $x, y_1, \ldots y_n$ and replace every propositional variable p_i by the equality $x \approx y_i$. In this way satisfiability is easily seen to be maintained. Hence we may and shall restrict to the satisfiability of CNFs.

2.1 Syntax

Let $\Sigma = (\mathsf{Fun}, \approx)$ be a signature, where $\mathsf{Fun} = \{f, g, h, \ldots\}$ is a set of *function symbols*.

For every function symbol its *arity* is defined, being a non-negative integer. We assume a set $\mathsf{Var} = \{x, y, z, \ldots\}$ of *variables*. The sets Var and Fun are pairwise disjoint.

The set Term of terms is inductively defined as follows.

- $x \in \mathsf{Var}$ is a term,
- $f(t_1, \ldots, t_n)$ is a term if $t_1, \ldots, t_n$ are terms, and $f \in \mathsf{Fun}$.

Symbols s, t, u denote terms.

A subterm of a term t is called *proper* if it is distinct from t. The set of subterms of a term t is denoted by $\mathsf{SubTerm}(t)$. The set of proper subterms of a term t is denoted by $\mathsf{SubTerm}_p(t)$.

The *depth* of a term t is denoted by $\mathsf{depth}(t)$ and inductively defined as follows.

- $\mathsf{depth}(x) = 1$ if $x \in \mathsf{Var}$,
- $\mathsf{depth}(f(t_1, \ldots, t_n)) = 1 + \max(\mathsf{depth}(t_1, \ldots, t_n))$.

An *atom* a is an equality of the form $s \approx t$, where $s, t \in \mathsf{Term}$. We consider $s \approx t$ and $t \approx s$ as the same atom. The set of atoms over the signature Σ is denoted by $\mathsf{At}(\Sigma, \mathsf{Var})$ or for simplicity by At.

A *literal* l is an atom or a negated atom. We say that l is a *positive literal* if it coincides with some atom A, otherwise it is called a *negative literal*. The set of all literals over the signature Σ is denoted by $\mathsf{Lit}(\Sigma, \mathsf{Var})$ or if it is not relevant by Lit. By $t \bowtie s$ is denoted either a literal $t \approx s$ or a literal $t \not\approx s$.

A *clause* C is a set of literals. The number of literals in a clause is called the *length* of the clause. The clause of length 0 is called the empty clause, and it is denoted by $\bot$. A non-empty clause C is called *positive* if it contains only positive literals, otherwise it is called *negative*. The set of clauses is denoted by Cls. A clause is called *unit* if it contains only one literal.

A *formula* ϕ is a set of clauses. The set of formulas is denoted by Cnf.

The set of positive clauses of length more than one contained in ϕ is called the *core* of ϕ, and it is denoted by $\mathsf{Core}(\phi)$.

In a CNF ϕ let

- $\mathsf{Var}(\phi)$ be the set of all constants in ϕ,
- $\mathsf{At}(\phi)$ be the set of all atoms in ϕ,
- $\mathsf{Lit}(\phi)$ be the set of all literals in ϕ,
- $\mathsf{MLit}(\phi)$ be the multiset of all literals contained in ϕ,
- $\mathsf{Cls}(\phi)$ be the set of all clauses in ϕ.

We define $\mathsf{Term}(\phi) = \{t \in \mathsf{Term} \mid \exists s \in \mathsf{Term} : (t \bowtie s) \in \mathsf{Lit}(\phi)\}$.
$\mathsf{SubTerm}(\phi) = \bigcup_{t \in \mathsf{Term}(\phi)} \mathsf{SubTerm}(t)$.

2.2   Semantics

Let At be a set of atoms.

We define an *interpretation* as a function

$$\mathsf{I} : \mathsf{At} \to \{\mathsf{true}, \mathsf{false}\}.$$

A literal l is true in I iff either l is an atom a and $\mathsf{I}(a) = \mathsf{true}$ or l is a negated atom $\neg a$ and $\mathsf{I}(a) = \mathsf{false}$. We write $\mathsf{I} \models l$, if a literal l is true in I.

We define an E-interpretation as one satisfying the following conditions.

- $\mathsf{I} \models t \approx t$;
- if $\mathsf{I} \models s \approx t$ then $\mathsf{I} \models t \approx s$;
- if $\mathsf{I} \models s \approx u$ and $\mathsf{I} \models u \approx t$ then $\mathsf{I} \models s \approx t$;
- if $\mathsf{I} \models s_i \approx t_i$ for all $i \in \{1, \ldots, n\}$ then $\mathsf{I} \models f(s_1, \ldots, s_n) \approx f(t_1, \ldots, t_n)$.

We write $\mathsf{I} \models \phi$ if a formula ϕ is true in I.

Definition 1. *A formula ϕ is called satisfiable if $\mathsf{I} \models \phi$ for some E-interpretation I. Otherwise ϕ is called unsatisfiable.*

By definition the empty clause $\bot$ is unsatisfiable.

We will use throughout the paper the following notations.

Let $s \notin \mathsf{SubTerm}(t)$. Then $\phi[t := s]$ denotes the formula that is obtained from ϕ by substituting recursively all occurrences of the term t by the term s till no occurrences of t is left.

Example 2. Let us consider $\phi = \{\{f(f(x)) \approx y\}, \{x \approx g(y)\}\}$.
    Then $\phi[f(x) := x] = \{\{x \approx y\}, \{x \approx g(y)\}\}$.

We define $\phi|_l = \{C - \{\neg l\} \mid C \in \phi, l \notin C\}$.

Example 3. Let us consider $\phi = \{\{x \approx f(y), z \approx g(z)\}, \{x \not\approx f(y), y \approx g(z)\}\}$.
    Then $\phi|_{x \approx f(y)} = \{\{y \approx g(z)\}\}$.

3 The DPLL Procedure

The DPLL procedure is a decision procedure for CNF formulae in propositional logic, and it is the basis of some of the most successful propositional satisfiability solvers to date. This is the fastest known algorithm for satisfiability testing that is not just sound, but also complete.

The basic DPLL procedure recursively implements the three rules: unit propagation, pure literal elimination and recursive splitting.

- *Unit propagation*: Given a unit clause $\{l\}$, remove all clauses that contain the literal l, including the clause itself and delete all occurrences of the literal $\neg l$ from all other clauses. Moreover, the literal l is assigned to true.
- *Pure literals elimination*: A pure literal is a literal that appears only in positive or only in negative form. Eliminate all clauses containing pure literals. Moreover, each pure literal is assigned to true.
- *Splitting*: Choose some literal $l \in \mathsf{Lit}(\phi)$. Now ϕ is unsatisfiable iff both $\phi \cup \{\{l\}\}$ and $\phi \cup \{\{\neg l\}\}$ are unsatisfiable. Selecting a literal for splitting is nondeterministic, so various heuristics can be employed.

The algorithm starts with a set of clauses and simplifies the set of clauses simultaneously. The stopping cases are:

- An empty set of clauses is satisfiable–in this case, the entire algorithm terminates with "satisfiable".
- A set containing an empty clause is not satisfiable–in this case, the algorithm backtracks and tries a different value for an instantiated variable.

The DPLL procedure returns a satisfying assignment, if one exists. It can be extended to return all satisfying assignments.

We have chosen the DPLL procedure to develop our procedure for the following reasons.

- DPLL is a simple and an efficient algorithm.
- Many current state-of-the-art solvers are based on DPLL.
- It can be extended to compute an interpretation.

4 The UIF-DPLL Calculus

UIF-DPLL can be used to decide the satisfiability of equality logic formulas with uninterpreted functions in conjunctive normal form. The main operations of the method are *unit propagation I*, *unit propagation II*, *tautology atom removing*, and *splitting* (recursive reduction to smaller problems). If the empty clause for all branches is derived then the procedure returns "unsatisfiable". Otherwise it returns "satisfiable".

The rules of the UIF-DPLL calculus are depicted in Figure 1.

Unit propagation I: $\dfrac{\{\{x \approx y\}\} \cup \phi}{\phi[x := y]}$ if $x, y \in \mathsf{Var}(\phi)$

Unit propagation II: $\dfrac{\{\{s \approx t\}\} \cup \phi}{\{\{s \approx t\}\} \cup \phi[t := s]}$ if $s, t \in \mathsf{Term}(\phi)$, $t \notin \mathsf{SubTerm}(s)$

Tautology atom removing: $\dfrac{\phi}{\phi|_{t \approx t}}$ if $(t \approx t) \in \mathsf{At}(\phi)$

Splitting: $\dfrac{\phi}{\{\{l\}\} \cup \phi \qquad \{\{\neg l\}\} \cup \phi}$ if $l \in \mathsf{Lit}(\mathsf{Core}(\phi))$

Fig. 1. The rules of the UIF-DPLL calculus

Definition 4. *Unit propagation I, unit propagation II and tautology atom removing rules are called reduction rules.*

The set of reduction rules of the UIF-DPLL is terminating. We will prove it in Section 4.1.

Definition 5. *A CNF ϕ is called reduced if none of the reduction rules of UIF-DPLL calculus is applicable.*

4.1   Termination of the Set of Reduction Rules

We will use the notation $\uplus$ for *disjoint union*, i.e., when we write $\phi \uplus \psi$ we are referring to the union $\phi \cup \psi$ and also asserting that $\phi \cap \psi = \emptyset$.

Definition 6. *A unit clause $\{s \approx t\}$ is called a non-propagated clause in $\phi \uplus \{\{s \approx t\}\}$ if $s, t \in \mathsf{Term}(\phi)$.*
   The set of all non-propagated clauses in ϕ is denoted by $\mathsf{NPCls}(\phi)$.

Example 7. Let us consider the formula

$$\phi : u_1 \approx f(x_1, y_1) \wedge u_2 \approx f(x_2, y_2) \wedge z \approx g(u_1, u_2) \wedge z \not\approx g(f(x_1, y_1), f(x_2, y_2)).$$

One can see that

$$\mathsf{NPCls}(\neg\phi) = \{\{u_1 \approx f(x_1, y_1)\}, \{u_2 \approx f(x_2, y_2)\}\}.$$

For each $\psi \in \mathsf{Cnf}$ we define $k(\phi) = |\mathsf{Term}(\phi)| + |\mathsf{MLit}(\phi)|$.

Definition 8. *We define a total order $\prec_1$ and a total order $\prec_2$ on CNFs as follows.*

$$\phi \prec_1 \psi \ \text{if} \ k(\phi) < k(\psi).$$

$$\phi \prec_2 \psi \ \text{if} \ \mathsf{NPCls}(\phi) < \mathsf{NPCls}(\psi).$$

Lemma 9. *(termination) Let $\phi \in$ Cnf. Then the set of reduction rules of the* UIF-DPLL *calculus is terminating.*

Proof. Let we have a CNF ϕ and let after applying an arbitrary reduction rule of the UIF-DPLL calculus we obtained a CNF ψ. One can check that either NPCls(ψ) < NPCls(ϕ) or $k(\psi) < k(\phi)$. Trivially, $\prec_1$ and $\prec_2$ are well-founded orders on CNFs. We obtain that the set of reduction rules of the UIF-DPLL calculus is terminating. □

5   The UIF-DPLL Procedure

In this section we introduce the UIF-DPLL procedure and prove its soundness and completeness.

The algorithm is implemented by the function UIF-DPLL() in Figure 2.

The UIF-DPLL procedure takes in input a EUF formula in conjunctive normal form and returns either "satisfiable" or "unsatisfiable". It invokes the function REDUCE.

REDUCE takes in input a CNF ϕ, applies the reduction rules of UIF-DPLL calculus till none of the rules is applicable, and returns a reduced CNF.

The function REDUCE(ϕ) is not uniquely defined as we will show with an example.

Example 10. Let us consider the formula

$$\phi = \{\{a \approx f(b)\}, \{a \approx g(c)\}, \{f(b) \approx h(a, c)\}\}.$$

We will apply unit propagation II rule on $a \approx f(b)$. We can replace a with $f(b)$. In this case we obtain

$$\phi' = \{\{a \approx f(b)\}, \{f(b) \approx g(c)\}, \{f(b) \approx h(f(b), c)\}\}.$$

The formula ϕ' is reduced.

We can also replace $f(b)$ with a. The result is the reduced formula

$$\phi'' = \{\{a \approx f(b)\}, \{a \approx g(c)\}, \{a \approx h(a, c)\}\}.$$

The UIF-DPLL procedure is done recursively, according the following steps.

- ϕ is replaced by a CNF REDUCE(ϕ) such that ϕ is satisfiable iff REDUCE(ϕ) is satisfiable.
- If $\perp \in \phi$, UIF-DPLL(ϕ) returns "unsatisfiable".
- If Core(ϕ) = $\emptyset$, where ϕ is reduced, then UIF-DPLL(ϕ) returns "satisfiable".
- If none of the above situations occurs, then ChooseLiteral(Core(ϕ)) returns a literal which occurs in Core(ϕ) according some heuristic criterion.

```
UIF-DPLL(φ) :
    begin
        φ := REDUCE(φ);
        if (⊥ ∈ φ)
            return "unsatisfiable";
        if (Core(φ) = ∅)
            return "satisfiable";
        l := ChooseLiteral(Core(φ));
        if (UIF-DPLL(φ ∪ {{l}}) is "satisfiable")
            return "satisfiable";
        else
            return UIF-DPLL(φ ∪ {{¬l}});
    end
```

Fig. 2. The UIF-DPLL procedure

Example 11. As am example we consider the formula raised during translation validation [9], where concrete functions replaced by uninterpreted function symbols.

$$\phi_0 : \{\{u_1 \approx f(x_1, y_1)\}, \{u_2 \approx f(x_2, y_2)\}, \{z \approx g(u_1, u_2)\},$$
$$\{z \not\approx g(f(x_1, y_1), f(x_2, y_2))\}\}.$$

After applying the unit propagation II rule, we obtain

$$\phi_1 = \{\{u_1 \approx f(x_1, y_1)\}, \{u_2 \approx f(x_2, y_2)\}, \{z \approx g(u_1, u_2)\},$$
$$\{z \not\approx g(u_1, f(x_2, y_2))\}\},$$
$$\phi_2 = \{\{u_1 \approx f(x_1, y_1)\}, \{u_2 \approx f(x_2, y_2)\}, \{z \approx g(u_1, u_2)\}, \{z \not\approx g(u_1, u_2)\}\},$$
$$\phi_3 = \{\{u_1 \approx f(x_1, y_1)\}, \{u_2 \approx f(x_2, y_2)\}, \{z \approx g(u_1, u_2)\}, \{z \not\approx z\}\}.$$

After applying tautology atom removing rule, we obtain

$$\phi_4 = \{\{u_1 \approx f(x_1, y_1)\}, \{u_2 \approx f(x_2, y_2)\}, \{z \approx g(u_1, u_2)\}, \bot\}.$$

Since $\bot \in \phi_4$ then ϕ_4 is unsatisfiable and therefore ϕ_0 is unsatisfiable.

5.1   Satisfiability Criterion

Let ϕ be a reduced CNF not containing the empty clause; $\text{Core}(\phi) = \emptyset$. We will give a proof that such CNF ϕ is satisfiable.

Let ϕ be a reduced CNF and $\text{Core}(\phi) = \emptyset$. Since $\text{Core}(\phi) = \emptyset$ then every clause of length more than one contains at least one negative literal. Let $\psi \in \text{Cnf}$ is obtained from ϕ by removing from all clauses of length more than one all literals except one negative literal. Then, trivially, if $I \models \psi$ for some E-interpretation I then $I \models \phi$, i.e. ϕ is satisfiable if ψ is satisfiable. It means that w.l.o.g. we can restrict ourself to the case when ϕ contains only unit clauses.

The set of CNFs containing only unit clauses is denoted by UCnf.

At first we introduce two binary relations on the set of terms contained in ϕ.

Definition 12. *Let $\phi \in$ UCnf. The binary relation $\sim_\phi$ is the smallest relation over* Term$(\phi) \times$ Term(ϕ) *such that:*

1. *$s \sim_\phi t$, if $\{s \approx t\} \in \phi$.*
2. *$\sim_\phi$ is reflexive, symmetric, and transitive.*

Definition 13. *The binary relation $\cong_\phi$ is the smallest relation over* SubTerm(ϕ) $\times$ SubTerm(ϕ) *such that:*

1. *$s \cong_\phi t$, if $\{s \approx t\} \in \phi$.*
2. *$f(s_1, \ldots, s_n) \cong_\phi f(t_1, \ldots, t_n)$, if $s_i \cong_\phi t_i$, $1 \leq i \leq n$, and $f(s_1, \ldots, s_n)$, $f(t_1, \ldots, t_n) \in$ SubTerm(ϕ).*
3. *$\cong_\phi$ is reflexive, symmetric, and transitive.*

Lemma 14. *Let $\phi \in$ UCnf be reduced. Then for each $s, t \in$ Term*

$$s \sim_\phi t \text{ if and only if } s \cong_\phi t.$$

Proof. ($\Rightarrow$) Suppose $s \sim_\phi t$. Then by Definitions 12 and 13, we obtain $s \cong_\phi t$.

($\Leftarrow$) Suppose $s \cong_\phi t$. By the lemma assumption ϕ is reduced. Then for each $\{s' \approx t'\} \in \phi$, either $s' \notin$ SubTerm$(\phi \backslash \{\{s' \approx t'\}\})$ or $t' \notin$ SubTerm$(\phi \backslash \{\{s' \approx t'\}\})$. We can conclude that the condition (2) never can be applied. Then $s \cong_\phi t$ implies $s \sim_\phi t$.   □

Lemma 15. *Let $\phi \in$ UCnf be reduced. If $\{s \not\approx t\} \in \phi$, where $s, t \in$ Term then $s \not\cong_\phi t$.*

Proof. Let us consider arbitrary $\{s \not\approx t\} \in \phi$, where $s, t \in$ Term. We will prove by contradiction that $s \not\sim_\phi t$.

Assume that $\{s \not\approx t\} \in \phi$ and $s \sim_\phi t$. If $s \sim_\phi t$ then one of the following holds.

- $s \equiv t$. Then $\{s \not\approx s\} \in \phi$. In this case a tautology atom removing rule can be applied. This contradicts that ϕ is reduced.
- there are $u_0, \ldots, u_n \in$ Term(ϕ) such that $s \equiv u_0$, $t \equiv u_n$ and

$$\{u_0 \approx u_1\}, \{u_1 \approx u_2\}, \ldots, \{u_{n-1} \approx u_n\} \in \phi.$$

This also contradicts that ϕ is reduced.

We can conclude that $s \not\sim_\phi t$. Then by Lemma 14, $s \not\cong_\phi t$.   □

Theorem 16. *Let $\phi \in$ UCnf. Then ϕ is unsatisfiable if and only if there exist $s, t \in$ Term(ϕ) such that*

$$\{s \not\approx t\} \in \phi \text{ and } s \cong_\phi t.$$

Proof. See [11].

Theorem 17. *(Satisfiability criterion) Let $\phi \in$ Cnf such that*

- ϕ *is reduced,*
- $\bot \notin \phi$,
- $\mathsf{Core}(\phi) = \emptyset$.

Then ϕ is satisfiable.

Proof. By the theorem assumption $\mathsf{Core}(\phi) = \emptyset$. Then every clause of length more than one contains at least one negative literal. Let $\psi \in$ Cnf is obtained from ϕ by removing from all clauses of length more than one all literals except one negative literal. Obtained ψ is reduced by construction. Since $\psi \in$ UCnf and ψ is reduced then by Lemma 15 and Theorem 16, ψ is satisfiable. If ψ is satisfiable then $I \models \psi$ for some E-interpretation I. One can easily see, that $I \models \phi$, i.e. ϕ is satisfiable. □

5.2   Soundness and Completeness of the UIF-DPLL Procedure

In this section we will prove that the UIF-DPLL procedure is sound and complete. One can see that both rules for unit propagation and the tautology atoms removing preserve (un)satisfiability of a formula.

Lemma 18. *Let $\phi \in$ Cnf. Then ϕ is satisfiable if and only if REDUCE(ϕ) is satisfiable.*

Proof. One can easily check that the rules of the UIF-DPLL calculus preserve (un)satisfiability. □

Theorem 19. *(Soundness and Completeness) A CNF ϕ is unsatisfiable if and only if the UIF-DPLL(ϕ) returns "unsatisfiable".*

Proof. ($\Rightarrow$) Let ϕ be unsatisfiable CNF.

Let $\bot \in \phi$. Then by definition of the function REDUCE(), $\bot \in$ REDUCE(ϕ), and the procedure returns "unsatisfiable".

Let $\bot \notin \phi$. We will give a proof by induction on $|\mathsf{MLit}(\mathsf{Core}(\phi))|$.

Base case. $|\mathsf{MLit}(\mathsf{Core}(\phi))| = 0$. Then one of the following holds.

- $\bot \in$ REDUCE(ϕ). Then the procedure returns "unsatisfiable".
- $\bot \notin$ REDUCE(ϕ). Since $|\mathsf{MLit}(\mathsf{Core}(\phi))| = 0$ then $\mathsf{Core}(\mathsf{REDUCE}(\phi)) = \emptyset$. By Theorem 17, we obtain that REDUCE(ϕ) is satisfiable. By Lemma 18, ϕ is also satisfiable. We obtain a contradiction with the assumption that ϕ is unsatisfiable. Then $\bot \in$ REDUCE(ϕ) and the procedure returns "unsatisfiable".

Inductive step. Let for every $\psi \in$ Cnf the procedure returns "unsatisfiable" if $|\mathsf{MLit}(\mathsf{Core}(\psi))| < n$.

Assume that $|\mathsf{MLit}(\mathsf{Core}(\phi))| = n$. We denote REDUCE($\phi$) by ψ. One can easily check that by definition of UIF-DPLL, for each $l \in \mathsf{Core}(\psi)$

$$|\mathsf{MLit}(\mathsf{Core}(\mathsf{REDUCE}(\psi \wedge l)))| < n,$$

$$|\mathsf{MLit}(\mathsf{Core}(\mathsf{REDUCE}(\psi \wedge \neg l)))| < n.$$

Since ϕ is unsatisfiable then by Lemma 18, ψ is also unsatisfiable. Then for each $l \in \mathsf{Lit}$, $\psi \wedge l$ and $\psi \wedge \neg l$ are unsatisfiable. By Lemma 18, $\mathsf{REDUCE}(\psi \wedge l)$ and $\mathsf{REDUCE}(\psi \wedge \neg l)$ are unsatisfiable. By induction hypothesis, the procedure returns "unsatisfiable" for $\mathsf{REDUCE}(\psi \wedge l)$ and $\mathsf{REDUCE}(\psi \wedge \neg l)$. By definition of UIF-DPLL, the procedure returns "unsatisfiable" for ϕ.

($\Leftarrow$) Let the procedure returns "unsatisfiable". We will give a proof by induction on the number of UIF-DPLL calls.

Base case. Let the procedure returns "unsatisfiable" after one call. Then $\bot \in \mathsf{REDUCE}(\phi)$. By Lemma 18, we obtain that ϕ is unsatisfiable.

Inductive step. Let ϕ be unsatisfiable if UIF-DPLL(ϕ) returns "unsatisfiable" after at most $n - 1$ calls. Assume that UIF-DPLL(ϕ) returns "unsatisfiable" after n calls. By definition of UIF-DPLL, UIF-DPLL$(\mathsf{REDUCE}(\phi) \wedge l)$ and UIF-DPLL$(\mathsf{REDUCE}(\phi) \wedge \neg l)$ returns "unsatisfiable" after at most $n - 1$ calls. Then by induction hypothesis, $\mathsf{REDUCE}(\phi) \wedge l$ and $\mathsf{REDUCE}(\phi) \wedge \neg l$ are unsatisfiable CNFs. We obtain that $\mathsf{REDUCE}(\phi)$ is unsatisfiable. By Lemma 18, ϕ is also unsatisfiable. $\qquad\square$

6   The Extended UIF-DPLL Calculus

In this section we will introduce an optimization technique which can be used as a preprocessing step.

Definition 20. *Let $t \in \mathsf{SubTerm}(\phi)$, $\mathsf{depth}(t) > 1$ and for each $s \in \mathsf{SubTerm}_p(t)$ and $u \in \mathsf{Term}(\phi)$, $(s \approx u) \notin \mathsf{Lit}(\phi)$. Then t is called reducible in ϕ.*

Example 21. Let us consider the formula from Example 11.

$$\neg\phi_0 : u_1 \approx f(x_1, y_1) \wedge u_2 \approx f(x_2, y_2) \wedge z \approx g(u_1, u_2) \wedge z \not\approx g(f(x_1, y_1), f(x_2, y_2)).$$

The terms $f(x_1, y_1)$ and $f(x_2, y_2)$ are reducible.

Definition 22. *A variable x is called fresh in ϕ if $x \notin \mathsf{Var}(\phi)$.*

6.1   The Term Reduction Rule

One may add one more rule to the UIF-DPLL calculus.

Term Reduction: $\dfrac{\phi}{\phi[t := x]}$ if t is reducible and x is fresh in ϕ

This rule can be applied as a preprocessing step. We will show it with an example.

Example 23. We consider the formula from Example 11.

$$\phi_0 : u_1 \approx f(x_1, y_1) \wedge u_2 \approx f(x_2, y_2) \wedge z \approx g(u_1, u_2) \wedge z \not\approx g(f(x_1, y_1), f(x_2, y_2)).$$

After applying the term reduction rule, we obtain

$$\phi_1 : u_1 \approx v_1 \wedge u_2 \approx v_2 \wedge z \approx g(u_1, u_2) \wedge z \not\approx g(v_1, v_2),$$

where a term $f(x_1, y_1)$ is replaced by a fresh variable v_1 and a term $f(x_2, y_2)$ is replaced by a fresh variable v_2.

6.2 The Optimized UIF-DPLL Procedure

The optimized UIF-DPLL procedure invokes the function REDUCE() which takes as an input a CNF ϕ and returns a reduced CNF which is obtained by applying the unit propagation I rule, the unit propagation II rule, the tautology atom removing rule and the term reduction rule.

Example 24. We consider the formula from Example 11.

$$\phi_0 : u_1 \approx f(x_1, y_1) \wedge u_2 \approx f(x_2, y_2) \wedge z \approx g(u_1, u_2) \wedge z \not\approx g(f(x_1, y_1), f(x_2, y_2)).$$

After applying the term reduction rule, we obtain

$$\phi_1 : u_1 \approx v_1 \wedge u_2 \approx v_2 \wedge z \approx g(u_1, u_2) \wedge z \not\approx g(v_1, v_2),$$

where a term $f(x_1, y_1)$ is replaced by a fresh variable v_1 and a term $f(x_2, y_2)$ is replaced by a fresh variable v_2.

After applying the unit propagation I rule, we obtain

$$\phi_2 : u_2 \approx v_2 \wedge z \approx g(u_1, u_2) \wedge z \not\approx g(u_1, v_2),$$

$$\phi_3 : z \approx g(u_1, u_2) \wedge z \not\approx g(u_1, u_2),$$

$$\phi_4 : z \approx g(u_1, u_2) \wedge z \not\approx z.$$

After applying tautology atoms removing rule, we obtain

$$\phi_5 : z \approx g(u_1, u_2) \wedge \bot.$$

6.3 Soundness and Completeness of the Optimized Procedure

In this section we will prove the soundness and completeness of the optimized procedure. An only difference the optimized UIF-DPLL comparing to the basic UIF-DPLL that it replaces all reducible terms by fresh variables.

Theorem 25. *Let t be reducible in ϕ; x be a fresh variable in ϕ. Then ϕ is satisfiable iff $\phi[t := x]$ is satisfiable.*

Proof. ($\Rightarrow$) Let ϕ be satisfiable. Then $I \models \phi$ for some E-interpretation I. Since t is a reducible term then by definition every proper subterm of t does not occur as a left side or a right side of any literal in ϕ. Then during the UIF-DPLL procedure no subterm of t is replaced by another term. In this case at least for one branch of UIF-DPLL(ϕ) the CNF satisfying Theorem 17 conditions is derived, where no subterm of t is replaced by another term. It means that for at least one branch of UIF-DPLL($\phi[t := x]$ the CNF satisfying Theorem 17 conditions is also derived. Then $I \models \phi[t := x]$. We can conclude that $\phi[t := x]$ is satisfiable.

($\Leftarrow$) Let $\phi[t := x]$ be satisfiable. Then $I \models \phi[t := x]$ for some E-interpretation I. In this case at least for one branch of UIF-DPLL($\phi[t := x]$), the CNF satisfying Theorem 17 conditions is derived. Let us say, the CNF ψ. Then $\psi[x := t]$ also satisfies the conditions of Theorem 17. Doing the backward substitution we obtain that at least for one branch of UIF-DPLL(ϕ) the CNF satisfying Theorem 17 conditions is derived. Then $I \models \phi$. We can conclude that ϕ is satisfiable. $\square$

Theorem 26. (*Soundness and Completeness*) *A CNF ϕ is unsatisfiable if and only if the optimized* UIF-DPLL(ϕ) *returns "unsatisfiable".*

Proof. It follows from Theorem 19 and Theorem 25.

$\square$

7   Conclusions and Future Work

We have presented a new approach for checking satisfiability of formulas in the EUF logic. A part of our method is a technique for reducing the size a formula that can be of interest itself. Our procedure can incorporate some optimization techniques developed by the SAT community for the DPLL method. We are going to implement our procedure and to compare it with existing techniques. We considered example, where after few steps we could prove the unsatisfiability of the formula. Traditional approach would lead at first to transformation of the formula to a propositional formula of bigger size and then applying a standard SAT checker. We can see from the considered example that our approach can be efficient for some formulas. Although at present we cannot make general conclusions about the efficiency of the procedure.

References

1. ACKERMANN, W. *Solvable cases of the decision problem.* Studies in Logic and the Foundations of Mathematics. North-Holland, Amsterdam, 1954.
2. BARRETT, C. W., DILL, D., AND LEVITT, J. Validity checking for combinations of theories with equality. In *Formal Methods in Computer-Aided Design (FMCAD'96)* (November 1996), M. Srivas and A. Camilleri, Eds., vol. 1166 of *LNCS*, Springer-Verlag, pp. 187–201.
3. BRYANT, R., AND VELEV, M. Boolean satisfiability with transitivity constraints. *ACM Transactions on Computational Logic 3*, 4 (October 2002), 604–627.

4. BURCH, J., AND DILL, D. Automated verification of pipelined microprocesoor control. In *Computer-Aided Verification (CAV'94)* (June 1994), D. Dill, Ed., vol. 818 of *LNCS*, Springer-Verlag, pp. 68–80.

5. DAVIS, M., LOGEMANN, G., AND LOVELAND, D. A machine program for theorem proving. *Communications of the Association for Computing Machinery 7* (July 1962), 394–397.

6. GOEL, A., SAJID, K., ZHOU, H., AZIZ, A., AND SINGHAL, V. BDD based procedures for a theory of equality with uninterpreted functions. In *Computer-Aided Verification (CAV'98)* (1998), A. J. Hu and M. Y. Vardi, Eds., vol. 1427 of *LNCS*, Springer-Verlag, pp. 244–255.

7. GROOTE, J., AND VAN DE POL, J. Equational binary decision diagrams. In *Logic for Programming and Reasoning (LPAR'2000)* (2000), M. Parigot and A. Voronkov, Eds., vol. 1955 of *LNAI*, pp. 161–178.

8. NIEUWENHUIS, R., AND OLIVERAS, A. Congruence closure with integer offsets. In *10h Int. Conf. Logic for Programming, Artif. Intell. and Reasoning (LPAR)* (2003), M. Vardi and A. Voronkov, Eds., LNAI 2850, pp. 78–90.

9. PNUELI, A., RODEH, Y., SHTRICHMAN, O., AND SIEGEL, M. Deciding equality formulas by small domains instantiations. In *Computer Aided Verification (CAV'99)* (1999), vol. 1633 of *LNCS*, Springer-Verlag, pp. 455–469.

10. RODEH, Y., AND SHTRICHMAN, O. Finite instantiations in equivalence logic with uninterpreted functions. In *Computer Aided Verification (CAV'01)* (July 2001), vol. 2102 of *LNCS*, Springer-Verlag, pp. 144–154.

11. SHOSTAK, R. An algorithm for reasoning about equality. *Communications of the ACM 21* (1978).

12. TSEITIN, G. On the complexity of derivation in propositional calculus. In *Studies in Constructive Mathematics and Mathematical Logic, Part 2*. Consultant Bureau, New York-London, 1968, pp. 115–125.

Generic Hermitian Quantifier Elimination

Andreas Dolzmann[1] and Lorenz A. Gilch[2]

[1] University of Passau, Passau, Germany
dolzmann@uni-passau.de
http://www.fmi.uni-passau.de/~dolzmann/
[2] Graz University of Technology, Graz, Austria
gilch@TUGraz.at
http://people.freenet.de/lorenzg/

Abstract. We present a new method for generic quantifier elimination that uses an extension of Hermitian quantifier elimination. By means of sample computations we show that this generic Hermitian quantifier elimination is, for instance, an important method for automated theorem proving in geometry.

1 Introduction

Ever since quantifier elimination by CAD has been implemented by Collins [1], quantifier elimination has become more and more important. This was reinforced especially by the development and implementation of the partial CAD [2] by Hong and later Brown [3]. Beside the approach used there, quantifier elimination by virtual substitution was published by Weispfenning [4, 5] and further developed and implemented by the first author together with Sturm in REDLOG [6]. The latter method can only be applied to degree restricted formulas. Although both methods are implemented highly efficiently, none is superior to the other one. Moreover sometimes the methods fail solving problems which seem to be solvable using quantifier elimination. Therefore it is necessary to develop and implement further quantifier elimination algorithms.

The quantifier elimination by real root counting was published by Weispfenning in 1998 [7], although he had already published in 1993 a technical report describing this method. The algorithm was first implemented by the first author in 1994 as a diploma thesis [8] in the computer algebra system MAS. Numerous new optimizations were developed by the authors. They were implemented by the second author [9] in a complete reimplementation of the method in the package REDLOG [6] of the computer algebra system REDUCE . The improved version of this quantifier elimination is called Hermitian quantifier elimination. The name "Hermitian" quantifier elimination was chosen to acknowledge the influence of Hermite's work in the area of real root counting.

Hermitian quantifier elimination has been proved to be a powerful tool for particular classes of elimination problems. In [10] the first author has used it for the automatic solution of a real algebraic implicitization problem by quantifier

B. Buchberger and J.A. Campbell (Eds.): AISC 2004, LNAI 3249, pp. 80–93, 2004.
© Springer-Verlag Berlin Heidelberg 2004

elimination. For this automatic solution he has used all three quantifier elimination methods, namely quantifier elimination by virtual substitution, Hermitian quantifier elimination, and quantifier elimination by partial cylindrical algebraic decomposition as well as the simplification methods described in [11].

The definition, development and implementation of new paradigms related to quantifier elimination algorithms have been very successful in the past. Extended quantifier elimination provides not only an equivalent formula but sample solutions. It can be applied e.g. in the area of generalized constraint solving yielding optimal solutions [12].

Generic quantifier elimination was introduced for the virtual substitution method by the first author together with Sturm and Weispfenning [13]. Let φ be an input formula with quantified variables $x_1, \ldots, x_n$ and parameters $u_1, \ldots, u_m$. Recall that regular quantifier elimination computes from φ a quantifier-free formula φ' such that for all real values $c_1, \ldots, c_m$ for the parameters $u_1, \ldots, u_m$ both φ and φ' are equivalent, i.e we have $\varphi(c_1, \ldots, c_m) \longleftrightarrow \varphi'(c_1, \ldots, c_m)$. In the case of generic quantifier elimination we compute additionally a conjunction Θ of non-trivial negated-equations, such that

$$\Theta \longrightarrow (\varphi \longleftrightarrow \varphi').$$

In other words, Θ restricts the parameter space. Note that Θ cannot become inconsistent, and moreover, the complement of the set described by Θ has a lower dimension than the complete parameter space. Thus it restricts our parameter space only slightly.

The idea behind the generic quantifier elimination is to add assumptions to Θ whenever this may either speed up the computation or may cause the algorithm to produce a shorter result formula φ'. The paradigm of generic quantifier elimination was introduced in [13] in the area of automated geometry proving. The key idea here is to express a geometric theorem as a quantified formula and then verify it by quantifier elimination. Regular quantifier elimination may fail due to lack of resources or if the theorem does not hold. In the latter case it may be false only for some degenerated situations, as for empty triangles or rectangles instead of arbitrary triangles. Generic quantifier elimination is in this area superior to the regular one for two reasons: The computations are in general much faster and the assumptions made in Θ may exclude degenerated situations in which the theorem is false. In the above cited paper, which is based on quantifier elimination by virtual substitution, it was heuristically shown that for this generic quantifier elimination in fact Θ contains mostly non-degeneracy conditions.

Meanwhile, using a generic projection operator, the concept of generic quantifier elimination was also successfully applied to quantifier elimination by partial cylindrical algebraic decomposition [14]. Seidl and Sturm study the general applicability of generic quantifier elimination in contrast to the regular one. As for regular quantifier elimination by cylindrical algebraic decomposition, this approach is successful mostly for problems containing only a few variables. This restricts the applicability of the generic projection operator to the area of automated theorem proving.

In this note we introduce a generic variant of Hermitian quantifier elimination and apply it in the area of automated theorem proving. This generic quantifier elimination is not degree restricted as the method based on virtual substitution, and it can handle in general, more variables than the method based on cylindrical algebraic decomposition. Hermitian quantifier elimination is, however, well suited for formulas containing many equations as in the case of automated geometric theorem proving. Nevertheless the generic Hermitian quantifier elimination is well suited in many other application areas, e.g., in physical applications in which a equation between two different values is always of no meaning.

The plan of the paper is as follows: In the next Section 2 we sketch the Hermitian quantifier elimination algorithm. In Section 3 we discuss three parts of the algorithm where the concept of generic quantifier elimination can be successfully applied. After describing the generic algorithm we show in Section 4 the scope of our method by computation examples. In the final Section 5 we conclude and summarize our results.

2 The Basic Algorithm

We want to eliminate the quantifiers from an arbitrary first-order formula in the language of ordered rings. In our discussion we restrict our attention to the main parts of the Hermitian quantifier elimination with some improvements. Given an arbitrary first-order formula, we first compute an equivalent prenex normal form of the form

$$Q_n x_{n1} \cdots Q_n x_{nm_n} \cdots Q_2 x_{21} \cdots Q_2 x_{2m_2} Q_1 x_{11} \cdots Q_1 x_{1m_1}(\psi), \quad Q_i \in \{\exists, \forall\},$$

with $Q_{i-1} \neq Q_i$ for $i \in \{2, \ldots, n\}$ and ψ quantifier-free.

Our elimination algorithm eliminates the quantifier blocks, block by block, beginning with the innermost one, i.e., we compute first a quantifier-free equivalent of

$$Q_1 x_{11} \cdots Q_1 x_{1m_1}(\psi).$$

Using the equivalence

$$\forall x_1 \ldots \forall x_n(\psi) \longleftrightarrow \neg \exists x_1 \ldots \exists x_n(\neg \psi),$$

we can obviously restrict our discussion to the case of one existential quantifier block, i.e. $Q_1 = \exists$. We can furthermore assume without lost of generality (for short w.l.o.g.) that ψ contains only atomic formulas of the form $t = 0$, $t > 0$ and $t \neq 0$ and that ψ is in disjunctive normal form. By applying the equivalence

$$\exists x_1 \ldots \exists x_n \left(\bigvee_{i=1}^{k} \psi_i \right) \longleftrightarrow \bigvee_{i=1}^{k} \exists x_1 \ldots \exists x_n(\psi_i)$$

we assume in the following that ψ is a conjunction of atomic formulas of the above form.

2.1   Preparation

We assume that our input formula has the following form

$$\exists x_1 \ldots \exists x_n \left(\bigwedge_{\hat{g} \in \hat{G}} (g = 0) \wedge \bigwedge_{h \in H} (h > 0) \wedge \bigwedge_{f \in F} (f \neq 0) \right),$$

where $\hat{G}$, H, and F are finite sets of polynomials in $\mathbb{Q}[u_1, \ldots, u_m][x_1, \ldots, x_n]$. We can obviously evaluate each variable-free atomic formula to a truth value making itself superfluous or the whole conjunction contradictive. Thus we can w.l.o.g. assume that each polynomial is an element of $\mathbb{Q}[u_1, \ldots, u_m][x_1, \ldots, x_n] \setminus \mathbb{Q}$.

For a polynomial $g \in \mathbb{Q}[u_1, \ldots, u_m][x_1, \ldots, x_n]$ and $(c_1, \ldots, c_m) \in \mathbb{R}^m$ we denote by $g(c_1, \ldots, c_m)$ the polynomial in $\mathbb{R}[x_1, \ldots, x_n]$ constructed from g by plugging in the c_i for u_i with $i \in \{1, \ldots, m\}$. We extend this notation in the natural manner to sets of polynomials.

If the set $\hat{G}$ is empty, we proceed with our quantifier elimination as described in Section 2.3. If $\hat{G}$ is not empty, we compute a Gröbner system [15] w.r.t. an arbitrary but fixed term order. This term order is also fixed for all subsequent computations in the following paragraphs.

The concept of Gröbner systems generalizes the concept of Gröbner bases to the parametric case. With the term "parametric case" we describe situations in which the coefficient of the polynomials are given parametric as polynomials in some variables, e.g. $mx + b$ is a univariate polynomial in x with the parametric coefficients m and b.

A Gröbner system S is a finite set of pairs (γ, G), called branches of the Gröbner system. Each branch consists of a quantifier-free formula γ in the u_1, $\ldots, u_m$ and a finite set of polynomials $\mathbb{Q}[u_1, \ldots, u_m][x_1, \ldots, x_n]$. For each $c \in \mathbb{R}^m$ there is one branch (γ, G) such that $\gamma(c)$ holds, we have $\mathrm{Id}(G(c)) = \mathrm{Id}(\hat{G}(c))$, and $G(c)$ is a Gröbner basis. In fact, all computations used for our algorithm can be performed parametrically using G.

Note, that for every (γ, G) and $c \in \mathbb{R}^m$ with $\gamma(c)$ we have that $\hat{G}(c)$, $G(c)$ and $\mathrm{Id}(G(c))$ have the same zeroes. By switching from

$$\bigwedge_{g \in \hat{G}} (g = 0) \quad \text{to} \quad \bigvee_{(\gamma, G) \in S} \gamma \wedge \bigwedge_{g \in G} (g = 0)$$

and interchanging the disjunction with the existential quantifier block it suffices to eliminate the quantifiers from

$$\gamma \wedge \exists x_1 \cdots \exists x_n \left(\bigwedge_{g \in G} g = 0 \wedge \bigwedge_{h \in H} h > 0 \bigwedge_{f \in F} f \neq 0 \right).$$

Let d be the dimension of $\mathrm{Id}(G(c))$ with $c \in \mathbb{R}^m$ and $\gamma(c)$. Note that this dimension is uniquely determined by γ. According to the dimension d we proceed as follows: If the ideal is zero dimensional, i.e., $d = 0$, we eliminate the complete block of existential quantifiers as described in the next Section 2.2. If the dimension is -1, i.e., the ideal is actually the entire polynomial ring, and thus there is obviously no zero of G, because 1 is member of the ideal. Our elimination result in this case is simply *false*. If the dimension is n, which is the number of main variables, we have to reformulate the problem and call our quantifier elimination recursively as described in Section 2.3. If, finally, the dimension is between 1 and $n - 1$ then we eliminate the quantifier block with two recursive calls of our Hermitian quantifier elimination as described in Section 2.4.

2.2   The Zero-Dimensional Case

We want to eliminate the quantifiers from

$$\gamma \wedge \exists x_1 \cdots \exists x_n \left(\bigwedge_{g \in G} g = 0 \wedge \bigwedge_{h \in H} h > 0 \wedge \bigwedge_{f \in F} f \neq 0 \right),$$

where for each c in $\mathbb{R}^m$ with $\gamma(c)$ we have that $G(c)$ is Gröbner basis of the zero-dimensional ideal $\mathrm{Id}\big(G(c)\big)$. For this we use a method originally developed for counting the real zeroes of $\mathrm{Id}\big(G(c)\big)$ w.r.t. the side conditions generated by H and F.

The result we use was found independently by Pedersen, Roy, Szpirglas [16] and Becker, Wörmann [17] generalizing a result of Hermite for the bivariate case. It was adapted to the parametric case including several side conditions by Weispfenning [7] and further extended by the first author [8].

For a moment, assume that $H = \{h_1, \ldots, h_r\}$ and $F = \{f_1, \ldots, f_s\}$. Let $E = \{1, 2\}^r$. For $e \in E$ define h^e by

$$h^e = \prod_{i=1}^{r} h_i^{e_i} \cdot \prod_{i=1}^{s} f_i^2.$$

For a univariate polynomial q define $Z_+(q)$ as the number of positive zeroes and $Z_-(q)$ as the number of negative zeroes, respectively, both counted with multiplicities.

Consider $R = \mathbb{Q}(u_1, \ldots, u_m)[x_1, \ldots, x_n]$ and let be $I = \mathrm{Id}(G)$ and $B = \{v_1, \ldots, v_b\}$ the reduced terms of G. Then $R(\underline{c})/I(\underline{c})$ is a $\mathbb{Q}$-algebra with basis $B(\underline{c})$ for each $\underline{c}$ with $\gamma(c)$. Note that each element in R can also be viewed as an element of R/I. For $q \in R$, the map

$$m_q : R/I \to R/I, \quad \text{defined by} \quad m_q(p) = q \cdot p$$

is linear. Using this definition we define for a polynomial $p \in R$ the $b \times b$ matrix $Q_p = (q_{ij})$ by

$$q_{ij} = \mathrm{trace}(m_{v_i v_j p}).$$

Finally let $\chi(Q_p)$ be the characteristic polynomial of Q_p.

Then we have for each $c \in \mathbb{R}^m$ with $\gamma(c)$, that

$$\left| \left\{ a \in \mathbb{R}^n \;\middle|\; \bigwedge_{g \in G} g(c)(a) = 0 \wedge \bigwedge_{h \in H} h(c)(a) > 0 \wedge \bigwedge_{f \in F} f(c)(a) \neq 0 \right\} \right|$$

equals $Z_+(\chi) - Z_-(\chi)$, where $\chi = \prod_{e \in E} \chi(Q_{h^e})$.

While the computations used so far are all uniform in $c \in \mathbb{R}^m$ with $\gamma(\tilde{c})$, we cannot uniformly count Z_+ or Z_- for χ. Note that $\chi(c)$ is of real type, i.e., there are no zeroes in $\mathbb{C} \setminus \mathbb{R}$. For those polynomials we can compute the number of positive and negative zeroes using Descartes rule of signs. It states that the positive real zeroes of a polynomial $\chi(c)$ of real type are exactly the number of sign changes in the list of coefficients of $\chi(c)$ ignoring 0. By considering $\chi(-c)$ one can also compute the negative zeroes using Descartes rule of signs.

Let $\chi = \sum_{i=0}^{l} a_i/b_i y^i$, where $a_i, b_i \in \mathbb{Q}[u_1, \ldots, u_m]$. For $\delta \in \{<, =, >\}^l$ let φ_δ be the formula

$$a_1 b_1 \; \delta_1 \; 0 \wedge \cdots \wedge a_l b_l \; \delta_l \; 0.$$

Using Descartes rule of signs we can now uniformly count the number Z_+^δ of positive and the number Z_-^δ of negative zeroes of $\chi(c)$ for all $c \in \mathbb{R}^m$ with $\gamma(c)$ and $\varphi_\delta(c)$.

Finally define φ by

$$\bigvee_{\delta \in \{<,=,>\}^l} \left\{ \varphi_\delta \;\middle|\; Z_+^\delta - Z_-^\delta = 0 \right\}.$$

Our formula φ states that the polynomial χ has exactly the same number of positive as of negative real zeroes. A quantifier-free formula with this property is called *type formula* for the polynomial χ. Recall from our discussion above that in this situation G has no zeroes which satisfy the given side conditions. Thus our final elimination result is $\gamma \wedge \neg \varphi$.

2.3   Constructing Equations

We enter this case of the Hermitian quantifier elimination if the input formula does not contain any equation or the dimension of $\mathrm{Id}(G(c))$ is n, i.e., $I(G) = \{0\}$ for $\underline{c}$ with $\gamma(\underline{c})$. In other words we consider the input formula

$$\exists x_1 \cdots \exists x_n \left(\bigwedge_{h \in H} h > 0 \wedge \bigwedge_{f \in F} f \neq 0 \right).$$

In this case, we can eliminate one quantifier, say x_n, and the other quantifiers of the considered block are eliminated by a recursive call of the Hermitian quantifier elimination.

Let h have a representation of the form $\sum_{k=0}^{d_h} a_{h,k} x_n^k$ where each $a_{h,k}$ is a polynomial in $\mathbb{Q}[u_1, \ldots, u_m, x_1, \ldots, x_{n-1}]$ with $a_{h,d_h} \neq 0$. Assume for a moment that $H = \{h_1, \ldots, h_r\}$ and let $D = \times_{k=1}^{r} \{0, \ldots, d_{h_k}\}$. For $\delta \in D$ we denote by δ_h the s-th element of δ such that $h_s = h$. Define $P = \{ (h_i, h_j) \mid 1 \leq i < j \leq r \} \subseteq H^2$.

For $h \in H$ and $d \in \{0, \ldots, d_h\}$ let Γ_d^h be the following formula:

$$\bigwedge_{k=d+1}^{d_h} a_{h,k} = 0 \wedge a_{h,d} \neq 0.$$

For fixed $h \in H$ the formulas Γ_d^h build a complete disjunctive case distinction of the degree of h under specifications of the parameters. Notice that it does not matter at this point, if Γ_d^h is equivalent to *false*. Let $\varrho_d(h) = \sum_{k=0}^{d} a_{h,k} x_n^k$. We have the equivalence

$$\exists x_n \Big(\bigwedge_{h \in H} h > 0 \wedge \bigwedge_{f \in F} f \neq 0 \Big) \longleftrightarrow \bigvee_{\delta \in D} \exists x_n \Big(\bigwedge_{h \in H} \Gamma_{\delta_h}^h \wedge \varrho_{\delta_h}(h) > 0 \wedge \bigwedge_{f \in F} f \neq 0 \Big).$$

For each δ we then in turn transform the formulas

$$\exists x_n \Big(\bigwedge_{h \in H} \Gamma_{\delta_h}^h \wedge \varrho_{\delta_h}(h) > 0 \wedge \bigwedge_{f \in F} f \neq 0 \Big)$$

separately to

$$\bigwedge_{f \in F} \bigvee_{i=0}^{d_f} a_{f,i} \neq 0 \wedge$$

$$\Big(\Big(\bigwedge_{h \in H} \Gamma_{\delta_h}^h \wedge a_{h,\delta_h} > 0 \Big) \vee$$

$$\Big(\bigwedge_{h \in H} \Gamma_{\delta_h}^h \wedge a_{h,\delta_h} < 0 \vee (\text{Even}(\delta_h) \wedge a_{h,\delta_h} > 0) \Big) \vee$$

$$\bigvee_{p \in Q} \exists x_n \Big(\bigwedge_{h \in H} \Gamma_{\delta_h}^h \wedge \varrho_{\delta_h}(h) > 0 \wedge \frac{\partial p}{\partial x_n} = 0 \Big) \vee$$

$$\bigvee_{(p,q) \in P} \exists x_n \Big(\bigwedge_{h \in H} \Gamma_{\delta_h}^h \wedge \varrho_{\delta_h}(h) > 0 \wedge (p - q) = 0 \Big) \Big),$$

where $Q = \{ h \in H \mid \delta_h \geq 2 \}$. The used predicate $\text{Even}(n)$ is true if and only if n is even. Thus we have shown how to trace back this case to the case with at least one existing equation in the input formula.

Let φ' denote the complete transformed input formula. Then we apply the Hermitian quantifier elimination to each quantified constituent and obtain, by eliminating x_n, a quantifier-free equivalent ψ'. Finally we apply the Hermitian quantifier elimination again recursively to

$$\exists x_1 \cdots \exists x_{n-1} (\psi')$$

obtaining a quantifier-free equivalent ψ. The final result of the elimination step is then

$$\gamma \wedge \psi.$$

2.4   Partial Elimination

We enter this case of the Hermitian quantifier elimination if the dimension of G is d with $d \in \{1, \ldots, n-1\}$. We compute a maximal strongly independent set Ξ [18]. Let w.l.o.g. be $\Xi = \{x_1, \ldots, x_k\}$. Then we apply recursively the Hermitian quantifier elimination to

$$\exists x_{k+1} \cdots \exists x_n \left(\bigwedge_{g \in G} g = 0 \wedge \bigwedge_{h \in H} h > 0 \wedge \bigwedge_{f \in F} f \neq 0 \right)$$

and obtain a quantifier-free formula ψ'. Then we apply our quantifier elimination procedure again recursively to

$$\exists x_1 \cdots \exists x_k (\psi')$$

yielding ψ. Our quantifier-free result is then $\gamma \wedge \psi$. This concludes the description of the Hermitian quantifier-elimination.

3 Generic Hermitian Quantifier Elimination

In this section we discuss our modifications to the algorithm for obtaining a generic quantifier elimination. As already mentioned in the introduction, a generic quantifier elimination computes for a first-order formula φ a quantifier-free formula φ' and a conjunction Θ of negated equations in the parameters u_1, $\ldots$, u_m such that

$$\Theta \longrightarrow (\varphi \longleftrightarrow \varphi').$$

Θ is called a theory. Recall from our discussion in the previous section that our quantifier elimination algorithm is recursive. In each recursive call we consider variables originally bound by quantifiers as additional parameters. Obviously we are not allowed to add assumptions about these additional parameters to Θ. To guarantee this restriction we denote by $v_1, \ldots, v_m$ the set of parameters of the input formula. In the discussion below we will always test whether an assumption is valid by checking whether it contains only variables from $\{v_1, \ldots, v_m\}$.

3.1 Generic Gröbner Systems

Our first and most prominent modification to the pure elimination algorithm is to compute in the preparation phase a generic Gröbner system instead of a regular one.

Let $<$ be a term order and let $p = c_1 t_1 + \cdots + c_d t_d$ be a polynomial in $\mathbb{Q}[u_1, \ldots, u_m][x_1, \ldots, x_n]$, where $c_1, \ldots, c_d \in \mathbb{Q}[u_1, \ldots, u_m]$ and $t_d > \cdots > t_1$ terms. Then the head term of p is c_d. For a given $c \in \mathbb{R}^m$ this may or may not be true for the polynomial $p(c)$. It depends on whether $c_d(c) \neq 0$ or not. During the construction of a Gröbner system we systematically construct a case distinction about some parameters of the occurring polynomials. In each case of this case distinction the head term of all polynomials is uniformly determined.

A generic Gröbner system allows us to exclude some cases by adding assumptions to Θ. In particular if c_d contains only parameters from $\{v_1, \ldots, v_m\}$ we add

$c_d \neq 0$ to Θ and assume in the following computation steps that the head term of p is t_d.

We denote by $\vdash$ a suitable heuristic to decide an implication: If $\gamma \vdash \alpha$, then we have the validity of $\gamma \longrightarrow \alpha$. Note that the construction of a Gröbner system requires that this heuristic can actually decide some implications.

The first algorithm extends a partial Gröbner system by an additional polynomial. Note that we assume that the theory Θ to be computed is globally available.

We use the following notations: $\mathrm{HC}(f)$ is the head or leading coefficient of f w.r.t. our fixed term order, $\mathrm{Red}(f)$ is the polynomial up to the head monomial, $\mathrm{Var}(f)$ is the set of variables actually occurring in f.

Algorithm 1 (extend) *Input: A partial system S, a branch (γ, G), and two polynomials h and h'. Output: An extended partial system.*

1   **if** $h' = 0$ **then**
2   return $S \cup \{(\gamma, G)\}$
3   **else if** $\mathrm{Var}(\mathrm{HC}(h')) \subseteq \{v_1, \ldots, v_m\}$ **then**
4       $\Theta := \Theta \wedge (\mathrm{HC}(h') \neq 0)$
5   return $S \cup \{(\gamma, G \cup \{h\})\}$
6   **else if** $\gamma \wedge \Theta \vdash \mathrm{HC}(h') \neq 0$ **then**
7   return $S \cup \{(\gamma, G \cup \{h\})\}$
8   **else if** $\gamma \wedge \Theta \vdash \mathrm{HC}(h') = 0$ **then**
9   return $\mathrm{extend}(S, (\gamma, G), h, \mathrm{Red}(h'))$
10 **else**
11      $S' := \{(\gamma \wedge \mathrm{HC}(h') \neq 0, G \cup \{h\})\}$
12   return $\mathrm{extend}(S', (\gamma \wedge \mathrm{HC}(h) = 0, G), h, \mathrm{Red}(h'))$
13 **fi**

This algorithm differs from the algorithm for regular Gröbner systems by accessing the theory Θ and by the lines 3 and 4 for generating new assumptions.

For computing a Gröbner system we start with computing an initial partial system S by calling the following algorithm Initialize with input $\hat{G}$.

Algorithm 2 (Initialize) *Input: A finite set H of polynomials. Output: A partial system.*

1   **begin**
2       $S := \{(\mathrm{true}, \emptyset)\}$
3   **for each** $h \in H$ **do**
4   **for each** $(\gamma, G) \in S$ **do**
5               $S := S \setminus \{(\gamma, G)\}$
6               $\mathrm{S} := \mathrm{extend}(S, (\gamma, G), h, h)$
7   **od**
8   **od**
9   **end**

For computing the Gröbner system from the partial system we proceed as follows: We select a branch (γ, G) of S, compute $S' = S \setminus \{(\gamma, G)\}$. Then we select g_1, g_2 from G such that the normal form h of the S-polynomial of g_1, g_2 is not 0. Finally we extend S' by (γ, G), h and h. This process is repeated until the normal form of all S-polynomials is 0.

As mentioned above the generic variant of the Gröbner system computation allows us to drop branches. Recall from the presentation of our quantifier elimination algorithm that we have to perform for each branch a separate quantifier elimination. If we are on the top-level of our quantifier-elimination algorithm we actually compute a Gröbner system containing one single branch, because the condition on line 3 of the algorithm "extend" is tautological in this situation. This reduces, in general, both the computation time and the size of the output formula dramatically. As a rule, observed from our sample computations, we compute only a few assumptions which can often be easily interpreted.

3.2 Generic Equation Construction

In Section 2.3 we have discussed how to construct an equation from a set of ordering relations. In this section we adapt this to the generic case.

Recall that we generate a complete case distinction about the highest coefficient of each $h \in H$. The size of this case distinction can be reduced by making appropriate assumptions as shown below.

For $h \in H$ let

$$n_h = \max\big(\{-1\} \cup \big\{ i \in \{0, \dots, d_h\} \mid \mathrm{Var}(a_{h,i}) \subseteq \{v_1, \dots, v_m\} \big\}\big).$$

For all n_h with $h \in H$ and $n_h \geq 0$ we add the assumption $a_{h,n_h} \neq 0$ to our theory Θ. Let finally $D' = \times_{k=1}^{r}\{\max(0, n_h), \dots, d_h\}$. Then we can proceed with the transformation described in Section 2.3 using D' instead of D. Note that $D' \subseteq D$ and often $D' \subsetneq D$.

3.3 Generic Type Formula Computation

In this section we discuss an approach to computing generic type formulas.

The type formula construction presented in Section 2.2 is a primitive version of the method used in our highly optimized Hermitian quantifier elimination. We actually compute a type formula τ_d for a polynomial $p = \sum_{i=0}^{d} c_i y^i$ of degree d recursively:

$$\tau_d(c_d, \dots, c_0) \equiv (c_0 = 0) \wedge \tau_{d-1} \vee \tau_d'(c_d, \dots, c_0).$$

The recursion basis are the simple type formulas up to the degree 3. The definition of τ_d' is similar to the definition of τ_d, but assumes a non-vanishing constant coefficient which implies the absence of the zero 0. The formula τ_d' is actually a disjunctive normal form. Each constituent has the following schema

$$c_{k_1} \varrho_{k_1} 0 \wedge \cdots \wedge c_{k_l} \varrho_{k_l} 0,$$

where $\{k_1, \dots, k_l\} \subseteq \{1, \dots, d\}$ and $\varrho_{k_j} \in \{<, >\}$.

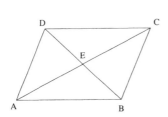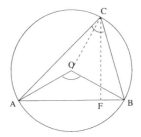

Fig. 1. Example 1 (left) and Example 2 (right)

For our generic type formula computation we cannot make use of our assumption for computing τ'_d. If $\mathrm{Var}(c_0) \subseteq \{v_1, \ldots, v_m\}$ we can however avoid the recursion by adding $c_0 \neq 0$ to Θ. This reduces the size of the output formula dramatically and if it occurs in the recursions it reduces the computation time, too. Our test computations have, however, shown that, in general, the assumptions made here are very complex and cannot be easily interpreted. For our application of automated theorem proving we have thus not analyzed this method further. This does not mean that generic type formulas are an irrelevant optimization for other application areas of generic quantifier elimination.

4   Examples

In this section we will apply our elimination algorithm to some automatic proofs of geometric theorems. We have implemented the algorithm in REDLOG 3.0, which is part of the current version 3.8 of the computer algebra system REDUCE.

For the geometric theorem proving we proceed here as described in [13]. Our examples will show the meaning of the assumptions which are created during the computations. In most cases, these assumptions can be interpreted as (necessary) non-degeneracy conditions, so they have a powerful geometric interpretation. Note that the constructed assumptions may not be a complete list of non-degeneracy conditions for the particular example. We will also show that generic Hermitian quantifier elimination will speed up the elimination procedure and will create smaller solution formulas than the regular Hermitian quantifier elimination. We explain in detail how to express a geometric theorem as a first-order formula by means of our first example.

Example 1. Given a parallelogram $ABCD$, let E be the intersection point of its diagonals. Then E is the midpoint of the diagonals (see Figure 1). This example was taken from [19]. By a suitable motion in $\mathbb{R}^2$ we can assume w.l.o.g.

$$A = (0,0), \quad B = (u_1, 0), \quad C = (u_2, u_3), \quad D = (x_2, x_1), \quad E = (x_4, x_3).$$

We now can describe the necessary properties to our statement by the following equations:

$$
\begin{array}{ll}
h_1 \equiv u_1 x_1 - u_1 u_3 = 0 & AB \| DC \\
h_2 \equiv u_3 x_2 - (u_2 - u_1)x_1 = 0 & DA \| CB \\
h_3 \equiv x_1 x_4 - (x_2 - u_1)x_3 - u_1 x_1 = 0 & E \in BD \\
h_4 \equiv u_3 x_4 - u_2 x_3 = 0 & E \in AC \\
g \equiv 2u_2 x_4 + 2u_3 x_3 - u_3^2 - u_2^2 = 0 & \text{Length}(AE) = \text{Length}(CE)
\end{array}
$$

The theorem can then be formulated as $\forall x_1 \forall x_2 \forall x_3 \forall x_4 (h_1 \wedge h_2 \wedge h_3 \wedge h_4 \rightarrow g)$. The application of our elimination algorithm leads in 30 ms to the result

$$
\Theta \equiv u_1 \neq 0 \wedge u_3 \neq 0, \quad \varphi' \equiv \text{true}.
$$

The theory Θ states that $ABCD$ is a proper parallelogram, i.e., opposite edges do not collapse. Non-generic Hermitian quantifier elimination yields in 170 ms a quantifier-free formula consisting of 24 atomic formulas.

Example 2. Let O be the center of the circumcircle of a triangle ABC. If O does not lie outside of ABC, then $\angle ACB = \angle AOB/2$. See Figure 1. This example was taken from [20].

W.l.o.g. we can assume the following coordinates

$$
A = (-u_1, 0), \quad B = (u_1, 0), \quad C = (u_2, u_3), \quad O = (0, x_1), \quad F = (0, u_3).
$$

We express the theorem as follows:

$$
\begin{aligned}
& \forall r \forall x_1 \forall t_1 \forall t_2 \forall t \forall t' \big(r^2 = u_1^2 + x_1^2 \wedge r^2 = u_2^2 + (u_3 - x_1)^2 \wedge \\
& u_3 t_1 = u_1 + u_2 \wedge u_3 t_2 = u_1 - u_2 \wedge (1 - t_1 t_2)t = t_1 + t_2 \wedge \\
& x_1 t' = u_1 \rightarrow t = t' \big).
\end{aligned}
$$

Generic Hermitian quantifier elimination on this formula leads in 10 ms to the result

$$
\Theta \equiv u_1^2 - u_2^2 - u_3^2 \neq 0 \wedge u_3 \neq 0, \quad \varphi \equiv \text{true}.
$$

We now take a closer look at Θ. $u_3 \neq 0$ ensures that not all points of the triangle lie on the x-axis. This is a necessary non-degeneracy condition. The assumption $u_1^2 - u_2^2 - u_3^2 \neq 0$ prevents that the midpoint of the circumcircle lies on the edge AB. We have proved this theorem if the constructed non-degeneracy assumptions hold. Actually the theorem holds for $u_3 \neq 0$, i.e., the second assumption is superfluous.

Example 3. (Feuerbach's Theorem) *The nine point circle of a triangle is tangent to the incircle and to each of the excircles of the triangle.* See [19] for a formulation of this problem. We get the following elimination result:

$$
\begin{aligned}
\Theta \equiv {}& u_1 u_3 + u_2^2 \neq 0 \wedge u_1 + 2u_2 + u_3 \neq 0 \wedge u_1 + u_2 \neq 0 \wedge u_1 - u_2 \neq 0 \\
& \wedge u_1 - 2u_2 + u_3 \neq 0 \wedge u_1 - u_3 \neq 0 \wedge u_1 \neq 0 \wedge u_2 + u_3 \neq 0 \wedge u_2 - u_3 \neq 0 \wedge \\
& u_2 \neq 0 \wedge u_3 \neq 0, \\
\varphi \equiv {}& u_1 - u_3 \neq 0.
\end{aligned}
$$

φ is obviously equivalent to *true* under the assumption of Θ. While we receive this result in 350 ms, regular Hermitian quantifier elimination cannot eliminate the quantifiers using 128MB.

Example 4. (M. Paterson's problem) *Erect three similar isosceles triangles A_1BC, AB_1C, and ABC_1 on the sides of a triangle ABC. Then AA_1, BB_1 and CC_1 are concurrent. How does the point of concurrency moves as the areas of the three similar triangles are varied between 0 and ∞.* This example is actually an example for theorem finding and not only theorem proving. See Chou [19] for a description of this problem. We get the following elimination result:

$$\Theta \equiv u_1u_2 - u_2x - u_3y \neq 0 \wedge u_2 - x \neq 0 \wedge u_2 \neq 0 \wedge u_3 - y \neq 0 \wedge y \neq 0,$$
$$\varphi' \equiv u_1^2u_2y + u_1^2u_3x - 2u_1^2xy + u_1u_2^2y - 2u_1u_2u_3x + 2u_1u_2xy - u_1u_3^2y$$
$$-u_1u_3x^2 + u_1u_3y^2 - 2u_2^2xy + 2u_2u_3x^2 - 2u_2u_3y^2 + 2u_3^2xy = 0 \vee u_1 = 0.$$

The result is obtained in 60 ms and describes a geometric locus. If one uses nongeneric Hermitian quantifier elimination for eliminating, the result is obtained in 2,8 seconds and consists of 295 atomic formulas.

5   Conclusions

We have presented a generic quantifier elimination method based on Hermitian quantifier elimination. For this purpose we have analyzed where making assumptions on parameters may support the algorithm: We compute generic Gröbner systems instead of regular ones reducing the practical complexity of our algorithm in all cases. In the special case that no equations occur in the input, we have additionally reduced the number of recursions needed.

By example computations we have shown that our generic Hermitian quantifier elimination can be successfully used for automatic theorem proving and theorem finding. In all examples the results are considerably shorter and the computation times are much faster than for regular Hermitian quantifier elimination.

References

1. Collins, G.E.: Quantifier elimination for the elementary theory of real closed fields by cylindrical algebraic decomposition. In Brakhage, H., ed.: Automata Theory and Formal Languages. 2nd GI Conference. Volume 33 of Lecture Notes in Computer Science. Springer-Verlag, Berlin, Heidelberg, New York (1975) 134–183
2. Collins, G.E., Hong, H.: Partial cylindrical algebraic decomposition for quantifier elimination. Journal of Symbolic Computation **12** (1991) 299–328
3. Brown, C.W.: Simplification of truth-invariant cylindrical algebraic decompositions. In Gloor, O., ed.: Proceedings of the 1998 International Symposium on Symbolic and Algebraic Computation (ISSAC 98), Rostock, Germany, ACM, ACM Press, New York (1998) 295–301

4. Weispfenning, V.: The complexity of linear problems in fields. Journal of Symbolic Computation **5** (1988) 3–27

5. Loos, R., Weispfenning, V.: Applying linear quantifier elimination. THE Computer Journal **36** (1993) 450–462 Special issue on computational quantifier elimination.

6. Dolzmann, A., Sturm, T.: Redlog: Computer algebra meets computer logic. ACM SIGSAM Bulletin **31** (1997) 2–9

7. Weispfenning, V.: A new approach to quantifier elimination for real algebra. In Caviness, B., Johnson, J., eds.: Quantifier Elimination and Cylindrical Algebraic Decomposition. Texts and Monographs in Symbolic Computation. Springer, Wien, New York (1998) 376–392

8. Dolzmann, A.: Reelle Quantorenelimination durch parametrisches Zählen von Nullstellen. Diploma thesis, Universität Passau, D-94030 Passau, Germany (1994)

9. Gilch, L.A.: Effiziente Hermitesche Quantorenelimination. Diploma thesis, Universität Passau, D-94030 Passau, Germany (2003)

10. Dolzmann, A.: Solving geometric problems with real quantifier elimination. In Gao, X.S., Wang, D., Yang, L., eds.: Automated Deduction in Geometry. Volume 1669 of Lecture Notes in Artificial Intelligence (Subseries of LNCS). Springer-Verlag, Berlin Heidelberg (1999) 14–29

11. Dolzmann, A., Sturm, T.: Simplification of quantifier-free formulae over ordered fields. Journal of Symbolic Computation **24** (1997) 209–231

12. Weispfenning, V.: Applying quantifier elimination to problems in simulation and optimization. Technical Report MIP-9607, FMI, Universität Passau, D-94030 Passau, Germany (1996) To appear in the Journal of Symbolic Computation.

13. Dolzmann, A., Sturm, T., Weispfenning, V.: A new approach for automatic theorem proving in real geometry. Journal of Automated Reasoning **21** (1998) 357–380

14. Seidl, A., Sturm, T.: A generic projection operator for partial cylindrical algebraic decomposition. In Sendra, R., ed.: Proceedings of the 2003 International Symposium on Symbolic and Algebraic Computation (ISSAC 03), Philadelphia, Pennsylvania. ACM Press, New York, NY (2003) 240–247

15. Weispfenning, V.: Comprehensive Gröbner bases. Journal of Symbolic Computation **14** (1992) 1–29

16. Pedersen, P., Roy, M.F., Szpirglas, A.: Counting real zeroes in the multivariate case. In Eysette, F., Galigo, A., eds.: Computational Algebraic Geometry. Volume 109 of Progress in Mathematics. Birkhäuser, Boston, Basel; Berlin (1993) 203–224 Proceedings of the MEGA 92.

17. Becker, E., Wörmann, T.: On the trace formula for quadratic forms. In Jacob, W.B., Lam, T.Y., Robson, R.O., eds.: Recent Advances in Real Algebraic Geometry and Quadratic Forms. Volume 155 of Contemporary Mathematics., American Mathematical Society, American Mathematical Society, Providence, Rhode Island (1994) 271–291 Proceedings of the RAGSQUAD Year, Berkeley, 1990–1991.

18. Kredel, H., Weispfenning, V.: Computing dimension and independent sets for polynomial ideals. Journal of Symbolic Computation **6** (1988) 231–247 Computational aspects of commutative algebra.

19. Chou, S.C.: Mechanical Geometry Theorem Proving. Mathematics and its applications. D. Reidel Publishing Company, Dordrecht, Boston, Lancaster, Tokyo (1988)

20. Sturm, T.: Real Quantifier Elimination in Geometry. Doctoral dissertation, Department of Mathematics and Computer Science. University of Passau, Germany, D-94030 Passau, Germany (1999)

Extending Finite Model Searching
with Congruence Closure Computation

Jian Zhang[1,*] and Hantao Zhang[2,**]

[1] Laboratory of Computer Science
Institute of Software, Chinese Academy of Sciences
Beijing 100080, China
zj@ios.ac.cn
[2] Department of Computer Science
University of Iowa Iowa City, IA 52242, USA
hzhang@cs.uiowa.edu

Abstract. The model generation problem, regarded as a special case of the Constraint Satisfaction Problem (CSP), has many applications in AI, computer science and mathematics. In this paper, we describe how to increase propagation of constraints by using the ground congruence closure algorithm. The experimental results show that using the congruence closure algorithm can reduce the search space for some benchmark problems.

1 Introduction

Compared to the research on propositional satisfiability problem, the satisfiability of first-order formulas has not received much attention. One reason is that the problem is undecidable in general. Since the early 1990's, several researchers have made serious attempts to solving the *finite* domain version of the problem. More specifically, the problem becomes deciding whether the formula is satisfiable in a given finite domain. Several model generation programs have been constructed [6, 3, 2, 10, 14, 16]. By *model generation* we mean, given a set of first order formulas as axioms, finding their models automatically. A *model* is an interpretation of the function and predicate symbols over some domain, which satisfies all the axioms. Model generation is very important to the automation of reasoning. For example, the existence of a model implies the satisfiability of an axiom set or the consistency of a theory. A suitable model can also serve as a counterexample which shows some conjecture does not follow from some premises. In this sense, model generation is complementary to classical theorem proving. Models help people understand a theory and can guide conventional theorem provers in finding proofs.

---

* Supported by the National Science Fund for Distinguished Young Scholars of China under grant 60125207. Part of the work was done while the first author was visiting the University of Iowa.
** Supported in part by NSF under grant CCR-0098093.

B. Buchberger and J.A. Campbell (Eds.): AISC 2004, LNAI 3249, pp. 94–102, 2004.

Some of the model generation methods are based on first-order reasoning (e.g., SATCHMO [6] and MGTP [2, 3]); some are based on constraint satisfaction (e.g., FINDER [10], FALCON [14] and SEM [16]); while others are based on the propositional logic (e.g., ModGen [4] and MACE [7]). These tools have been used to solve a number of challenging problems in discrete mathematics [11, 7, 14]. Despite such successes, there is still space for improvement on the performance of the tools.

In this paper, we study how to improve the performance of finite model searchers by more powerful reasoning mechanisms. Specifically, we propose to incorporate congruence closure computation into the search procedure of the model generation tools.

2   Model Generation as Constraint Satisfaction

The finite model generation problem studied in this paper is stated as follows. Given a set of first order clauses and a non-empty finite domain, find an interpretation of all the function symbols and predicate symbols appearing in the clauses such that all the clauses are true under this interpretation. Such an interpretation is called a *model*. Here we assume that all the input formulas are clauses. Each variable in a clause is (implicitly) universally quantified.

Without loss of generality, we assume that an n-element domain is the set D_n = { $0, 1, \ldots, n-1$ }. The Boolean domain is { FALSE, TRUE }. If the arity of each function/predicate symbol is at most 2, a finite model can be conveniently represented by a set of *multiplication tables*, one for each function/predicate. For example, a 3-element model of the clause $f(x, x) = x$ is like the following:

f	0	1	2
0	0	1	0
1	1	1	0
2	0	1	2

Here f is a binary function symbol and its interpretation is given by the above 2-dimensional matrix. Each entry in the matrix is called a *cell*.

In this paper, we treat the problem as a constraint satisfaction problem (CSP), which has been studied by many researchers in Artificial Intelligence. The variables of the CSP are the *cell* terms (i.e., ground terms like $f(0, 0)$, $f(0, 1)$, etc.). The domain of each variable is D_n (except for predicates, whose domain is the Boolean domain). The constraints are the set of ground instances of the input clauses, denoted by Ψ. The goal is to find a set of assignments to the cells (e.g., $f(0, 1) = 2$) such that all the ground clauses hold.

Theoretically speaking, any approach of constraint satisfaction can be used for finite model generation. For instance, a simple backtracking algorithm can always find a finite model (if it exists). Of course, a brute-force search procedure is too inefficient to be of any practical use. There are many other search procedures and heuristics proposed in the AI literature, e.g., forward checking and lookahead. See [5] for a good survey.

In this paper, we will solve the finite model generation problem using backtrack search. The basic idea of such a search procedure is roughly like the following: repeatedly extend a partial model (denoted by *Pmod*) until it becomes a complete model (in which every cell gets a value). Initially *Pmod* is empty. *Pmod* is extended by selecting an unassigned cell and trying to find a value for it (from its domain). When no value is appropriate for the cell, backtracking is needed and *Pmod* becomes smaller.

The execution of the search procedure can be represented as a search tree. Each node of the tree corresponds to a partial model, and each edge corresponds to assigning a value to some cell by the heuristic. We define the *level* of a node as usual, i.e., the level of the root is 0 and the level of the children of a level n node is $n + 1$.

3   Congruence Closure for Constraint Propagation

The efficiency of the search procedure depends on many factors. One factor is how we can perform reasoning to obtain useful information when a certain number of cells are assigned values. That is, we have to address the following issue: how can we implement constraint propagation and consistency checking efficiently?

This issue may be trivial for some constraint satisfaction algorithms because the constraints they accept are often assumed to be unary or binary. It is true that n-ary constraints can be converted into an equivalent set of binary constraints; but this conversion usually entails the introduction of new variables and constraints, and hence an increase in problem size. This issue is particularly important to model generation because in this case, the constraints are represented by complicated formulas. Experience tells us that a careful implementation can improve the performance of a program significantly.

In [15], a number of inference rules are described in detail. They are quite effective on many problems. In this paper, we discuss another inference rule, namely *congruence closure*, which can be quite useful for equational problems.

3.1   Congruence Closure

An *equivalence relation* is a reflexive, symmetric and transitive binary relation. A relation $\sim$ on the terms is *monotonic* if $f(s_1, \ldots, s_n) \sim f(t_1, \ldots, t_n)$ whenever f is an n-ary function symbol and for every i ($1 \leq i \leq n$), $s_i \sim t_i$. A *congruence relation* is a monotonic equivalence relation. A set of ground equations, E, defines a relation among the ground terms. The *congruence generated* by E, denoted by E^*, is the smallest congruence relation containing E.

There are several algorithms for computing the congruence generated by a set of ground equations, called *congruence closure* algorithms, e.g., [1,8]. The Nelson-Oppen algorithm [8] represents terms by vertices in a directed graph. It uses UNION and FIND to operate on the partition of the vertices, to obtain the congruence relation.

A congruence closure algorithm can deduce some useful information for finite model searching. One example is given in [8]. From the following two equations:

```
f(f(f(a))) = a
f(f(f(f(f(a))))) = a
```

we can deduce that `f(a) = a`. Thus we need not "guess" a value for `f(a)`, if the above two equations are in the input clauses. Without using the congruence closure, we may have to try unnecessary assignments such as `f(a) = b`, `f(a) = c`, etc., before or after the assignment of `f(a) = a`.

3.2   An Example

Let us look at a problem in the combinatorial logic. A fragment of combinatory logic is an equational system defined by some equational axioms. We are interested in the fragment { B, N1 }, whose axioms are:

$$a(a(a(\mathrm{B}, x), y), z) = a(x, a(y, z))$$
$$a(a(a(\mathrm{N1}, x), y), z) = a(a(a(x, y), y), z)$$

Here B and N1 are constants, while the variables x, y and z are universally quantified. The *strong fixed point property* holds for a fragment of combinatory logic if there exists a combinator y such that for all combinators x, $a(y, x) = a(x, a(y, x))$. In other words, the fragment has the strong fixed point property if the formula $\varphi\colon \exists y \forall x\, [a(y, x) = a(x, a(y, x))]$ is a logical consequence of the equational axioms.

In [13], it is shown that the fragment { B, N1 } does not have the strong fixed point property, because there is a counterexample of size 5. It is a model of the following formulas:

```
(BN1-1)   a(a(a(0,x),y),z) = a(x,a(y,z)).
(BN1-2)   a(a(a(1,x),y),z) = a(a(a(x,y),y),z).
(BN1-3)   a(y,f(y)) != a(f(y),a(y,f(y))).
```

The last formula is the negation of the formula φ, where f is a Skolem function and != means "not equal". In the first two formulas, we assume that B is 0 and N1 is 1. That is, B and N1 take different values. (But it is also possible to generate a counterexample in which B = N1.)

Search by SEM. When using the standard version of SEM [16], the first few steps of the search tree are the following:

(1) Choose the cell `a(0,0)` and assign the value 0 to it.
(2) Choose the cell `f(0)` and assign the value 1 to it. Note that `f(0)` cannot be 0; otherwise the formula (BN1-3) does not hold, because we already have `a(0,0) = 0`.

(3) Choose the cell a(0,1) and assign the value 1 to it. After the first two choices, SEM deduces that a(0,1) cannot be 0. Otherwise, suppose a(0,1) = 0. Let x = 1 and y = z = 0 in the formula (BN1-1), we have a(a(a(0,1),0),0) = a(1,a(0,0)), which is simplified to 0 = a(1,0). Then let y = 0 in the formula (BN1-3), we shall have 0 ≠ 0.

(4) Choose the cell a(1,1) and assign the value 2 to it.

After the first three steps, SEM will also deduce that a(1,1) ≠ 1.

The Effect of Congruence Closure Computation. If we apply the congruence closure algorithm to the ground instances of the formulas (BN1-1) and (BN1-2) and the first three assignments

$$a(0,0) = 0; \quad f(0) = 1; \quad a(0,1) = 1; \qquad (P1)$$

we shall get the conclusion a(1,1) = 1. This contradicts with the inequality a(1,1) ≠ 1 which is deduced by SEM. Thus if we extend SEM's reasoning mechanism with the congruence closure algorithm, we know that, at step 3, assigning the value 1 to the cell a(0,1) will lead to a dead end.

Now suppose we try a(0,1) = 2 next. For this new branch, we add the following three equations to Ψ:

$$a(0,0) = 0; \quad f(0) = 1; \quad a(0,1) = 2. \qquad (P2)$$

After computing the congruence closure, we can get these equations:

$$a(0,0) = a(1,0) = a(2,0) = 0;$$
$$f(0) = 1;$$
$$a(0,1) = a(0,2) = a(1,2) = a(2,2) = 2.$$

However, with the old version of SEM, after the three assignments (P2), we only deduce a(0,2) = 2 as a new cell assignment. In other words, the partial model consists of these four assignments:

$$a(0,0) = 0; \quad f(0) = 1; \quad a(0,1) = 2; \quad a(0,2) = 2.$$

SEM can also deduce some negative assignments, e.g., a(1,2) ≠ 2. This contradicts with what we get from the closure computation algorithm. Thus, we have come to a dead end again. We don't need to go further and expand the search tree (as the old version of SEM does). The difference is illustrated by the following two search trees:

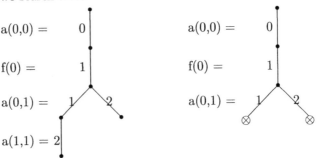

The left tree describes the search process of old SEM, while the right one describes the effect of congruence closure computation on the search process. We see that some branches of the search tree can be eliminated if we combine congruence closure computation with existing reasoning methods of SEM.

4   The Extended Search Algorithm

The model generation process can be described by the recursive procedure in Fig. 1, which searches for every possible model. It can be easily modified to search for only one model. The procedure uses the following parameters:

- $Pmod = \{(ce, e) \mid e \in Dom(ce)\}$: assignments (cells and their assigned values), where $Dom(ce)$ is the domain of values for ce;
- $\mathcal{D} = \{(ce, D) \mid D \subset Dom(ce)\}$: unassigned cells and their possible values;
- Ψ: constraints (i.e. the clauses).

Initially $Pmod$ is empty, and $\mathcal{D}$ contains $(ce, Dom(ce))$ for every cell ce.

```
proc search(Pmod, D, Ψ)
{
    if D = ∅ then /* a model is found */
        { print(Pmod); return; }
    choose and delete (ce_i, D_i) from D;
    if D_i = ∅ then return; /* no model */
    for e ∈ D_i do
    {
        (Pmod', D', Ψ') := propa(Pmod ∪ {(ce_i, e)}, D, Ψ);
        if Ψ' is not FALSE /* no contradiction found */
            then   search(Pmod', D', Ψ');
    }
}
```

Fig. 1. The abstract search procedure

The procedure propa($Pmod, \mathcal{D}, \Psi$) propagates assignment $Pmod$ in Ψ: it simplifies Ψ and may force some variables in $\mathcal{D}$ to be assigned. The procedure propa($Pmod, \mathcal{D}, \Psi$) is essentially a closure operation (with respect to a set of sound inference rules). It repeatedly modifies $Pmod, \mathcal{D}$, and Ψ until no further changes can be made. When it exits, it returns the modified triple $(Pmod, \mathcal{D}, \Psi)$. The basic steps of this procedure can be described as follows.

(1) For each new assignment (ce, e) in $Pmod$, replace the occurrence of ce in Ψ by e.
(2) If there exists an empty clause in Ψ (i.e., each of its literals becomes FALSE during the propagation), replace Ψ by FALSE, and exit from the procedure. Otherwise, for every unit clause in Ψ (i.e. all but one of its literals become FALSE), examine the remaining literal l.

  - If l is a Boolean cell term ce, and $(ce, D) \in \mathcal{D}$, then delete (ce, D) from $\mathcal{D}$ and add (ce, TRUE) to $Pmod$; similarly, if l is the negation of a Boolean cell term in $\mathcal{D}$, i.e. $\neg ce$, delete (ce, D) from $\mathcal{D}$ and add (ce, FALSE) to $Pmod$.
  - If l is of the form $\text{EQ}(ce, e)$ (or $\text{EQ}(e, ce)$), and $(ce, D) \in \mathcal{D}$, delete (ce, D) from $\mathcal{D}$ and add (ce, e) to $Pmod$; similarly, if l is of the form $\neg\text{EQ}(ce, e)$ (or $\neg\text{EQ}(e, ce)$), and $(ce, D) \in \mathcal{D}$, then delete e from D.
(3) For each pair $(ce, D) \in \mathcal{D}$, if $D = \emptyset$, then replace Ψ by FALSE, and exit from the procedure; if $D = \{e\}$ (i.e. $|D| = 1$), then delete the pair (ce, D) from $\mathcal{D}$, and add the assignment (ce, e) to $Pmod$.
(4) Let $E^* = \text{groundCC}(Pmod \cup E(\Psi))$, where $E(X)$ is the set of all equations in X and $\text{groundCC}(Q)$ returns the consequences of the congruence closure of ground equations Q. If E^* and $Pmod$ are inconsistent, replace Ψ by FALSE and exit from the procedure; otherwise extend $Pmod$ with all cell assignments in E^*.

The last item in the above list is an addition to the original constraint propagation procedure implemented in SEM. Basically, the extended search algorithm tries to deduce useful information using congruence closure computation, from all the equations in Ψ and all the cell assignments in $Pmod$.

When will E^* and $Pmod$ be inconsistent? Firstly, if E^* puts two domain elements in the same equivalence class (e.g., 1 = 2, or TRUE = FALSE), we get an inconsistency. Secondly, if E^* contains $ce = v$ but v is not in $Dom(ce)$, then it is also inconsistent. When consistent, we extend the current partial assignment $Pmod$ with all cell assignments in E^*, i.e. equations like $\text{a}(0,1) = 2$.

5   Implementation and Experiments

We have extended SEM [16] with the Nelson-Oppen algorithm [8] for computing congruences. We call the new version SEMc.

The implementation of the congruence closure algorithm is straightforward. No advanced data structures are used. As a result, the efficiency is not so good. But this still allows us to experiment with the new search procedure.

Since congruence computation may not offer new information at all nodes of the search tree, we add a control parameter (denoted by Lvl) to SEMc. It means that congruence is only computed at nodes whose levels are less than Lvl.

We have experimented with SEMc on some model generation problems. Table 1 compares the performances of SEMc with that of SEM. The results were obtained on a Dell Optiplex GX270 (Pentium 4, 2.8 GHz, 2G memory), running RedHat Linux. In the table, the running times are given in seconds. A "round" refers to the number of times when we try to find an appropriate value for a selected cell. This can be done either when extending a partial solution (i.e., trying to assign a value for a new cell) or during backtracking (i.e., trying to assign a new value to a cell which already has a value).

Here BN1 refers to the combinatorial logic problem described in Section 3. The problem g1x has the following two clauses:

Table 1. Performance comparison

Problem	Size	Satisfiable	SEM		SEMc	
			Round	Time	Round	Time
BN1	4	No	888	0.00	502	0.22
BN1	5	Yes	34	0.00	26	0.02
g1x	4	Yes	11803	0.05	675	0.11
BOO033-1	5	Yes	29	0.00	27	0.05
LCL137-1	6	Yes	609	0.01	569	0.03
LCL137-1	8	Yes	5662	0.09	5466	0.15
ROB012-1	3	No	983694	2.55	982652	2.69
ROB015-1	3	No	1004638	2.25	1001420	2.35

```
f(z,f(g(f(y,z)),f(g(f(y,g(f(y,x)))),y))) = x
f(0,g(0)) != f(1,g(1))
```

The existence of a model implies that the first equation is not a single axiom for group theory. The other problems are from TPTP [12].

For some problems, there are non-unit clauses as well as negated equations. Thus the use of congruence computation does not make much difference. For problems like g1x where complex equations dominate in the input, the new version of SEM can reduce the number of branches greatly (by 17 times).

In the above experiments, *Lvl* is set to 10 (an arbitrary value). Without the restriction, the overhead of computing congruence closure might be significant.

6   Concluding Remarks

Many factors can affect the efficiency of a backtrack procedure. In the context of finite model searching, these include the heuristics for choosing the next cell and the inference rules for deducing new information from existing assignments. In this paper, we have demonstrated that adding congruence computation as a new inference rule can reduce the number of branches of the search tree, especially when most clauses are complex equations.

The current implementation is not so efficient because we have implemented only the original Nelson-Oppen algorithm [8] for computing congruences. With more efficient data structures and algorithms (e.g. [9]), we expect that the running time of SEMc can be reduced. Another way of improvement is that the congruence should be computed incrementally. In the current version of SEMc, each time the congruence is computed, the algorithm starts from all ground instances of the input equations. However, during the search, many of them can be neglected (when both the left-hand side and the right-hand side are reduced to the same value).

Acknowledgements

We are grateful to the anonymous reviewers for their detailed comments and suggestions.

References

1. P.J. Downey, R. Sethi and R.E. Tarjan, Variations on the common subexpression problem, *J. ACM* 27(4): 758–771, 1980.
2. M. Fujita, J. Slaney and F. Bennett, Automatic generation of some results in finite algebra, *Proc. 13th IJCAI*, 52–57, 1993.
3. R. Hasegawa, M. Koshimura and H. Fujita, MGTP: A parallel theorem prover based on lazy model generation, *Proc. CADE-11*, LNAI 607, 776–780, 1992.
4. S. Kim and H. Zhang, ModGen: Theorem proving by model generation, *Proc. 12th AAAI*, 162–167, 1994.
5. V. Kumar, Algorithms for constraint satisfaction problems: A survey, *AI Magazine* 13(1): 32–44, 1992.
6. R. Manthey and F. Bry, SATCHMO: A theorem prover implemented in Prolog, *Proc. CADE-9*, LNCS 310, 415–434, 1988.
7. W. McCune, A Davis-Putnam program and its application to finite first-order model search: Quasigroup existence problems. Technical Report ANL/MCS-TM-194, Argonne National Laboratory, 1994.
8. G. Nelson and D.C. Oppen, Fast decision procedures based on congruence closure, *J. ACM* 27(2): 356–364, 1980.
9. R. Nieuwenhuis and A. Oliveras, Congruence closure with integer offsets, *Proc. 10th LPAR*, LNAI 2850, 78–90, 2003.
10. J. Slaney, FINDER: Finite domain enumerator – system description, *Proc. CADE-12*, LNCS 814, 798–801, 1994.
11. J. Slaney, M. Fujita and M. Stickel, Automated reasoning and exhaustive search: Quasigroup existence problems, *Computers and Mathematics with Applications* 29(2): 115–132, 1995.
12. G. Sutcliffe and C. Suttner, The TPTP problem library for automated theorem proving. http://www.cs.miami.edu/~tptp/
13. J. Zhang, Problems on the generation of finite models, *Proc. CADE-12*, LNCS 814, 753–757, 1994.
14. J. Zhang, Constructing finite algebras with FALCON, *J. Automated Reasoning* 17(1): 1–22, 1996.
15. J. Zhang and H. Zhang, Constraint propagation in model generation, *Proc. Int'l Conf. on Principles and Practice of Constraint Programming*, LNCS 976, 398–414, 1995.
16. J. Zhang and H. Zhang, SEM: A system for enumerating models, *Proc. 14th IJCAI*, 298–303, 1995.

On the Combination
of Congruence Closure and Completion

Christelle Scharff[1,*] and Leo Bachmair[2]

[1] Department of Computer Science, Pace University, NY, USA
cscharff@pace.edu
[2] Department of Computer Science, SUNY Stony Brook, NY, USA
leo@cs.sunysb.edu

Abstract. We present a graph-based method for constructing a congruence closure of a given set of ground equalities that combines the key ideas of two well-known approaches, completion and abstract congruence closure, in a natural way by relying on a specialized and optimized version of the more general, but less efficient, SOUR graphs. This approach allows for efficient implementations and a visual presentation that better illuminates the basic ideas underlying the construction of congruence closures and clarifies the role of original and extended signatures and the impact of rewrite techniques for ordering equalities.

1 Introduction

Theories presented by finite sets of ground (i.e., variable-free) equalities are known to be decidable. A variety of different methods for solving word problems for ground equational theories have been proposed, including algorithms based on the computation of a congruence closure of a given relation. Efficient congruence closure algorithms have been described in [5, 8, 10, 12]. These algorithms typically depend on sophisticated, graph-based data structures for representing terms and congruence relations.

A different approach to dealing with ground equational theories is represented by term rewriting [1, 4], especially the completion method [7] for transforming a given set of equalities into a convergent set of directed rules that defines unique normal forms for equal terms and hence provides a decision procedure for the word problem of the underlying equational theory. Completion itself is a semi-decision procedure but under certain reasonable assumptions about the strategy used to transform equalities, is guaranteed to terminate if the input is a set of ground equalities. Completion methods are not as efficient as congruence closure, though an efficient ground completion method has been described by [13], who obtains an $O(n \, log(n))$ algorithm that cleverly uses congruence closure to transform a given set of ground equalities into a convergent ground rewrite system. Standard completion is quadratic in the worst case [11].

---

* This work is supported by the National Science Foundation under grant ITR-0326540.

We combine the two approaches in a novel way, different from [6], and present an efficient graph-based method that combines the key ideas of completion and abstract congruence closure [2, 3]. Our approach employs a specialized version of the SOUR graphs that were developed for general completion [9]. In SOUR graphs the vertices represent terms and the edges carry information about sub-term relations between terms (S), rewrite rules (R), unifiability of terms (U) and order relations between terms (O). In the application to congruence closure we consider only ground terms and hence do not need unification edges. Moreover, general term orders are too restrictive as well as too expensive to maintain, and hence we also dispense with order edges and manipulate rewrite edges in a differ-ent way, based on the explicit use of edges (E) representing unordered equalities. Thus, our modifications amount to what might be called "SER graphs." This modified, and simplified, graph structure provides a suitable basis for computing congruence closures. We represent terms and equalities by a directed graph that supports full structure sharing. The vertices of the graph represent terms, or more generally equivalence classes of terms; edges represent the subterm struc-ture of the given set of terms, as well as equalities and rewrite rules.

Some of the graph transformation rules we use are simpler versions of SOUR graph rules and, in logical terms, correspond to combinations of critical pair computations and term simplifications. We also include an explicit "merge rule" that is well-known from congruence closure algorithms, but only implicitly used in SOUR graphs. Exhaustive application of these transformation rules termi-nates, is sound in that the equational theory represented over Σ-terms does not change, and complete in that the final rewrite system over the extended signature is convergent.

The main difference between our approach and the abstract congruence clo-sure framework of [3] is that the graph-based formalism naturally supports a term representation with full structure sharing, which can not be directly de-scribed in an abstract, inference-based framework, though the effect of graph transformation rules can be indirectly simulated by suitable combinations of inference rules.

The efficiency of our method crucially depends on the use of a simple ordering (that needs to be defined only on the set of constants extending the original term signature), rather than a full term ordering. The corresponding disadvantage is that we do not obtain a convergent rewrite system over the original signature, but only over an extended signature. However, we may obtain a convergent rewrite system on the original signature by further transforming the graph in a way reminiscent of the compression and selection rules of [2, 3].

We believe that our approach allows for a visual presentation that better illuminates the basic ideas underlying the construction of congruence closures. In particular, it clarifies the role of original and extended signatures and the impact of rewrite techniques for ordering equalities. It should also be suitable for educational purposes.

The graph-based construction of a congruence closure is described in Sec-tion 3. In Section 4 we show how to obtain a convergent rewrite system over the original signature. Section 5 contains examples. In Section 6 we discuss the influence of the ordering on the efficiency of our method and present complexity results. A full version of this article with proofs and more examples is available at: `http://www.csis.pace.edu/~scharff/CC`.

2 Preliminaries

We assume the reader is familiar with standard terminology of equational logic and rewriting. Key definitions are included below, more details can be found in [1, 4]. In the following, let Σ and $\mathcal{K}$ be disjoint sets of function symbols. We call Σ the (basic) *signature*, and $\Sigma \cup \mathcal{K}$ the *extended signature*. The elements of $\mathcal{K}$ are assumed to be constants, and are denoted by subscripted letters c_i ($i \geq 0$). **Flat Ground Terms.** The height H of a term is recursively defined by: if t is a variable, then $H(t) = 0$, otherwise $H(t) = H(f(t_1, \ldots, t_n)) = 1 + max\{H(t_1), \ldots, H(t_n)\}$. A term is said to be *flat* if its height is 2 at most. We will consider flat ground terms, i.e., variable-free terms t with $H(t) \leq 2$.

D-Rules, C-Rules, C-Equalities. A *D-rule* on $\Sigma \cup \mathcal{K}$ is a rewrite rule $f(c_1, \ldots, c_n) \rightarrow c_0$, where $f \in \Sigma$ is a function symbol of arity n and $c_0, c_1, \ldots, c_n$ are constants of $\mathcal{K}$. A *C-rule* on $\Sigma \cup \mathcal{K}$ (respectively, a *C-equality*) is a rule $c_0 \rightarrow c_1$ (respectively, an equality $c_0 \approx c_1$), where c_0 and c_1 are constants of $\mathcal{K}$.

The constants in $\mathcal{K}$ will essentially serve as names for equivalence classes of terms. Thus, an equation $c_i \approx c_j$ indicates that c_i and c_j are two names for the same equivalence class. A constant c_i is said to *represent* a term $t \in \mathcal{T}(\Sigma \cup \mathcal{K})$ via a rewrite system R, if $t \leftrightarrow^*_R c_i$.

Abstract Congruence Closure: A ground rewrite system $R = D \cup C$ of D and C-rules on $\Sigma \cup \mathcal{K}$ is called an *abstract congruence closure* if: (i) each constant $c_0 \in \mathcal{K}$ represents some term $t \in \mathcal{T}(\Sigma)$ via R and (ii) R is convergent. If E is a set of ground equalities over $\mathcal{T}(\Sigma \cup \mathcal{K})$ and R an abstract congruence closure such that for all terms s and t in $\mathcal{T}(\Sigma)$, $s \leftrightarrow^*_E t$ if, and only if, there exists a term u with $s \rightarrow^*_R u \leftarrow^*_R t$, then R is called an *abstract congruence closure of E*. That is, the word problem for an equational theory E can be decided by rewriting to normal form using the rules of an abstract congruence closure R of E.

3 Graph-Based Congruence Closure

We describe a graph-based method for computing an abstract congruence closure of a given set of ground equalities E over a signature Σ. First a directed acyclic graph (DAG) is constructed that represents the set E, as well as terms occurring in E. In addition, each vertex of this initial graph is labeled by a distinct constant c_i of $\mathcal{K}$. Next various graph transformation rules are applied that represent equational inferences with the given equalities. Specifically, there are four *mandatory* rules (*Orient*, *SR*, *RRout*, and *Merge*) and one optional rule (*RRin*). Exhaustive application of these rules, or *saturation*, will under certain reasonable assumptions yield an abstract congruence closure. The vertices of each (initial and transformed) graph represent equivalence classes of terms.

Transformed graphs need not be acyclic, but will conform to full structure sharing in that different vertices represent different terms (or equivalence classes). The transformation rules crucially depend on an ordering $\succ$ on $\mathcal{K}$ [1].

3.1   Initial Graphs

We consider directed graphs where each vertex v is labeled by (i) a function symbol of Σ, denoted by $Symbol(v)$, and (ii) a constant of $\mathcal{K}$, denoted by $Constant(v)$. In addition, edges are classified as *equality*, *rewrite*, or *subterm* edges. We write $u -_E v$ and $u \to_R v$ to denote equality and rewrite edges (between vertices u and v), respectively. Subterm edges are also labeled by an index, and we write $u \to_S^i v$. Informally, this subterm edge indicates that v represents the i-th subterm of the term represented by u.

An *initial* graph $DAG(E)$ represents a set of equalities E as well as the subterm structure of terms in E. It is characterized by the following conditions: (i) If $Symbol(v)$ is a constant, then v has no outgoing subterm edges; and (ii) if $Symbol(v)$ is a function symbol of arity n, then there is exactly one edge of the form $v \to_S^i v_i$, for each i with $1 \le i \le n$. (That is, the number of outgoing vertices from v reflects the arity of $Symbol(v)$.)

The term $Term(v)$ *represented* by a vertex v is recursively defined as follows: If $Symbol(v)$ is a constant, then $Term(v) = Symbol(v)$; if $Symbol(v)$ is a function symbol of arity n, then $Term(v) = Symbol(v)(Term(v_1), \ldots, Term(v_n))$, where $v \to_S^i v_i$, for $1 \le i \le n$. Evidently, $Term(v)$ is a term over signature Σ. We require that distinct vertices of $DAG(E)$ represent different terms. Moreover, we insist that $DAG(E)$ contain no rewrite edges and that each equality edge $u -_E v$ correspond to an equality $s \approx t$ of E (with u and v representing s and t, respectively), and vice versa.

The vertices of the graph $DAG(E)$ also represent flat terms over the extended signature $\Sigma \cup \mathcal{K}$. More specifically, if $Symbol(v)$ is a constant, then $ExtTerm(v) = Constant(v)$, and if $Symbol(v)$ is a function symbol of arity n, then $ExtTerm(v) = Symbol(v)(Constant(v_1), \ldots, Constant(v_n))$, where $v \to_S^i v_i$, for $1 \le i \le n$.

We should point out that the labels $Constant(v)$ allow us to dispense with the extension rule of abstract congruence closure [2,3]. The initial graph $DAG(E)$ contains only subterm and equality edges. Rewrite edges are introduced during graph transformations. The term representation schema for transformed graphs is also more complex, see Section 4.

3.2   Graph Transformation

We define the graph transformations by rules. The first rule, *Orient*, can be used to replace an equality edge, $v -_E w$, by a rewrite edge, $v \to_R w$, provided $Constant(v) \succ Constant(w)$. If the ordering $\succ$ is total, then every equality edge

---

[1] An ordering is an irreflexive and transitive relation on terms. An ordering $\succ$ is total, if for any two distinct terms s and t, $s \succ t$ or $t \succ s$.

for which $Constant(v) \neq Constant(w)$ can be replaced by a rewrite edge (one way or the other).

The ordering $\succ$ needs to be defined on constants in $\mathcal{K}$ only, not on terms over $\Sigma \cup \mathcal{K}$. Term orderings, as employed in SOUR-graphs, are inherently more restrictive and may be detrimental to the efficiency of the congruence closure construction. For instance, well-founded term orderings must be compatible with the subterm relation, whereas we may choose an ordering with $Constant(v) \succ Constant(w)$ for efficiency reasons, even though $Term(v)$ may be a subterm of $Term(w)$.

The *SR* rule replaces one subterm edge by another one. In logical terms it represents the simplification of a subterm by rewriting, or in fact the simultaneous simplification of all occurrences of a subterm, if the graph presentation encodes full structure sharing for terms.

The *RRout* and *RRin* each replace one rewrite edge by another. They correspond to certain equational inferences with the underlying rewrite rules (namely, critical pair computations and compositions, which for ground terms are also simplifications). The *RRin* rule is useful for efficiency reasons, though one can always obtain a congruence closure without it. If the rule is applied exhaustively, the resulting congruence closure will be a right-reduced rewrite system over the extended signature.

The *Merge* rule collapses two vertices that represent the same term over the extended signature into a single vertex. It ensures closure under congruence and full structure sharing.

The graph transformation rules are formally defined as pairs of tuples of the form $(E_s, E_e, E_r, V, \mathcal{K}, KC) \rightarrow (E'_s, E'_e, E'_r, V', \mathcal{K}', KC')$, where the individual components specify a graph, an extended signature, and an ordering on new constants, before and after rule application. Specifically,

- the first three components describe the sets of subterm, equality, and rewrite edges, respectively;
- the fourth component describes the set of vertices;
- the fifth component describes the extension of the original signature Σ [2];
and
- the last component describes the (partial) ordering on constants. Specifically, KC is a set of "ordering constraints" of the form $\{c_i \succ c_j \mid c_i, c_j \in \mathcal{K}\}$. (A set of such constraints is considered *satisfiable* if there is an irreflexive, transitive relation on $\mathcal{K}$ that meets all of them.)

The specific conditions for the various rules are shown in Figures 1 and 2. For example, if two vertices v and w represent the same flat term (over the extended signature), then the merge rule can be used to delete one of the two vertices, say v, and all its outgoing subterm edges. All other edges that were incident on v need to be redirected to w, with the proviso that outgoing rewrite edges have to be changed to equality edges.

---

[2] We have $\mathcal{K}' \subseteq \mathcal{K}$, which is different from abstract congruence closure [2, 3], where new constants can be introduced.

Construction of a congruence closure starts from an initial tuple $(E_s, E_e, E_r, V, \mathcal{K}, \emptyset)$, the first five components of which are derived from $DAG(E)$ (so that E_r is empty)[3]. Transformation rules can then be applied non-deterministically. The rules SR, $RRout$ and $RRin$ are only applied if they result in a *new* edge.

There are various possible strategies for orienting equality edges. One may start with a fixed, total ordering on constants, or else construct an ordering "on the fly." Different strategies may result in different saturated graphs, see Example 2. The choice of the ordering is crucial for efficiency reasons, as discussed in section 6. The ordering also prevents the creation of cycles of with equality or rewrite edges, though the SR rule may introduce self-loops or cycles involving subterm edges. (Such a cycle would indicated that a term is equivalent to one of its subterms in the given equational theory.) See section 3.3 and example 2 for more details.

Definition 1. *We say that a graph G is* saturated *if it contains only subterm and rewrite edges and no further graph transformation rules can be applied.*

3.3   Extraction of Rules

We can extract D-rules, C-rules and C-equalities from the initial and transformed graphs as follows. A vertex v_0 with $ExtTerm(v_0) = t$ and $Constant(v_0) = c_0$ induces a D-rule $t \rightarrow c_0$. An equality edge $v_1 -_E v_2$ induces a C-equality $c_1 \approx c_2$, where $Constant(v_1) = c_1$ and $Constant(v_2) = c_2$. A rewrite edge $v_1 \rightarrow_R v_2$ induces a C-rule $c_1 \approx c_2$, where $Constant(v_1) = c_1$ and $Constant(v_2) = c_2$.

With each tuple $(E_s, E_e, E_r, V, \mathcal{K}, KC)$ we associate a triple $(\mathcal{K}, Ex, Rx)$, where Ex is the set of C-equalities and Rx is the set of D and C-rules extracted from the graph G specified by the first four components of the given tuple. Thus, with the initial graph we associate a triple $(\mathcal{K}_0, Ex_0, Rx_0)$, where Rx_0 is empty and Ex_0 represents the same equational theory over Σ-terms as the given set of equations E. The goal is to obtain a triple $(\mathcal{K}_n, Ex_n, Rx_n)$, where Ex_n is empty and Rx_n is an abstract congruence closure of E.

3.4   Correctness

The following can be established:

- Exhaustive application of the graph transformation rules is *sound* in that the equational theory represented over Σ-terms does not change.
- Exhaustive application of the graph transformation rules *terminates*. This can be proved by assigning a suitable weight to graphs that decreases with each application of a transformation rule.
- Exhaustive application of the rules is *complete* in that the rewrite system that can be extracted from the final graph is convergent and an abstract congruence closure for E. (If the optional $RRin$ rule has been applied exhaustively, the final rewrite system over the extended signature is right-reduced.)

---

[3] We essentially begin with the same graph as the Nelson-Oppen procedure, as described in the abstract congruence closure framework [2, 3].

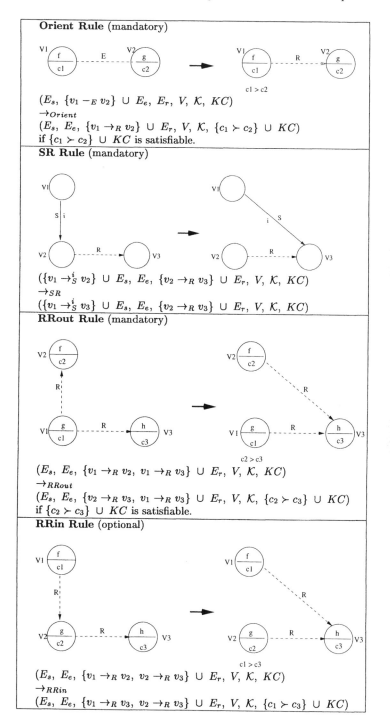

Fig. 1. *Orient, SR, RRout* and *RRin* graph transformation rules

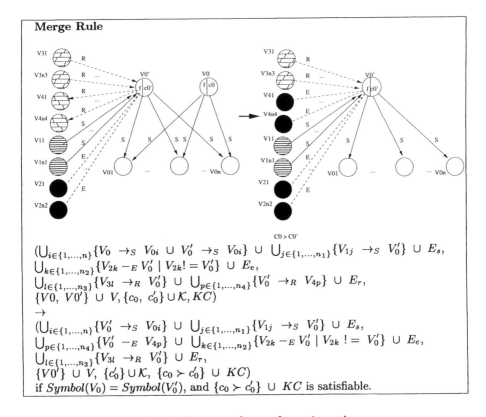

Fig. 2. *Merge* graph transformation rule

4   Rewrite System over the Original Signature

In this section we explain how to obtain a convergent rewrite system over the original signature Σ (independent from the ordering on constants of $\mathcal{K}$) from a graph G saturated with respect to *Orient, SR, RRout, RRin* and *Merge*. Basically, at this point, constructing the convergent rewrite system over Σ from G consists of eliminating the constants of $\mathcal{K}$ from G. Indeed, the constants of $\mathcal{K}$ are "names" of equivalence classes, and the same equivalence class may have several "names." There are two methods. The first method works on the convergent rewrite system over $\Sigma \cup \mathcal{K}$ extracted from G. Redundant constants (constants appearing on the left-hand side of a C rules) are eliminating by applications of *compression* rules and non-redundant constants are eliminated by *selection* rules as described in [2,3]. The second method, that we propose in this article, permits us to work directly and only on the graph G, by transforming the graph in a way reminiscent of the compression and selection rules. We define the notion of redundant constants on a saturated graph with respect to *Orient, SR, RRout, RRin* and *Merge*, and introduce three mandatory graph transformation rules: *Compression, Selection 1* and *Selection2*. These inference rules eliminate constants from $\mathcal{K}$, and redirect R and S edges. They also remove cycles of S edges from the graph. Their exhaustive nondeterministic application produces a DAG representing a convergent rewrite system over the original signature. The graph data structure permits us to visualize in parallel what happens on the original and extended signatures. For example, the *Selection 1* rule redirects R edges that were oriented in the "wrong" way during the abstract congruence process. Indeed, if we need to redirect an R edge, it means that the ordering on the constants of $\mathcal{K}$ that was used or constructed "on the fly" is in contradiction with any ordering on terms of $\mathcal{T}(\Sigma)$.

4.1   Graph-Based Inference Rules

Definition 2. *A constant c_0 of $\mathcal{K}$ labeling a vertex with an incident outgoing R edge is called* redundant.

The *Compression, Selection 1* and *Selection 2* rules are provided in Figure 3. Redundant constants of $\mathcal{K}$ are eliminated by the *Compression* rule. Both *Selection 1* and *Selection 2* implement the original selection rule of [2, 3]. *Selection 1* is implemented by finding an R edge from a vertex v representing a term t of $\mathcal{T}(\Sigma)$ (i.e. all the vertices reachable from v following S edges are labeled by constants of Σ only) to a vertex labeled by a non redundant constant c_0 of $\mathcal{K}$ on the graph, inversing the direction of this R edge, redirecting all the incoming S and R edges incident to c_0 to v, and eliminating c_0 from the graph. Hence, it consists of picking a representative term t over Σ for the equivalence class of c_0. *Selection 2* is described as follows. If a vertex v is labeled by a constant c_0 of $\mathcal{K}$, and all the vertices reachable from v following S edges are labeled by constants of Σ only, then the constant c_0 of $\mathcal{K}$ is eliminated from the graph. A particular case of this rule occurs when a vertex is labeled by a constant c_0 of $\mathcal{K}$ and a constant c of Σ.

4.2   Correctness

When applying *Compression, Selection 1* and *Selection 2*, the graph is in its maximal structure sharing form. It can be represented by a state $(\mathcal{K}, R)$, where $\mathcal{K}$ is the set of constants disjoint from Σ labeling the vertices of the graph, and R is the set of rewrite rules read from the graph over $\Sigma \cup \mathcal{K}$. We use the symbol $\vdash$ to denote the one-step transformation relation on states induced by *Compression, Selection 1*, and *Selection 2*. A *derivation* is a sequence of states $(\mathcal{K}_0, R_0) \vdash (\mathcal{K}_1, R_1) \vdash \ldots$, and $\mathcal{K}_i \subseteq \mathcal{K}_j$ for $i \geq j \geq 0$. We call a state $(\mathcal{K}, R)$ *final*, if no *mandatory* transformation rules (*Compression, Selection 1, Selection 2*) are applicable to this state. We prove that the *Compression, Selection 1* and *Selection 2* are sound in that the equational theory represented over Σ-terms does not change. If there is a constant of $\mathcal{K}$ labeling a vertex of the graph, then either *Compression, Selection 1* or *Selection 2* can be applied, and the exhaustive application of *Compression, Selection 1* and *Selection 2* terminates. The termination is easily shown because the application of *Compression, Selection 1* or

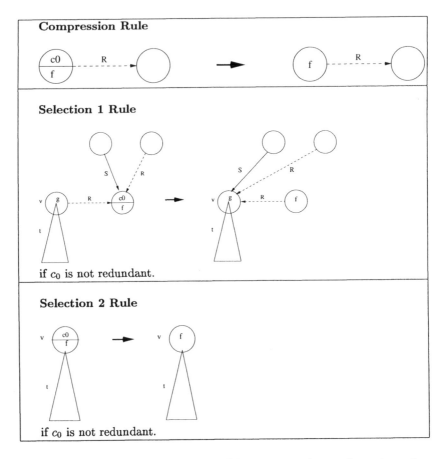

Fig. 3. *Compression*, *Selection 1* and *Selection 2* graph transformation rules

Selection 2 reduces the number of constants of $\mathcal{K}$ by one. The final state of a derivation describes a DAG labeled by constants of Σ only, that does not contain cycles of S edges, and, represents a convergent rewrite system over Σ.

5   Examples

Example 1. Figure 4 a) presents the construction of $DAG(E)$, where $E = \{f(f(f(a))) \approx a,\ f(f(a)) \approx a,\ g(c,c) \approx f(a),\ g(c,h(a)) \approx g(c,c),\ c \approx h(a),\ b \approx m(f(a))\}$. $\mathcal{K} = \{c_1, \ldots, c_{10}\}$. We apply all the following mandatory transformations on $DAG(E)$ in a certain order constructing the order on the constants of $\mathcal{K}$ "on the fly." *Orient* orients the E edge between c_1 and c_4 into an R edge from c_4 to c_1 ($c_4 \succ c_1$), and the E edge between c_1 and c_3 into an R edge from c_1 to c_3 ($c_1 \succ c_3$). We can apply an SR transformation that replaces the S edge from c_2 to c_3 by an S edge from c_2 to c_1. Let $c_2 \succ c_4$. We merge c_2 and c_4; c_2 is removed from the graph, the S edges between c_2 and c_3 are removed,

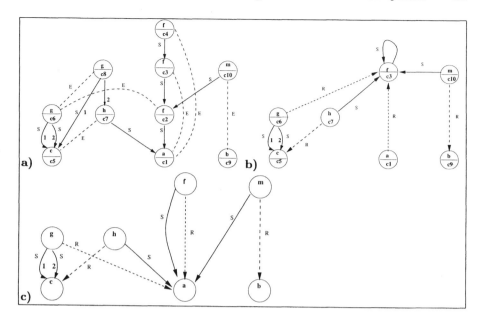

Fig. 4. a) Initial DAG for $E = \{f(f(f(a))) \approx a,\ f(f(a)) \approx a,\ g(c,c) \approx f(a),\ g(c,h(a)) \approx g(c,c),\ c \approx h(a),\ b \approx m(f(a))\}$, b) Saturated graph for E on the extended signature, c) DAG on the original signature

the E edge from c_6 to c_2 is replaced by an E edge from c_6 to c_3, and a self-loop composed of an S edge is added from c_3 to itself. c_4 and c_3 can be merged, and c_4 is removed from the graph $(c_4 \succ c_3)$. The S edge between c_4 and c_3 is removed from the graph, and the R edge from c_4 to c_1 is simply removed, because there is already an R edge from c_1 to c_3. We orient the E edge between c_{10} and c_9 from c_{10} to c_9 $(c_{10} \succ c_9)$, the E edge between c_7 and c_5 from c_7 to c_5 $(c_7 \succ c_5)$, and the E edge between c_6 and c_3 from c_6 to c_3 $(c_6 \succ c_3)$, and the E edge between c_8 and c_6 from c_8 to c_6 $(c_8 \succ c_6)$. We can apply an SR transformation that adds an S edge from c_8 to c_5, and removes the S edge from c_8 to c_7. c_8 and c_6 are merged, and c_8 is removed from the graph. The R edge between c_8 and c_6 is removed from the graph. We obtain the saturated graph labeled by $\{c_1, c_3, c_5, c_6, c_7, c_9, c_{10}\}$ such that $\{c_6 \succ c_3, c_7 \succ c_5, c_1 \succ c_3, c_{10} \succ c_9\}$ presented in Figure 4 b). The convergent rewrite system over the extended signature is: $\{c_7 \rightarrow c_5,\ c_6 \rightarrow c_3,\ c_1 \rightarrow c_3,\ c_{10} \rightarrow c_9,\ c \rightarrow c_5,\ a \rightarrow c_1,\ b \rightarrow c_9,\ h(c_3) \rightarrow c_7,\ g(c_5, c_5) \rightarrow c_6,\ f(c_3) \rightarrow c_3,\ m(c_3) \rightarrow c_{10}\}$. By applying *Compression* for the redundant constants c_{10}, c_7, c_6 and c_1, *Selection 2* to eliminate c_5 and c_9, and *Selection 1* to eliminate c_3, we obtain $\{h(a) \rightarrow c,\ g(c,c) \rightarrow a,\ f(a) \rightarrow a,\ m(a) \rightarrow b\}$, the convergent rewrite system over Σ that we read from 4 c).

Example 2. Figure 5 presents the constructions of a) $DAG(E)$ where $E = \{f(a,b) \approx a\}$, and its saturated counter-parts following two strategies. In b),

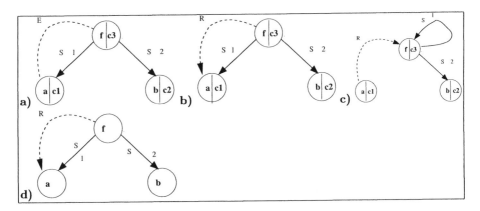

Fig. 5. a) Initial DAG for $E = \{f(a, b) \approx a\}$, b) Saturated graph for E with $c_3 \succ c_1$, c) Saturated graph for E with $c_1 \succ c_3$, and d) DAG on the original signature

$c_3 \succ c_1$ and in c), $c_1 \succ c_3$. In c), there is a self-loop composed of an S edge. The two saturated graphs with respect to *Orient*, *SR*, *RRout*, *RRin* and *Merge* are labeled by $\{c_1, c_2, c_3\}$. In b), the rewrite system on the extended signature is $\{f(c_1, c_2) \rightarrow c_3, c_3 \rightarrow c_1, a \rightarrow c_1, b \rightarrow c_2\}$, and in c), it is $\{f(c_3, c_2) \rightarrow c_3, c_1 \rightarrow c_3, a \rightarrow c_1, b \rightarrow c_2\}$. Both saturated graphs generate the DAG d) on the original signature representing $\{f(a, b) \rightarrow a\}$.

6   Complexity Results and Implementation

Our algorithms to obtain rewrite systems over extended and original signatures use only polynomial space to store the rewrite systems. Let n be the number of vertices of the initial DAG. During the abstract congruence process, there can only be n^2 subterm edges, and $n(n-1)/2$ equality and rewrite edges, because an edge can only be added once to a graph, and there are never two equality or rewrite edges between the same vertices (because of the ordering on the constants of $\mathcal{K}$.) Moreover, the number of vertices can only decrease (as a result of merging).

In comparing our graph-based approach with the logic-based approach of [2, 3], we find that graphs support full structure sharing and consequently our approach will tend to lead to fewer applications of transformation rules than corresponding applications of logical inference rules in the standard method. This by itself does not imply that our approach will be more efficient as full structure sharing depends on systematic application of the *Merge* rule, which can be expensive.

Efficient implementations of congruence closure require specialized data structures. Also, abstract congruence closure efficiency is parametrized by the choice or construction "on the fly" of the ordering on the constants of $\mathcal{K}$. In our case, E edges are oriented into R edges using this ordering. There exist naive ways to choose the ordering. For example, we can use a total and linear ordering on the constants of $\mathcal{K}$. There is a tradeoff between the effort spent in constructing an ordering, the time spent in comparing two constants, and the lengths of derivations (number of graph transformations). The graph data structure permits us to understand the influence of the ordering on the length of any derivation, and to construct "on the fly" orderings that control the number of inferences. It suggests new efficient procedures for constructing efficient abstract congruence closures. Indeed, the number of times we apply *SR* and *RRout* depends on how we apply *Orient*. Applying an *RRout* rule can create an *SR* configuration; so we would orient an E edge in the direction that creates the less *SR* and *RRout* configurations. *Merge* configurations can be created after the application of an *SR* rule.

When constructing the ordering, we do not allow backtracking, so we are only interested in *feasible* orderings, i.e. orderings that will produce *unfailing* derivations, i.e. derivations terminating with a saturated graph (as defined in definition 1). Starting from an initial state representing the initial DAG, the maximal derivation is in $O(n\delta)$, where n is the number of vertices of the initial DAG, and δ is the depth of the ordering (i.e. the longest chain $c_0 \succ c_1 \succ \ldots \succ c_\delta$.) So, any maximal derivation starting from an initial state is bounded by a quadratic upper bound. Indeed, any total and linear order is feasible and can be used. There exits a feasible ordering with smaller depth that can be computed "on the fly," and produces a maximal derivation of length $n \, log(n)$ [2,3]. This ordering is based on orienting an E edge from v_0 to v_1 if the number of elements of the equivalence class of $Constant(v_0)$ is less than or equal to the number of elements of the equivalence class of the $Constant(v_1)$. This can be applied only if all *RRout* and *RRin* inference rules have been processed for v_0 and v_1. The number of elements in an equivalence class of a constant $Constant(v)$ is computed by counting the number of incoming R edges in a vertex v.

A first implementation of our method is available for online experimentation at: `http://www.csis.pace.edu/~scharff/CC`. It is written in java, and uses java servlets, and XML. The implemented system contains (i) a parsing component that also transforms a given set of equalities into a DAG, (ii) a graph component that manages the graph, and (iii) an inferences component that deals with the application of the transformation rules. In particular, the strategy for application of rules is coded in XML, and therefore, is modifiable. An efficient strategy applies simplifications before orienting equalities (one at a time):

$(Orient \, . \, ((SR \, . \, Merge)^*)^* \, . \, (RRout \, . \, RRin \, . \, ((SR \, . \, Merge)^*)^*)^*)^*$ [4]

The ordering that is used in the implementation is a total and linear ordering. Given a signature and a set of equalities, the system displays the initial C equalities and D rules, the convergent rewrite system over the extended signature and the convergent rewrite system over the original signature. Some statistics results are also provided: the number of inferences of each type, and the processing times.

---

[4] * means that the strategy is applied 0 or more times. $X \, . \, Y$ means that Y is applied after X.

7 Conclusion and Future Work

We have presented a new graph-based method for constructing a congruence closure of a given set of ground equalities. The method combines the key ideas of two approaches, completion and standard congruence closure, in a natural way by relying on a data structure, called "SER graphs," that represents a specialized and optimized version of the more general, but less efficient, SOUR-graphs. We believe that our approach allows for efficient implementations and a visual presentation that better illuminates the basic ideas underlying the construction of congruence closures. In particular it clarifies the role of original and extended signatures and the impact of rewrite techniques for ordering equalities. Our approach should therefore be suitable for educational purposes.

A first implementation of our method is available. The method we described processes all equalities at once during construction of the initial graph. It is relatively straightforward to devise a more flexible, incremental approach, in which equalities are processed one at a time. Once the first equality is represented by a graph, transformation rules are applied until a "partial" congruence closure has been obtained. Then the next equation is processed by extending the current graph to represent any new subterms, followed by another round of graph transformations. This process continues until all equations have been processed. An advantage of the incremental approach is that simplifying graph transformations can be applied earlier. We expect to make implementations of both incremental and non-incremental available.

Acknowledgments

The authors would like to thank Eugene Kipnis, student at Pace University for the implementation of the work presented in this article, and the anonymous reviewers for their comments.

References

1. Baader, F., Nipkow, T.: Term rewriting and all that. Cambridge University Press (1998).
2. Bachmair, L., Ramakrishnan, I.V., Tiwari, A., Vigneron, L.: Congruence Closure Modulo Associativity and Commutativity. 3rd Intl Workshop FroCoS 2000, Kirchner, H. and Ringeissen, C. Editors, Lecture Notes in Artificial Intelligence, Springer-Verlag, **1794** (2000) 245–259.
3. Bachmair, L., Tiwari, A., Vigneron, L.: Abstract congruence closure. J. of Automated Reasoning, Kluwer Academic Publishers **31(2)** (2003) 129–168.
4. Dershowitz, N., Jouannaud, J.-P.: Handbook of Theoretical Computer Science, Chapter 6: Rewrite Systems. Elsevier Science Publishers **B** (1990) 244–320.
5. Downey, P. J., Sethi, R.,Tarjan, R. E.: Variations on the common subexpression problem. Journal of the ACM **27(4)** (1980) 758–771.
6. Kapur, D.L: Shostak's congruence closure as completion. Proceedings of the 8th International Conference on Rewriting Techniques and Applications, H. Comon Editor, Lecture Notes in Computer Science, Springer-Verlag **1232** (1997).

7. Knuth, D. E., Bendix, P. B.: Simple word problems in universal algebras. Computational Problems in Abstract Algebra, Pergamon Press, Oxford (1970) 263–297.
8. Kozen, D.: Complexity of Finitely Presented Algebras. PhD. thesis. Cornell University (1977).
9. Lynch, C., Strogova, P.: SOUR graphs for efficient completion. Journal of Discrete Mathematics and Theoretical Computer Science **2(1)** (1998) 1–25.
10. Nelson, C. G., Oppen, D. C.: Fast Decision Procedures based on Congruence Closure. Journal of the ACM **27(2)** (1980) 356–364.
11. Plaisted, D.A., Sattler-Klein, A.: Proof lengths for equational completion. Journal of Inf. Comput., Academic Press, Inc. **125(2)** (1996) 154–170.
12. Shostak, R.E.: Deciding Combinations of Theories. Journal of the ACM **31** (1984) 1–12.
13. Snyder, W.: A Fast Algorithm for Generating Reduced Ground Rewriting Systems from a Set of Ground Equations. Journal of Symbolic Computation **15** (1993) 415–450.

Combination of Nonlinear Terms
in Interval Constraint Satisfaction Techniques

Laurent Granvilliers[*] and Mina Ouabiba

LINA, University of Nantes
2, rue de la Houssinière, BP 92208 44322 Nantes Cedex 3, France
{Laurent.Granvilliers,Mina.Ouabiba}@lina.univ-nantes.fr

Abstract. Nonlinear constraint systems can be solved by combining consistency techniques and search. In this approach, the search space is reduced using local reasoning on constraints. However, local computations may lead to slow convergences. In order to handle this problem, we introduce a symbolic technique to combine nonlinear constraints. Such redundant constraints are further simplified according to the precision of interval computations. As a consequence, constraint reasoning becomes tighter and the solving process faster. The efficiency of this approach is shown using experimental results from a prototype.

Keywords: Interval arithmetic, numerical constraint, local consistency, symbolic algorithm, redundant constraint.

1 Introduction

The problem of solving a conjunction – a system – of nonlinear constraints over the real numbers is uncomputable in general [3]. Only approximations may be computed using machine numbers. Interval arithmetic [14] provides a set of operations over interval numbers. In this framework, every real quantity is enclosed by an interval. The general interval-based algorithm for solving nonlinear constraint systems is a bisection process. The search space, defined by the Cartesian product of variable domains, is recursively bisected and reduced until a desired precision is reached. Unfortunately the problem of approximating a constraint system using interval numbers is intractable in general.

Local consistency techniques are tractable algorithms that may accelerate the bisection algorithm [13,12]. Basically, the search space may be reduced using constraint projections on variables. Given a variable occurring in a constraint the projection is the set of values that may be assigned to the variable such that the constraint can be satisfied. These values are said to be consistent. It follows that the complementary set (the inconsistent values) can be removed from the variable domain. Such domain reductions have to be iterated until

---

[*] This work has been partially supported by a 2003–04 Procope project funded by the French Ministry of Education, the French Ministry of Foreign Affairs, and the German Academic Exchange Service (DAAD).

B. Buchberger and J.A. Campbell (Eds.): AISC 2004, LNAI 3249, pp. 118–131, 2004.

reaching quiescence. Unfortunately, these techniques may be weak, since some constraint projection may be less precise than the projection of the solution set.

The locality problem of projection-based reasonings can be handled in several ways, one of which being the symbolic combination of constraints [1]. The purpose is to eliminate variables in order to concentrate global information of the system in a smaller set of variables. As a consequence, local reasoning over the new constraints may derive global information. In this paper, we propose to improve consistency reasonings by means of redundant constraints [16], as shown in the following example (which is very simple for the sake of comprehension).

Example 1. Let x and y be two variables lying in the real interval $[-10, 10]$. Consider the following constraint systems:

$$P = \begin{cases} xy = 1 \\ xy = 2 \end{cases} \qquad P' = \begin{cases} xy = 1 \\ xy \sin(y) = 2 \end{cases}$$

It is worth noticing that variable domains cannot be reduced using projections, *e.g.*, for every value of x (take $x = 2$), there exists a value of y such that $xy = 1$ (here $y = 0.5$). The same conclusions are obtained for y and the second constraint. However, these problems are not satisfiable, which can be easily verified. The term xy is equal to two terms in P, whose evaluations are not compatible ($1 \neq 2$). As a consequence, P is not satisfiable. The division of the left-hand terms of equations from P' must be equal to the division of right-hand terms. Now simplify the result to derive $\sin(y) = 2$. Since a sine cannot be greater than 1, then P' is not satisfiable.

Example 1 shows that the combination of terms introduces more global reasonings in the process of solving nonlinear constraint systems. The first, well-known, idea is to share projections on terms, not only on variables. The second idea is to combine terms using symbolic transformations. This approach can be seen as a form of microscopic redundant computations, which may work for equations or inequalities. Since nonlinear constraints cannot be simplified in general, only well-chosen terms have to be combined.

In the following, we will focus on box consistency [2], the local consistency technique implemented in Numerica [17]. Box consistency is a basis for approximating constraint projections using interval numbers. The first contribution of this paper is the generalization of the definition of box consistency to process projections on terms. The second contribution is the combination technique used to generate redundant constraints that may improve box consistency. An implementation has been done in RealPaver [6]. Encouraging experimental results have been obtained on a set of known problems from different application domains.

The rest of this paper is organized as follows. Section 2 presents the basics related to numerical constraints and interval constraint solving. The main contributions are introduced in Section 3. The experimental results are discussed in Section 4. Last, Section 5 summarizes the contributions and points out some directions for future research.

2   Preliminaries

2.1   Formulas

Let $\Sigma_\mathbb{R}$ be the usual real-based structure where the set of real numbers $\mathbb{R}$ is associated with arithmetic operations, elementary functions and equality and inequality relations. Formulas are written using a first-order language L defined by an infinite countable set of variables $\{x_1, x_2, \dots\}$ and a set of symbols with arities, namely constants, functions and relations. In the following, we will use the same symbols to denote the elements of $\Sigma_\mathbb{R}$ and the symbols of L.

A *term* is a syntactic expression built by induction from variables, constants and function symbols of L. A *constraint* is an atomic formula of the form $f \bowtie g$ where f and g are terms and $\bowtie$ is a relation symbol. A *formula* is defined by induction from constraints, connectives and quantifiers. In the following, formulas will be written with the variables $x_1, \dots, x_n$, the natural n being arbitrarily large. Given a syntactic element S, the set of variables occurring in S is denoted by $\mathrm{Var}(S)$, and the notation $S(x_1, \dots, x_n)$ will mean that the sets $\mathrm{Var}(S)$ and $\{x_1, \dots, x_n\}$ are identical. The number of occurrences of any variable x_i in S will be denoted by $\mathrm{mult}(S, x_i)$. Given two terms f and g and a term h occurring in f, let $f[h \leftarrow g]$ denote the term obtained from f by replacing all the occurrences of h with g. Let $\mathbb{T}$ be the set of terms.

Evaluating a term $f(x_1, \dots, x_n)$ in the domain $\mathbb{R}$ consists in assigning a value $a_i \in \mathbb{R}$ to each variable x_i and applying the functions of $\Sigma_\mathbb{R}$ that correspond to the symbols. The satisfaction of a constraint $C(x_1, \dots, x_n)$ requires evaluating terms and verifying the relation. If the relation is verified then the tuple $(a_1, \dots, a_n)$ is a *solution of C* in the domain $\mathbb{R}$. Let $f_\mathbb{R}$ and $C_\mathbb{R}$ respectively denote the mapping from $\mathbb{R}^n$ to $\mathbb{R}$ corresponding to the evaluation of f and the relation of $\mathbb{R}^n$ defined by the set of solutions of C in $\mathbb{R}$.

A conjunction of constraints is a formula $F(x_1, \dots, x_n)$ of the form $C_1 \wedge \cdots \wedge C_m$. The satisfaction of F consists in assigning a value $a_i \in \mathbb{R}$ to each variable x_i and verifying that all constraints C_j are satisfied. If the formula is satisfied then the tuple $(a_1, \dots, a_n)$ is a *solution of F* in the domain $\mathbb{R}$. Let $F_\mathbb{R}$ denote the relation of $\mathbb{R}^n$ defined by the set of solutions of F in $\mathbb{R}$. In the following, only conjunctions of constraints will be considered, and a solution in $\mathbb{R}$ will simply be called *solution*.

2.2   Interval Arithmetic

Interval arithmetic [14,9] can be used to conservatively enclose the solutions of formulas by Cartesian products of intervals. Interval computations are done in an interval-based structure $\Sigma_\mathbb{I}$. This structure can be defined by an homomorphism μ from $\Sigma_\mathbb{R}$ to $\Sigma_\mathbb{I}$, as follows. The notations $f_\mathbb{R}$, $C_\mathbb{R}$ and $F_\mathbb{R}$ used with the domain $\mathbb{R}$ will simply be changed into $f_\mathbb{I}$, $C_\mathbb{I}$ and $F_\mathbb{I}$.

Constants. Consider the set of floating-point numbers $\mathbb{F}$ defined by the IEEE standard [10] and the set of intervals $\mathbb{I}$ whose bounds are in $\mathbb{F}$. Let $[\underline{I}, \overline{I}]$ denote the interval I defined by the set $\{a \in \mathbb{R} \mid \underline{I} \leqslant a \leqslant \overline{I}\}$. The standard defines two rounding operations that associate to each real number a the closest floating-point numbers $\lfloor a \rfloor$ and $\lceil a \rceil$ such that $\lfloor a \rfloor \leqslant a \leqslant \lceil a \rceil$. These operations enable the computation of the convex hull of every set of real numbers A, defined by the interval: $\Box A = [\lfloor \inf A \rfloor, \lceil \sup A \rceil]$. Then μ maps every real number a to the interval $\Box\{a\}$. The membership, union and intersection operations in $\mathbb{I}$ are naturally extended to interval vectors. It is worth noticing that an interval vector $D = (D_1, \ldots, D_n)$ is a machine representation of the box $D_1 \times \cdots \times D_n$. In the following, interval vectors and associated Cartesian products will simply be called *boxes*.

Terms. Interval arithmetic is a reliable extension of real arithmetic in the sense that every interval computation results in a superset of the result of the corresponding real computation. Arithmetic operations and elementary functions are implemented by computation over interval bounds according to monotonicity properties. For instance, given two intervals $I = [a, b]$ and $J = [c, d]$, μ maps the addition and subtraction operations and the exponential function to the following interval functions:

$$\begin{cases} I + J = [\lfloor a + c \rfloor, \lceil b + d \rceil] \\ I - J = [\lfloor a - d \rfloor, \lceil b - c \rceil] \\ \exp(I) = [\lfloor \exp(a) \rfloor, \lceil \exp(b) \rceil] \end{cases}$$

It is worth noticing that term evaluation in $\Sigma_{\mathbb{I}}$ exactly corresponds to the application of an interval function called *natural form*. The main result from interval arithmetic is the following inclusion theorem:

Theorem 1. *Consider a term $f(x_1, \ldots, x_n)$. Then for all $D \in \mathbb{I}^n$ the following property holds:*

$$\{f_{\mathbb{R}}(a_1, \ldots, a_n) \mid a_1 \in D_1, \ldots, a_n \in D_n\} \subseteq f_{\mathbb{I}}(D_1, \ldots, D_n).$$

Informally speaking, the evaluation of $f_{\mathbb{I}}$ is a superset of the range of $f_{\mathbb{R}}$ on the domain D.

Constraints. The interval versions of relations originate from an existential extension of real relations. For instance, given two intervals $I = [a, b]$ and $J = [c, d]$, μ maps equations and inequalities to the following predicates:

$$\begin{cases} I = J \iff \exists u \in I \ \exists v \in J : u = v \iff \max(a, c) \leqslant \min(b, d) \\ I \leqslant J \iff \exists u \in I \ \exists v \in J : u \leqslant v \iff a \leqslant d \end{cases}$$

Constraint satisfaction in $\Sigma_{\mathbb{I}}$ uses natural forms of terms and existential extensions of relations. The aim is to compute reliable approximations of constraint solutions, which leads to the following theorem:

Theorem 2. *Consider a constraint $C(x_1, \ldots, x_n)$. Then for all $D \in \mathbb{I}^n$ we have:*

$$\exists a_1 \in D_1 \ \ldots \ \exists a_n \in D_n : (a_1, \ldots, a_n) \in C_{\mathbb{R}} \implies (D_1, \ldots, D_n) \in C_{\mathbb{I}}.$$

Informally speaking, each box containing a solution of C in $\mathbb{R}$ must be included in the interval relation. As a consequence, the approximation does not lose any solution. Methods based on constraint satisfaction in $\Sigma_{\mathbb{I}}$ are said to be *complete*.

Notations. Given a box D and a term f (resp. a constraint C) such that each of its variables has a domain in D, let $D_{|f}$ (resp. $D_{|C}$) be the restriction of D to the variables of f (resp. C). In the following, we will simply write $f_{\mathbb{I}}(D)$ $(D \in C_{\mathbb{I}})$ for $f_{\mathbb{I}}(D_{|f})$ (resp. $D_{|C} \in C_{\mathbb{I}}$).

2.3   Numeric Constraints

A *numerical constraint satisfaction problem* (NCSP) is given by a conjunction of constraints $F(x_1, \ldots, x_n)$ of the form:

$$C_1 \wedge \cdots \wedge C_m \wedge x_1 \geqslant a_1 \wedge x_1 \leqslant b_1 \wedge \cdots \wedge x_n \geqslant a_n \wedge x_n \leqslant b_n$$

NCSPs can be solved by a bisection algorithm, which maintains a set of boxes. The initial box is given by the variable domains. Boxes are processed by splitting and narrowing operations. A splitting step transforms a box into a set of smaller boxes whose union is equivalent. A narrowing operation reduces a box by removing facets that are solution-free. The output is a set of boxes whose union is a superset of the set of solutions of the NCSP.

2.4   Narrowing

A basic narrowing procedure consists in using constraint satisfaction in the interval structure. Given a box, if at least one constraint is not satisfied then the whole NCSP cannot be satisfied. By Theorem 2, it follows that it cannot be satisfied in the real structure. As a consequence, the box can be rejected. Unfortunately this method is weak since a free-solution space needs to be completely isolated by splitting before being eliminated.

A more powerful approach is based on constraint projections. In the constraint programming literature [12], projections have been defined to determine the allowed values of one variable in order to satisfy one constraint in the real domain. More precisely, given a constraint $C(x_1, \ldots, x_n)$, a box $D \in \mathbb{I}^n$ and a variable x_i occurring in C, the *projection* of C on x_i in D is the set of real numbers

$$\begin{aligned}
\Pi_{x_i}(C, D) = \{a_i \in D_i \mid &\exists a_1 \in D_1, \ldots, \exists a_{i-1} \in D_{i-1}, \\
&\exists a_{i+1} \in D_{i+1}, \ldots, \exists a_n \in D_n : \\
&(a_1, \ldots, a_n) \in C_{\mathbb{R}}\}.
\end{aligned}$$

The definition of projections is naturally extended to conjunctions of constraints by considering that all constraints have to be satisfied. In most solving engines of existing solvers, only one constraint is used during a narrowing operation. That leads to the *locality problem*, which is the main concern of our work. Given a box D, a variable x_i and two constraints C and C' the following inclusion holds, though the equality is not verified in general:

$$\Pi_{x_i}(C \wedge C', D) \subseteq \Pi_{x_i}(C, D) \cap \Pi_{x_i}(C', D) \tag{1}$$

Example 2. Consider the NCSP defined by two constraints $C : x_1 + x_2 = 0$ and $C' : x_1 - x_2 = 0$ and the box $D = [-1, 1]^2$. The projections over x_1 are as follows:

$$\Pi_{x_1}(C \wedge C', D) = [0, 0] \subseteq [-1, 1] = \Pi_{x_1}(C, D) = \Pi_{x_1}(C', D)$$

The problem of processing existential quantifications in the definition of projections is intractable. A main idea is then to implement constraint satisfaction in the interval domain, which leads to the narrowing procedure called box consistency [2]. In the following, the restriction of the original definition to natural forms is given:

Definition 1 (Box consistency). *Consider a constraint $C(x_1, \ldots, x_n)$, a box $D \in \mathbb{I}^n$ and a natural i that belongs to the set $\{1, \ldots, n\}$. The domain D_i is said to be* box consistent *wrt. C if we have:*

$$D_i = \Box\{a \in D_i \mid (D_1, \ldots, D_{i-1}, \Box\{a\}, D_{i+1}, \ldots, D_n) \in C_\mathbb{I}\}$$

The corresponding narrowing procedure allows one to reduce the domain D_i if the equality is not verified, using the following operation:

$$\theta : (x_i, C, D) \mapsto \Box\{a \in D_i \mid (D_1, \ldots, D_{i-1}, \Box\{a\}, D_{i+1}, \ldots, D_n) \in C_\mathbb{I}\}$$

This operation computes an approximation of the projection of C on x_i in D. The narrowing procedure is complete, since the computed domain encloses the corresponding projection, that is:

$$\Pi_{x_i}(C, D) \subseteq \theta(x_i, C, D) \tag{2}$$

However the resulting interval may be arbitrarily large wrt. the projection. This is closely related to the *dependency problem* of interval arithmetic, which leads to weak interval computations [8].

3   Symbolic-Numeric Techniques

Box consistency and projections have been defined wrt. the variables. Noticing that variables are instances of terms leads to generalize these notions.

Definition 2 (Box consistency on terms). *Let $C(x_1, \ldots, x_n)$ be a constraint. Consider a box $D \in \mathbb{I}^n$ and a term f occurring in C. Let x_{n+1} be a fresh variable and let the constraint C' be $C[f \leftarrow x_{n+1}]$. The domain $f_\mathbb{I}(D)$ is said to be box consistent wrt. C if we have:*

$$f_\mathbb{I}(D) = \Box\{a \in f_\mathbb{I}(D) \mid (D_1, \ldots, D_n, \Box\{a\}) \in C'_\mathbb{I}\}$$

If the domain of term f is not box consistent then it may be reduced by means of the narrowing operation $\theta(f, C, D)$. By Theorem 1 the reduced domain must be a superset of the range of $f_{\mathbb{R}}$ on D.

Box consistency is a method for bounding the range of a term over a domain. Given a term f and an interval I, let $f \in I$ denote a *projection constraint*, which means that the range of $f_{\mathbb{R}}$ over the current domain must be included in I. We will assume in this section that there exist efficient algorithms able to compute such an interval I. An effective implementation will be described in Section 4.

In the following projection constraints will be considered as first-class constraints. The purpose of our work is to combine projection constraints in order to handle the locality problem for the variables occurring in f.

3.1   Term Sharing

The first idea, which may be considered as a well-known result, is to share terms in NCSPs. It suffices to represent the set of constraint expressions as a Directed Acyclic Graph. As a consequence, intersections of domains do not only happen at variable nodes but also at non-variable term nodes. The following lemma shows that projections of different constraints on the same term just need to be intersected to reduce its domain of possible values.

Lemma 1. *Let f be a term and let I and J be two intervals. Then we have:*

$$f \in I \wedge f \in J \implies f \in I \cap J$$

This technique is a first approach to handle the locality problem. As shown in the introduction and in the following example, which corresponds to the use of a redundant equation. However the redundant equation has not to be represented since the reductions are directly obtained in the DAG.

Example 3. Consider the conjunction of constraint

$$\cos(xy) = 2\sin(x) \ \wedge \ \cos(xy) + \cos(y) = 0$$

in the box $[-10, 10]^2$. The application of a bisection algorithm at the maximal machine precision leads to the generation of about $1.5 \cdot 10^4$ boxes and to the evaluation of $3 \cdot 10^6$ interval operations. If the term $\cos(xy)$ is shared then the number of boxes decreases to $1.5 \cdot 10^3$ and the number of evaluations is $5 \cdot 10^4$. The gain is more than one order of magnitude on this problem. In fact sharing the term $\cos(xy)$ is equivalent to using the redundant constraint $2\sin(x) + \cos(y) = 0$.

3.2   Term Combination

The second idea is to mimic elimination procedures for linear and polynomial systems of equations. The goal is to derive redundant constraints which may improve the precision of consistency computations and the computation time. These remarks led us to the following motivations: combination of terms which can be simplified and reuse of already computed intervals for terms. The notion of combination function is well-known for linear terms or S-polynomials [5,3]. We give now a general definition to cope with nonlinear combinations.

Definition 3. *(Combination function). Let $\phi(u,v)$ be a term in which u and v occur only once. The underlying combination function φ is defined as follows:*

$$\varphi: \begin{cases} \mathbb{T}^2 \to \mathbb{T} \\ (f,g) \mapsto \phi[u \leftarrow f, v \leftarrow g] \end{cases}$$

In the following every function φ will be supposed to be associated with a term ϕ, even if ϕ is not mentioned. These functions are used to process projection constraints of the form $f \in I$. The combination method is described in the following lemma, the result being a redundant constraint.

Lemma 2. *Let f and g be two terms and let I and J be two intervals. Given a combination function φ, the following relation holds:*

$$f \in I \wedge g \in J \implies \varphi(f,g) \in \phi_{\mathbb{I}}(I,J)$$

The previous lemma is interesting since it shows how to introduce a redundant constraint. Terms f and g are combined by φ and the evaluation of $\phi_{\mathbb{I}}$ on (I,J) gives a range of the redundant term $\varphi(f,g)$. However, the following lemma shows that this redundancy is useless for box consistency computations.

Lemma 3. *Let D be a box and let φ be a combination function. Let $C : f \in I$ and $C' : g \in J$ be two projection constraints such that $f_{\mathbb{I}}(D)$ is box consistent wrt. C and $g_{\mathbb{I}}(D)$ is box consistent wrt. C'. If C'' is the redundant constraint defined by $\varphi(f,g) \in \phi_{\mathbb{I}}(I,J)$ then $f_{\mathbb{I}}(D)$ and $g_{\mathbb{I}}(D)$ are box consistent wrt. C''.*

The previous lemma shows that the redundant constraint does not allow to reduce the domain of f since it is already box consistent. As a consequence, the redundancy is useless. The next step introduces a simplification process.

3.3   Term Simplification

Simplification procedures are generally associated with two properties. Firstly, the equivalence of terms in the domain $\mathbb{R}$ must be preserved. Secondly, rewritten terms have to be adapted for further computations. The first property is the basis of the following definition.

Definition 4. *(Simplification function). A simplification function ψ is a function on terms such that for all $f(x_1, \ldots, x_n) \in \mathbb{T}$, the set of variables occurring in $\psi(f)$ is included in $\{x_1, \ldots, x_n\}$ and the following formula is true in the real structure:*

$$\forall x_1 \ldots \forall x_n \ f = \psi(f)$$

The main idea is to simplify left-hand terms of redundant projection constraints. The following lemma shows that the redundancy property (Lemma 2) is preserved by simplification.

Lemma 4. *Let $f \in I$ and $g \in J$ be two projection constraints. Given a combination function φ and a simplification function ψ, the following relation holds:*

$$f \in I \wedge g \in J \implies \psi(\varphi(f,g)) \in \phi_{\mathbb{I}}(I,J)$$

The following lemma provides a sufficient condition for defining efficient simplification functions. The aim is to derive more precise interval functions. Given two terms f and g, g is said to be *more precise* than f if for all $D \in \mathbb{I}^n$, $g_\mathbb{I}(D) \subseteq f_\mathbb{I}(D)$ holds. This relation defines a partial ordering $\preccurlyeq$ on terms (here we have $g \preccurlyeq f$).

Lemma 5. *Let $f(x_1, \ldots, x_n)$ be a term and let g be a term resulting from a simplification of f. Let x_i be a variable occurring in f and g. Suppose that g is more precise than f. Then for every interval I and every box D the following result holds:*

$$\theta(x_i, g \in I, D) \subseteq \theta(x_i, f \in I, D)$$

The dependency problem of interval arithmetic underlines the weakness of interval evaluation wrt. multiple occurrences of variables [14]. More precisely, two occurrences of one variable are processed as two different variables. As a consequence, there is no elimination of values as in the domain of real numbers. In this paper, we define the simplification procedure as an application in sequence of elementary rules supposed to handle the dependency problem. A non exhaustive set of rules is given below[1], where f, g and h are terms:

$$\begin{cases} \exp: & e^f/e^g & \rightarrow & e^{f-g} \\ \lin: & fg + fh & \rightarrow & f(g+h) \\ \operatorname{div}: & fg/fh & \rightarrow & g/h \end{cases}$$

It can be shown that each rule $f \rightarrow g$ is such that $g \preccurlyeq f$. In particular the second one looks like the sub-distributivity property of interval arithmetic. Moreover, every sequence of application of rules is finite. It suffices to remark that the number of symbols in rewritten terms strictly decreases.

Example 4. Consider the NCSP $x_2 \exp(-2.5x_1) = 0.85 \wedge x_2 \exp(-1.5x_1) = 1.25$ given the box $[0, 1] \times [0, 10]$. Constraint propagation leads to slow convergence. This behavior is illustrated in Figure 1 where x_2 is expressed as a function of x_1. The initial box is first reduced at the upper bound of x_2 using g_2: the eliminated box contains no solution of the second constraint, *i.e.*, no point of g_2. Then the upper bound of x_1 is contracted using g_1, and so on.

The two constraints can be combined and simplified as follows:

$$\frac{x_2 \exp(-2.5x_1)}{x_2 \exp(-1.5x_1)} = \frac{0.85}{1.25} \xrightarrow{div} \frac{\exp(-2.5x_1)}{\exp(-1.5x_1)} = \frac{0.85}{1.25} \xrightarrow{exp}$$

$$\exp(-2.5x_1 + 1.5x_1) = \frac{0.85}{1.25} \xrightarrow{lin} \exp(-x_1) = \frac{0.85}{1.25}$$

If the redundant constraint is added in the NCSP, a gain of more than one order of magnitude is obtained in the solving process using box consistency. The result is the box $[0.38566248, 0.38566249] \times [2.2291877, 2.2291878]$.

---

[1] Due to a lack of space, only the rules that are used in the experimentations are given.

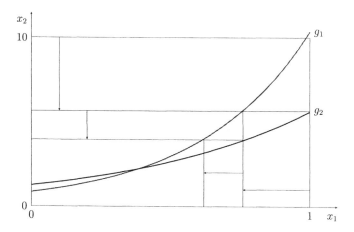

Fig. 1. Slow convergence in constraint propagation.

4   Implementation and Results

An implementation has been realized in `RealPaver`. In the following we give details of the encoding of constraint projections. In order to illustrate the effects of our approach in consistency reasonings, a set of experimental results is also discussed.

4.1   Implementation Details

First of all let us note that every constraint can be expressed as a projection constraint: $f = 0$ corresponds to $f \in [0,0]$ and $f \leqslant 0$ to $f \in [-\infty, 0]$. Projections on terms are computed by the `HC4revise` algorithm, which is a chain rule enforcing elementary constraint inversion steps [7].

Example 5. Let f be a term and let C be the constraint $(f + y) - 1 = 0$. Given the box $[-10, 10]^2$ let us implement `HC4revise` to compute an interval I such that the projection constraint $f \in I$ holds. Constraint C is rewritten as the projection constraint $(f + y) - 1 \in [0,0]$. The first inversion step eliminates the minus operation: $f + y \in [0,0] + 1 = [1,1]$. The second inversion step removes the plus operation: $f \in [1,1] - y$ and y is replaced with its domain: $f \in [-9, 11]$. Now, given the constraint $f \in [-9, 11]$ box consistency may be enforced over f.

Given two projection constraints $f \in I$ and $g \in J$ every redundant constraint $\psi(\varphi(f, g)) \in \phi_{\mathbb{I}}(I, J)$ is represented as follows. Term $\psi(\varphi(f, g))$ is represented in explicit form in the DAG. $\phi_{\mathbb{I}}(I, J)$ is implemented as an interval function taking as input the intervals I and J available at the root nodes of f and g in the DAG. This function is located in the root node of term $\psi(\varphi(f, g))$. This way $\phi_{\mathbb{I}}$ is evaluated as fast as possible and memory consumption is kept in reasonable bounds. If the expression of $\psi(\varphi(f, g))$ is simple then the computation of box consistency may be precise and cheap.

4.2   Experimental Results

This section reports the results of the term sharing and the term combination procedures. For the experimental tests, we have used some known problems from the numerical analysis and interval analysis communities. All the problems were solved by a bisection algorithm with a precision of output boxes of 10^{-10}. The experiments have been conducted on a Linux/PC Pentium III 933MHz. The description of test problems follows. Every variable whose domain is not explicitly mentioned lies in the interval $[-100, 100]$.

1. **Brown's almost linear problem** [11]:

$$\begin{cases} x_i + \sum_{j=1}^{n} x_j = n+1 & 1 \leqslant i \leqslant n-1 \\ \prod_{j=1}^{n} x_j = 1 \end{cases}$$

   In the system, the sum of variables can be shared and represented only once.

2. **Extended Wood Problem** [15] $(1 \leqslant i \leqslant n)$:

$$\begin{cases} -200x_i(x_{i+1} - x_i^2) - (1 - x_i) = 0 & \mod (i, 4) = 1 \\ 200(x_i - x_{i-1}^2) + 20(x_i - 1) + 19.8(x_{i+2} - 1) = 0 & \mod (i, 4) = 2 \\ -180x_i(x_{i+1} - x_i^2) - (1 - x_i) = 0 & \mod (i, 4) = 3 \\ 180(x_i - x_{i-1}^2) + 20.2(x_i - 1) + 19.8(x_{i-2} - 1) = 0 & \mod (i, 4) = 0 \end{cases}$$

   Terms $(x_{i+1} - x_i^2)$ and $(x_i - x_{i-1}^2)$ can be shared in the first two equations and in the last two constraints.

3. **Circuit design problem** [7] $(1 \leqslant i \leqslant 4)$:

$$\begin{cases} x_1 x_3 = x_2 x_4 \\ (1 - x_1 x_2)x_3(\exp(x_5(a_{1i} - a_{3i}x_7 - a_{5i}x_8)) - 1) = a_{5i} - a_{4i}x_2 \\ (1 - x_1 x_2)x_4(\exp(x_6(a_{1i} - a_{2i} - a_{3i}x_7 + a_{4i}x_9)) - 1) = a_{5i}x_1 - a_{4i} \end{cases}$$

   The symbols a_{ij} are known coefficients. In this system, it can be observed that the term $(1 - x_1 x_2)$ occurs in all but one constraint.

4. **Product problem**:

$$\prod_{j=1, j \neq i}^{n} x_j = i, \quad 1 \leqslant i \leqslant n$$

   Two consecutive constraints i and $i+1$ can be divided and simplified into $x_{i+1}/x_i = i/(i+1)$.

5. **Extended product problem**:

$$x_i + \prod_{j=1}^{n} x_j = i, \quad 1 \leqslant i \leqslant n$$

   The products from two consecutive constraints i and $i+1$ can be eliminated, which corresponds to the generation of the redundant constraint $i - x_i = (i+1) - x_{i+1}$.

6. **Parameter estimation problem**: Parameter estimation is the problem of determining the parameters of a model given experimental data. Let us consider the following model using exponential sums:

$$y(t) = x_1 \exp(-x_2 t) + x_3 \exp(-x_4 t) + x_5 \exp(-x_6 t)$$

In the real-world a series of measures (t_i, y_i) is known and the aim is to determine the x_i's such that every measure verifies the model. In the following let us compute the data by simulation. The series of timings is fixed as $t = (0, 1, 4, 9, 16, 25, 36, 49, 69, 81, 100)$. The exact parameter values are defined by $(10, 1, -5, 0.1, 1, 0.01)$ and the y_i's are computed as $y(t_i)$. Now let us try to compute the parameters values in the box

$$[-10^3, 10^3] \times [0.5, 2] \times [-10^3, 10^3] \times [0.05, 0.5] \times [-10^3, 10^3] \times [0, 0.05].$$

The main idea is to combine terms from the same columns. Given the projection constraints $x_k \exp(-x_{k+1} t_i) \in I$ and $x_k \exp(-x_{k+1} t_j) \in J$, the following redundant constraint is derived:

$$\frac{x_k \exp(-x_{k+1} t_i)}{x_k \exp(-x_{k+1} t_j)} \in \frac{I}{J} \xrightarrow{div} \frac{\exp(-x_{k+1} t_i)}{\exp(-x_{k+1} t_j)} \in \frac{I}{J} \xrightarrow{exp}$$

$$\exp(x_{k+1} t_j - x_{k+1} t_i) \in \frac{I}{J} \xrightarrow{lin} \exp((t_j - t_i) x_{k+1}) \in \frac{I}{J}$$

The redundant constraint may be used to reduce further the domain of x_{k+1}. Since x_{k+1} occurs once the reduction may be computed by a direct interval expression $x_{k+1} \leftarrow (t_j - t_i)^{-1} \cdot \log(I/J)$. Furthermore the term $(t_j - t_i)^{-1}$ can be evaluated only once. As a consequence the reduction is very cheap since it needs evaluating three interval operations.

Table 1 summarizes the results. In the table, Name denotes the problem and n stands for the number of constraints. The next columns present the computation time in seconds and the number of boxes of the solving process for the classical bisection algorithm, the bisection algorithm using term sharing and combination, and the improvement on the number of boxes. A "?" stands for problems that cannot be solved in less than one hour.

Term sharing transforms Brown's system as a gentle problem for consistency techniques. There is clearly a great interest in sharing complex terms occurring in many constraints. The improvement for Wood and Transistor is less impressive since only small terms are shared among a subset of constraints. However, the improvement is still more than one order of magnitude.

The product and extended product problems are efficiently solved using term combination. In this case, the constraints are simplified enough to greatly improve constraint processing. The improvement is smaller for the estimation problem since the constraints to be combined are complex, and only small parts of them are combined and simplified. Our approach is clearly very efficient for problems with many small constraints having similar expressions. For the other problems, it remains a technique of choice since the process of evaluating the new constraints is cheap.

Table 1. Experimental results.

Benchmark		Classical method		New method		Ratio
Name	n	Time	Box	Time	Box	Box
Brown	4	1	1 018	0	76	13
Brown	5	2	12 322	0	94	131
Brown	6	43	183 427	0	67	2 737
Brown	7	1 012	3 519 097	0	88	39 989
Wood	4	1	3 688	0	325	11
Wood	8	256	452 590	3	4 747	95
Wood	12	?	?	42	45 751	?
Transistor	9	1 061	236 833	130	23 887	9
Product	3	1	52	0	31	1
Product	5	459	5 698 714	0	37	154 019
Product	7	?	?	0	40	?
Extended Product	3	1	3 217	0	16	201
Extended Product	4	1 346	19 315 438	0	19	101 660
Extended Product	5	?	?	0	19	?
Estimation	11	6	11 581	3	5 467	2

4.3   On Strategies

A critical component of the algorithm is the strategy for combining terms. The first approach is to tune the method according to the problem structure, typically exponential sums. This allows one to develop so-called global constraints, namely complex constraints associated with specific and efficient algorithms.

The second approach is to tackle the general problem. In our implementation, only consecutive constraints are combined in order to introduce a reasonable number of new constraints and patterns are determined wrt. specific tree-representations of constraints. These limitations have to be relaxed with the aim of controling the combinatorial explosion.

5   Conclusion

In this paper, we have introduced a general framework for improving consistency techniques using redundant constraints. Redundancies are obtained by combination of terms and simplification according to the precision of interval computations. In particular, the well-known problem of processing exponential sums arising in mathematical modeling of dynamic systems has been efficiently handled. In this case the simplification process follows from the property of the exponential function $e^a e^b = e^{a+b}$.

The first issue is to design an efficient combination strategy for the general problem. For this purpose, the study of term rewriting engines will be useful [4]. The second issue is to develop new global constraints, e.g., for geometric problems modeled by trigonometric functions.

Acknowledgements

Frédéric Goualard is gratefully acknowledged for his comments on a previous version of this paper.

References

1. F. Benhamou and L. Granvilliers. Automatic Generation of Numerical Redundancies for Non-Linear Constraint Solving. *Reliable Computing*, 3(3):335–344, 1997.
2. F. Benhamou, D. McAllester, and P. Van Hentenryck. CLP(Intervals) Revisited. In M. Bruynooghe, editor, *Proceedings of International Symposium on Logic Programming*, pages 124–138, Ithaca, USA, 1994. MIT Press.
3. A. Bockmayr and V. Weispfenning. *Handbook of Automated Reasoning*, chapter Solving Numerical Constraints, pages 753–842. Elsevier Science Publishers, 2001.
4. P. Borovansky, C. Kirchner, H. Kirchner, and P.-E. Moreau. ELAN from a rewriting logic point of view. *Theoretical Computer Science*, 285(2):155–185, 2002.
5. B. Buchberger. Gröbner Bases: an Algorithmic Method in Polynomial Ideal Theory. In N. K. Bose, editor, *Multidimensional Systems Theory*, pages 184–232. D. Reidel Publishing Company, 1985.
6. L. Granvilliers. On the Combination of Interval Constraint Solvers. *Reliable Computing*, 7(6):467–483, 2001.
7. L. Granvilliers and F. Benhamou. Progress in the Solving of a Circuit Design Problem. *Journal of Global Optimization*, 20(2):155–168, 2001.
8. L. Granvilliers, E. Monfroy, and F. Benhamou. Symbolic-Interval Cooperation in Constraint Programming. In G. Villard, editor, *Proceedings of International Symposium on Symbolic and Algebraic Computation*, pages 150–166. ACM Press, 2001.
9. T. J. Hickey, Q. Ju, and M. H. Van Emden. Interval Arithmetic: From Principles to Implementation. *Journal of the Association for Computing Machinery*, 48(5):1038–1068, 2001.
10. Institute of Electrical and Electronics Engineers. *IEEE Standard for Binary Floating-Point Arithmetic*, 1985. IEEE Std 754-1985, Reaffirmed 1990.
11. R. B. Kearfott and M. Novoa. Intbis: A Portable Interval Newton/Bisection Package. *ACM Transactions on Mathematical Software*, 2(16):152–157, 1990.
12. O. Lhomme. Consistency Techniques for Numeric CSPs. In R. Bajcsy, editor, *Proceedings of International Joint Conference on Artificial Intelligence*, pages 232–238, Chambéry, France, 1993. Morgan Kaufman.
13. A. K. Mackworth. Consistency in Networks of Relations. *Artificial Intelligence*, 8(1):99–118, 1977.
14. R. E. Moore. *Interval Analysis*. Prentice-Hall, Englewood Cliffs, NJ, 1966.
15. J. J. Moré, B. S. Garbow, and K. E. Hillstrom. Testing Unconstrained Optimization Software. *ACM Transactions on Mathematical Software*, 7(1):17–41, 1981.
16. M. H. Van Emden. Algorithmic Power from Declarative Use of Redundant Constraints. *Constraints*, 4(4):363–381, 1999.
17. P. Van Hentenryck, L. Michel, and Y. Deville. *Numerica: a Modeling Language for Global Optimization*. MIT Press, 1997.

Proving and Constraint Solving
in Computational Origami

Tetsuo Ida[1], Dorin Ţepeneu[1], Bruno Buchberger[2], and Judit Robu[3]

[1] Department of Computer Science
University of Tsukuba, Tennoudai 1-1-1, Tsukuba 305-8573, Japan
{ida,dorinte}@score.cs.tsukuba.ac.jp
[2] Research Institute for Symbolic Computation
Johannes Kepler University, A 4232 Schloss Hagenberg, Austria
Bruno.Buchberger@risc.uni-linz.ac.at
[3] Department of Computer Science
Babes-Bolyai University, Mihail Kogalniceanu No.1, Cluj-Napoca 3400, Romania
robu@cs.ubbcluj.ro

Abstract. Origami (paper folding) has a long tradition in Japan's culture and education. We are developing a computational origami system, based on symbolic computation system Mathematica, for performing and reasoning about origami on the computer. This system is based on the implementation of the six fundamental origami folding steps (origami axioms) formulated by Huzita. In this paper, we show how our system performs origami folds by constraint solving, visualizes each step of origami construction, and automatically proves general theorems on the result of origami construction using algebraic methods. We illustrate this by a simple example of trisecting an angle by origami. The trisection of an angle is known to be impossible by means of a ruler and a compass. The entire process of computational origami shows nontrivial combination of symbolic constraint solving, theorem proving and graphical processing.

1 Introduction

Origami is a Japanese traditional art of paper folding. The word origami[1] is a combined word coming from ori (fold) and kami (paper). For several centuries, origami has been popular among Japanese common people as an art, as playing toys, and teaching material for children. Over a decade, origami is now also receiving wide interest among mathematicians, mathematics educators and computer scientists, as well as origami artists, as origami poses interesting fundamental geometrical questions.

We are proposing computational origami, as part of a discipline of origami science (also coined *origamics* by Haga, a pioneer in Japanese origami research

---

[1] Traditionally, the word origami is used to represent a folding paper, the act of paper folding or the art of origami. As such, we also use the word origami in a flexible way depending on the context.

B. Buchberger and J.A. Campbell (Eds.): AISC 2004, LNAI 3249, pp. 132–142, 2004.
© Springer-Verlag Berlin Heidelberg 2004

[7]). We believe that the rigor of paper folding and the beauty of origami art-works enhance greatly when the paper folding is supported by a computer. In our earlier work, computational origami performs paper folding by solving both symbolically and numerically certain geometrical constraints, followed by the visualization of origami by computer graphics tools. In this paper we extend the system for proving the correctness of the origami constructions, as proposed in our previous paper [3].

Origami is easy to practice. Origami is made even easier with a computer: we can construct an origami by calling a sequence of origami folding functions on the Mathematica Notebook [12] or by the interaction with our system, running in the computing server, using the standard web browser. The constructed final result, as well as the results of the intermediate steps, can be visualized and manipulated. Moreover, in cases where a proof is needed, the system will produce the proof of the correctness of the construction. Namely, for a given sequence of origami construction steps and a given property, the system proves that the resulting shape will satisfy the property (or disprove the property).

The rest of the paper is organized as follows. In sections 2 and 3 we present a formal treatment of origami construction. In section 4 we discuss a method for trisecting an angle by origami, which cannot be made using the traditional ruler-and-compass method. The construction is followed by the proof of the correctness of the construction in section 5. In section 6 we summarize our contributions and point out some directions for future research.

2   Preliminaries

In this section we summarize basic notions and notations that are used to explain the principles of origami construction. We assume the basic notions from elementary Euclidean geometry.

In our formalism origami is defined as a structure $O = \langle \chi, \mathcal{R} \rangle$, where χ is the set of faces of the origami, and $\mathcal{R}$ is a relation on χ. Let $\mathcal{O}$ be the set of origamis. Origami construction is a finite sequence of origamis $O_0, O_1, ..., O_n$ with O_0 an initial origami (usually square paper) and $O_{i+1} = g_i(O_i)$ for some function g_i defined by a particular fold. $\mathcal{R}$ is the combination of overlay and neighborhood relations on faces, but in this paper we will not elaborate it further.

A point P is said to be *on origami* $O = \langle \chi, \mathcal{R} \rangle$ if there exists a face $X \in \chi$ such that P is on X. Likewise, a line l is said to *pass through origami* $O = \langle \chi, \mathcal{R} \rangle$ if there exists a face $X \in \chi$ such that l passes through the interior of X. We denote by $\mathcal{L}$ the set of lines, and by $\mathcal{P}$ the set of points. The set of points on O is denoted by $\mathcal{P}_O$, and the set of lines on O by $\mathcal{L}_O$. We abuse the notation $\cap$ and $\in$, to denote by $P \cap O \neq \phi$ (resp. $l \cap O \neq \phi$) the property that P (resp. l) is on O, and by $P \in f$ the property that point P is on line f.

Furthermore, 'sym' is the function which computes the symmetric point of a given point with respect to a given line, 'dist' is the function which computes the distance between two parallel lines, and 'bisect' is the function which computes the bisector(s) of two lines.

3   Principles of Origami Construction

An origami is to be folded along a specified line on the origami called *fold line*. The line segment of a fold line on the origami is called a *crease*, since the consecutive operation of a fold and an unfold along the same fold line makes a crease on the origami.

A fold line can be determined by the points it passes through or by the points (and/or lines) it brings together. As in Euclidean geometry, by specifying points, lines, and their configuration, we have the following six basic fold operations called *origami axioms* of Huzita [9, 8]. It is known that Huzita's origami Axiom set is more powerful than the ruler-and-compass method in Euclidean geometry [6]. Origami can construct objects that are impossible by the ruler-and-compass method [4]. One of them is trisecting an angle, which we will show in this paper, as our example of origami proving and solving.

3.1   Origami Axioms

Huzita's origami axioms are described in terms of the following fold operations:

(O1) Given two points P and Q, we can make a fold along the crease passing through them.

(O2) Given two points P and Q, we can make a fold to bring one of the points onto the other.

(O3) Given two lines m and n, we can make a fold to superpose the two lines.

(O4) Given a point P and a line m, we can make a fold along the crease that is perpendicular to m and passes through P.

(O5) Given two points P and Q and a line m, either we can make a fold along the crease that passes through Q, such that the fold superposes P onto m, or we can determine that the fold is impossible.

(O6) Given two points P and Q and two lines m and n, either we can make a fold along the crease, such that the fold superposes P and m, and Q and n, simultaneously, or we can determine that the fold is impossible.

The operational meaning of these axioms is the following: finding crease(s) and folding the origami along the crease.

Let us first formalize the process of finding the creases. Let O be an origami, P and Q be points, and f, m and n be lines. The above fold operations can be stated by the following logical formulas. They are the basis for the implementation of the computational origami system.

Axiom 1: $\displaystyle\forall_{P,Q \in \mathcal{P}_O} \left(\exists_{f \in \mathcal{L}_O} (P \in f \wedge Q \in f) \right)$

Axiom 2: $\displaystyle\forall_{P,Q \in \mathcal{P}_O} \left(\exists_{f \in \mathcal{L}_O} (\mathrm{sym}(P, f) \equiv Q) \right)$

Axiom 3: $\displaystyle\forall_{m,n \in \mathcal{L}_O} \left(\exists_{f \in \mathcal{L}_O} \left((m \parallel n) \wedge (f \parallel m) \wedge (\mathrm{dist}(f, m) = \mathrm{dist}(f, n)) \right) \vee \left(\neg (m \parallel n) \wedge (f \equiv \mathrm{bisect}(m, n)) \right) \right)$

Axiom 4: $\displaystyle \forall_{P \in \mathcal{P}_O, m \in \mathcal{L}_O} \left(\exists_{f \in \mathcal{L}_O} \left((f \perp m) \wedge (P \in f) \right) \right)$

Axiom 5: $\displaystyle \forall_{P, Q \in \mathcal{P}_O, m \in \mathcal{L}_O} \left(\Phi_1 \left(P, Q, m \right) \Rightarrow \exists_{f \in \mathcal{L}_O} \left((Q \in f) \wedge (\mathrm{sym}(P, f) \in m) \right) \right)$

Axiom 6: $\displaystyle \forall_{P, Q \in \mathcal{P}_O, m, n \in \mathcal{L}_O} \Big(\Phi_2 \left(P, Q, m, n \right) \Rightarrow$

$$\exists_{f \in \mathcal{L}_O} \left((\mathrm{sym}(P, f) \in m) \wedge (\mathrm{sym}(Q, f) \in n) \right) \Big)$$

In Axioms 5 and 6 we have to define the constraints that ensure the existence of the fold line on the origami. Conditions Φ_1 and Φ_2 could be explicitly given as boolean combinations of formulas of the form $A = 0$ or $A > 0$, where the A's are polynomials in the coordinates of P, Q, m (and n).

All the axioms have existential sub-formulas, hence the essence of an origami construction is finding concrete terms for the existentially quantified variable f. Our computational origami system returns the solution both symbolically and numerically, depending on the input.

4   Trisecting an Angle

We give an example of trisecting an angle in our system. This example shows a nontrivial use of Axiom 6. The method of construction is due to H. Abe as described in [6, 5]. In the following, all the operations are performed by *Mathematica* function calls. Optional parameters can be specified by *"keyword → value"*.

Steps 1 and 2: First, we define a square origami paper, whose corners are designated by the points A, B, C and D. The size may be arbitrary, but for our example, let us fix it to 100 by 100. The new origami figure is created with two differently colored surfaces: a light-gray front and a dark-gray back. We then introduce an arbitrary point, say E at the coordinate $(30, 100)$, assuming A is at $(0, 0)$.

```
NewOrigami[Square[100, MarkPoints → {'A','B','C','D'}],
    FigureCaption → 'Step '];
PutPoint['E', Point[30, 100]];
```

Our problem is to trisect the angle $\angle EAB$. The method consists of the following seven steps (steps 3-9) of folds and unfolds.

Steps 3 and 4: We make a fold to bring point A to point D, to obtain the perpendicular bisector of segment $\overline{AD}$. This is the application of (O2). The points F and G are automatically generated by the system. We unfold the origami and obtain the crease $\overline{FG}$.

```
FoldBring[A, D];
Unfold[];
```

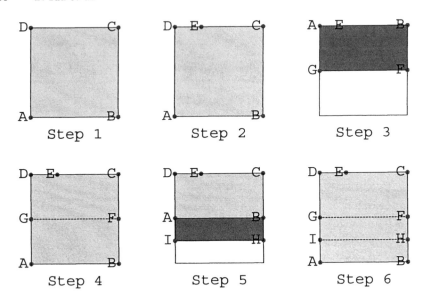

Fig. 1. Trisecting an angle by origami, steps 1-6.

Steps 5 and 6: Likewise we obtain the crease $\overline{IH}$.

```
FoldBring[A, G];
Unfold[];
```

Steps 7 and 8: Step 7 is the crucial step of the construction. We will superpose point G and the line that is the extension of the segment $\overline{AE}$, and to
superpose point A and the line that is the extension of the segment $\overline{IH}$, simultaneously. This is possible by (O6) and is realized by the call of function
FoldBrBr. There are three candidate fold lines to make these superpositions
possible. The system responds with the query of "Specify the line number"
together with the fold lines on the origami image. We reply with the call of
FoldBrBr with the additional parameter 3, which tells the system that we
choose the line number 3. This is the fold line that we are primarily interested in. However, the other two fold lines are also solutions (which trisect
different angles).

```
FoldBrBr[G, AE, A, IH];
FoldBrBr[G, AE, A, IH, 3];
```

Steps 9 and 10: We will duplicate the points A and I on the other face that is
below the face that A and I are on, and unfold the origami. The duplicated
points appear as L and J for A and I, respectively. These names are automatically generated. Finally, we see that the segments $\overline{AJ}$ and $\overline{AL}$ trisect
the angle $\angle EAB$.

```
DupPoint[{'I', 'A'}]; Unfold[];
ShowOrigamiSegment[{{{A, E}}, {{A, J}}, {{A, L}}}];
```

Specify the line number.

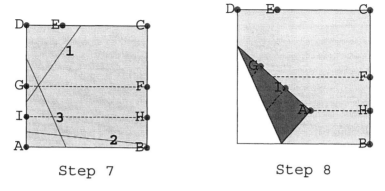

Step 7   Step 8

Fig. 2. Trisecting an angle by origami, steps 7-8.

Although it is not obvious to see that equational solving is performed, our system solves a set of polynomial equations, up to the third degree. In the case of the folds applying (O1) - (O4), the system computes a solution using the well known mathematical formulas of elementary geometry rather than proving the existential formulas of Axioms 1 to 4. In the case of FoldBrBr, the system solves a cubic equation. This explains why we have (at most) 3 possible fold lines at step 7.

5   Proof of the Correctness of the Trisection Construction

We now prove the following theorem with our system:

Theorem 1. *The origami construction in section 4 trisects an angle.*

5.1   Proof Method

In this simple example, the correctness of the trisection construction could be easily verified either by geometric reasoning or by a sequence of simplification steps of the algebraic equations representing geometric constrains. However, for proceeding towards a general (and completely automatic) proving method for origami theorems, we formulate the proving steps in a more general setting by showing that the above theorem would be proved if we could show that

$$\tan \angle BAL = \tan \angle LAJ = \tan \angle JAE \tag{1}$$

in step 10 of Fig. 3.

A general proof procedure is as follows:

(i) We first translate the equality (1) into the algebraic form. This is done after we fix the coordinate system (in our case Cartesian system).

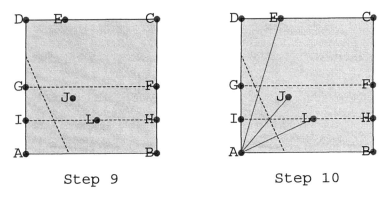

Step 9   Step 10

Fig. 3. Trisecting an angle by origami, steps 9-10.

(ii) We already observed that all the folding steps are formulated in Axioms 1 - 6. These axioms are easily transcribed in terms of polynomial constraints, once the representations of lines and points are fixed. The functions sym, bisect, and dist, and relations $\|$ and $\perp$ are easily defined.

(iii) We use the Gröbner bases method. We collect all the premise equalities and inequalities $C = \{c_1, ..., c_n\}$ obtained at step (ii), and the conclusion equalities and inequalities $D = \{d_1, ..., d_m\}$ obtained at step (i). Let M be the boolean combinations of the equalities and inequalities of the form $\neg (c_1 \wedge ... \wedge c_n) \vee (d_1 \wedge ... \wedge d_m)$, i.e. $C \Rightarrow D$. We prove $\forall M$ by refutation. The decision algorithm, roughly, proceeds as follows:

(a) Bring M into a conjunctive normal form and distribute $\forall$ over the conjunctive parts. Treat each of the parts

$$\underset{x,y,z,...}{\forall} P$$

separately. Note that P is a disjunction

$$E_1 = 0 \vee ... \vee E_k = 0 \vee N_1 \neq 0 \vee ... \vee N_l \neq 0$$

of equalities and inequalities.

(b) Then

$$\underset{x,y,z,...}{\forall} (E_1 = 0 \vee ... \vee E_k = 0 \vee N_1 \neq 0 \vee ... \vee N_l \neq 0)$$

is transformed into

$$\neg \underset{x,y,z,...}{\exists} (E_1 \neq 0 \wedge ... \wedge E_k \neq 0 \wedge N_1 = 0 \wedge ... \wedge N_l = 0)$$

and further on to

$$\neg \underset{x,y,z,...,\xi_1,...,\xi_k}{\exists} (E_1 \xi_1 - 1 = 0 \wedge ... \wedge E_k \xi_k - 1 = 0 \wedge N_1 = 0 \wedge ... \wedge N_l = 0)$$

with new variables $\xi_1, ..., \xi_k$ ("Rabinovich trick").

(c) Now, our question becomes a question on the solvability of a system of polynomial equalities, which can be decided by computing the reduced Gröbner basis of $\{E_1 \xi_1 - 1, ..., E_k \xi_k - 1, N_1, ..., N_l\}$. Namely, one of the fundamental theorems of Gröbner bases theory tells us that this Gröbner basis will be {1} iff the system is unsolvable (i.e. has no common zeros), which was first proved in [1].

5.2   Automated Proof

The proof steps would in general require laborious work of symbol manipulation. We are developing software for performing such transformations completely automatically. The software is an extension of the geometrical theorem prover [11] based on *Theorema* [2].

The following piece of programs will do all the necessary transformations and the proof outlined above. The program will be self-explanatory, as we use the names similar to the functions for folding used before.

The proposition to be proved is the following:

```
Proposition['Trisection', any[A,B,C,D,E,F,G,H,I,J,K,L,M,N],
   neworigami[A,B,C,D] ∧ pon[E,line[D,C]] ∧
   foldBring[A,D,crease[G on line[A,D], F on line[B,C]]] ∧
   foldBring[A,G,crease[I on line[A,G], H on line[B,F]]] ∧
   foldBrBr[L,A,line[H,I],N,G,line[A,E],
      crease[K on line[C,D], M on line[A,B]]] ∧
   symmetricPoint[J, I, line[M, K]] ⇒
      equalzero[tanof[B, A, A, L] - tanof[L, A, A, J]] ∧
   equalzero[tanof[B, A, A, L] - tanof[J, A, A, E]]]

KnowledgeBase['C1',any[A,B],{{A,{0,0}},{B,{100,0}}}]
```

To display graphically the geometrical constraints among the involved points and lines, we call function `Simplify`.

```
Simplify[Proposition['Trisection'],
   by → GraphicSimplifier, using → KnowledgeBase['C1']]
```

The following is the output of the prover.
We have to prove:

(Proposition(Trisection))

$$\forall_{A,B,C,D,E,F,G,H,I,J,K,L,M,N} (\text{neworigami}[A, B, C, D] \wedge \text{pon}[E, line[D, C]] \wedge$$
$$\text{foldBring}[A, D, \text{crease}[G \text{ on } line[A, D], F \text{ on } line[B, C]]] \wedge$$
$$\text{foldBring}[A, G, \text{crease}[I \text{ on } line[A, G], H \text{ on } line[B, F]]] \wedge$$
$$\text{foldBrBr}[L, A, line[H, I], N, G, line[A, E],$$
$$\text{crease}[K \text{ on } line[C, D], M \text{ on } line[A, B]]] \wedge$$
$$\text{symmetricPoint}[J, I, line[M, K]] \Rightarrow$$
$$\text{equalzero}[\text{tanof}[B, A, A, L] - \text{tanof}[L, A, A, J]] \wedge$$
$$\text{equalzero}[\text{tanof}[B, A, A, L] - \text{tanof}[J, A, A, E]])$$

with no assumptions.

To prove the above statement we use the Gröbner bases method. First we have to transform the problem into algebraic form.

To transform the geometric problem into an algebraic form we choose an orthogonal coordinate system.

Let us have the origin in point A, and points $\{B, M\}$ and $\{G, D, I\}$ on the two axes.

Using this coordinate system we have the following points:

$$\{\{A, 0, 0\}, \{B, u_1, 0\}, \{D, 0, x_1\}, \{E, x_2, u_2\}, \{\alpha_G, 0, x_3\}, \{G, 0, x_4\},$$
$$\{\alpha_I, 0, x_5\}, \{I, 0, x_6\}, \{M, x_7, 0\}, \{C, x_8, x_9\}, \{F, x_{10}, x_{11}\},$$
$$\{H, x_{12}, x_{13}\}, \{L, x_{14}, x_{15}\}, \{\alpha_L, x_{16}, x_{17}\}, \{K, x_{18}, x_{19}\},$$
$$\{\alpha_N, x_{20}, x_{21}\}, \{N, x_{22}, x_{23}\}, \{\alpha_J, x_{24}, x_{25}\}, \{J, x_{26}, x_{27}\}\},$$

where α_X is a variable generated internally to create point X.

The algebraic form[2] of the given construction is:

$$\forall_{x_0, \ldots, x_{27}} \ ((-1) + u_1 x_0 == 0 \wedge -u_1 + -x_1 == 0 \wedge -u_1^2 + u_1 x_8 == 0 \wedge$$
$$-x_1^2 + x_1 x_9 == 0 \wedge -x_1 x_2 + -u_2 x_8 + x_1 x_8 + x_2 x_9 == 0 \wedge$$
$$-x_1 + 2x_3 == 0 \wedge x_1 x_3 + -x_1 x_4 == 0 \wedge$$
$$-u_1 x_9 + x_9 x_{10} + u_1 x_{11} + -x_8 x_{11} == 0 \wedge$$
$$x_1 x_3 + -x_1 x_{11} == 0 \wedge -x_4 + 2x_5 == 0 \wedge x_4 x_5 + -x_4 x_6 == 0 \wedge$$
$$-u_1 x_{11} + x_{11} x_{12} + u_1 x_{13} + -x_{10} x_{13} == 0 \wedge x_4 x_5 + -x_4 x_{13} == 0 \wedge$$
$$-x_6 x_{12} + x_6 x_{14} + -x_{13} x_{14} + x_{12} x_{15} == 0 \wedge -x_{14} + 2x_{16} == 0 \wedge$$
$$-x_{15} + 2x_{17} == 0 \wedge -x_{14} x_{16} + -x_{15} x_{17} + x_{14} x_{18} + x_{15} x_{19} == 0 \wedge$$
$$-x_1 x_8 + x_1 x_{18} + -x_9 x_{18} + x_8 x_{19} == 0 \wedge$$
$$x_7 x_{14} + -x_{14} x_{16} + -x_{15} x_{17} == 0 \wedge$$
$$x_7 x_{19} + -x_{19} x_{20} + -x_7 x_{21} + x_{18} x_{21} == 0 \wedge$$
$$-x_4 x_{19} + -x_7 x_{20} + x_{18} x_{20} + x_{19} x_{21} == 0 \wedge 2x_{20} + -x_{22} == 0 \wedge$$
$$-x_4 + 2x_{21} + -x_{23} == 0 \wedge -u_2 x_{22} + x_2 x_{23} == 0 \wedge$$
$$-x_7 x_{19} + x_{19} x_{24} + x_7 x_{25} + -x_{18} x_{25} == 0 \wedge$$
$$x_6 x_{19} + x_7 x_{24} + -x_{18} x_{24} + -x_{19} x_{25} == 0 \wedge$$
$$2x_{24} + -x_{26} == 0 \wedge -x_6 + 2x_{25} + -x_{27} == 0 \Rightarrow$$
$$u_2 x_{14} x_{26} + -x_2 x_{15} x_{26} + -x_2 x_{14} x_{27} + -u_2 x_{15} x_{27} == 0 \wedge$$
$$-2x_{14} x_{15} x_{26} + x_{14}^2 x_{27} + -x_{15}^2 x_{27} == 0)$$

The further output of the proof is omitted here; the proof proceeds as outlined in step (iii).

Namely,

1. The above problem is decomposed into two independent problems.
2. The individual problems are separately proved.
3. For each problem, the reduced Gröbner bases are computed.
4. Since the result of the computation is $\{1\}$ for each individual problem, the proposition is generally true.

---

[2] Notation $x_0, \ldots, x_{27}$ represents the full sequence of consecutive variables from x_0 to x_{27}.

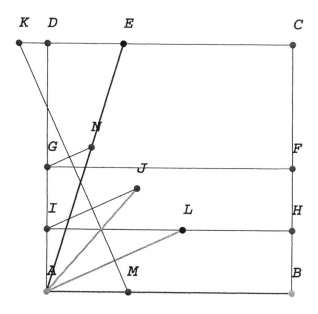

Fig. 4. Graphical output of the call of `Simplify`.

6   Conclusion

We have shown the computer origami construction with the example of trisecting an angle. Combining the origami simulation software[10], the implementation of the Gröbner basis algorithm [1], the implementation of the decision algorithm in *Theorema* [2], and a new tool [11] in *Theorema*, which translates the descriptions of origami construction into the corresponding polynomial equalities, we are able to offer a coherent tool for computational origami that can

- simulate arbitrary origami sequences both algebraically and graphically,
- translate conjectures about properties of the results of origami sequences into statements in the form of universally quantified boolean combinations of polynomial equalities,
- decide the truth of such conjectures and produce a proof or refutation of the conjecture fully automatically.

We are now working to integrate those software tools into a coherent system, each component working independently but coordinating each other on the Internet as a symbolic computation grid.

As a next step of our research we plan to study "Origami Solving Problem", which asks for finding a sequence of origami steps that will lead to an origami object with a desired property. However, it is clear that this problem is analogous to the problem of finding geometric objects with desired properties using only a ruler and a compass. Note, however, that the two problems – origami construction and the ruler-and-compass construction – are not equivalent, as we have seen. For further development of origami construction, in analogy to the ruler-and-compass construction problem, Galois theory suggests itself as the main approach to solving the origami construction problem.

Acknowledgements

Sponsored by Austrian FWF (Österreichischer Fonds zur Förderung der Wissenschaftlichen Forschung), Project 1302, in the frame of the SFB (Special Research Area) 013 "Scientific Computing"and RICAM (Radon Institute for Applied and Computational Mathematics, Austrian Academy of Science, Linz).

References

1. Buchberger, B.: Ein algorithmisches Kriterium für die Lösbarkeit eines algebraischen Gleichungssystems (An Algorithmical Criterion for the Solvability of Algebraic Systems of Equations). Aequationes mathematicae 4/3, 1970, pp. 374-383. (English translation in: Buchberger, B., and Winkler, F. (eds.), Gröbner Bases and Applications, Proceedings of the International Conference "33 Years of Gröbner Bases", 1998, Research Institute for Symbolic Computation, Johannes Kepler University, Austria, London Mathematical Society Lecture Note Series, Vol. 251, Cambridge University Press, 1998, pp. 535 -545.)
2. Buchberger, B., Dupre, C., Jebelean, T., Kriftner, F., Nakagawa, K., Vasaru, D., Windsteiger, W.: The Theorema Project: A Progress Report, In Symbolic Computation and Automated Reasoning (Proceedings of CALCULEMUS 2000, Symposium on the Integration of Symbolic Computation and Mechanized Reasoning, August 6-7, 2000, St. Andrews, Scotland), Kerber, M. and Kohlhase, M. (eds.), A.K. Peters Publishing Co., Natick, Massachusetts, pp. 98-113.
3. Buchberger, B. and Ida, T.: Origami Theorem Proving, SFB Scientific Computing Technical Report 2003-23-Oct, Research Institute for Symbolic Computation, Johannes Kepler University Linz, Austria, 2003.
4. Chen, T. L.: Proof of the Impossibility of Trisecting an Angle with Euclidean Tools, Mathematics Magazine Vol. 39, pp. 239-241, 1966.
5. Fushimi, K.: Science of Origami, a supplement to Saiensu, Oct. 1980, p. 8 (in Japanese).
6. Geretschläger, R.: Geometric Constructions in Origami (in Japanese, translation by Hidetoshi Fukagawa), Morikita Publishing Co., 2002.
7. Haga, K.: Origamics Part I: Fold a Square Piece of Paper and Make Geometrical Figures (in Japanese), Nihon Hyoronsha, 1999.
8. Hull, T.: Origami and Geometric Constructions, available online at http://web.merrimack.edu/ thull/geoconst.html, 1997.
9. Huzita, H.: Axiomatic Development of Origami Geometry, Proceedings of the First International Meeting of Origami Science and Technology, pp. 143-158, 1989.
10. Ida, T., Marin, M. and Takahashi, H.: Constraint Functional Logic Programming for Origami Construction, Proceedings of the First Asian Symposium on Programming Languages and Systems (APLAS2003), Lecture Notes in Computer Science, Vol. 2895, pp. 73-88, 2003.
11. Robu, J.: Automated Geometric Theorem Proving, PhD Thesis, Research Institute for Symbolic Computation, Johannes Kepler University Linz, Austria, 2002.
12. Wolfram, S.: The Mathematica Book, 5th ed., Wolfram Media, 2003.

An Evolutionary Local Search Method for Incremental Satisfiability

Mohamed El Bachir Menaï

Artificial Intelligence Laboratory, University of Paris 8
2 rue de la liberté, 93526 Saint Denis, France
menai@ai.univ-paris8.fr

Abstract. Incremental satisfiability problem (ISAT) is considered as a generalisation of the Boolean satisfiability problem (SAT). It involves checking whether satisfiability is maintained when new clauses are added to an initial satisfiable set of clauses. Since stochastic local search algorithms have been proved highly efficient for SAT, it is valuable to investigate their application to solve ISAT. Extremal Optimization is a simple heuristic local search method inspired by the dynamics of living systems with evolving complexity and their tendency to self-organize to reach optimal adaptation. It has only one free parameter and had proved competitive with the more elaborate stochastic local search methods on many hard optimization problems such as MAXSAT problem. In this paper, we propose a novel Extremal Optimization based method for solving ISAT. We provide experimental results on ISAT instances and compare them against the results of conventional SAT algorithm. The promising results obtained indicate the suitability of this method for ISAT.

Keywords: Incremental Satisfiability, Stochastic Local Search, Extremal Optimization, Self-Organized Criticality

1 Introduction

The Boolean satisfiability problem (SAT) is one of the most important discrete constraint satisfaction problems. It is defined as follows. Let *Bool* denotes the Boolean domain $\{0,1\}$ and $X = \{x_1, x_2,...,x_n\}$ be a set of Boolean variables. The set of literals over X is $L = \{x, \bar{x} / x \in X\}$. A clause C on X is a disjunction of literals. A SAT instance is a conjunction of clauses. An assignment of n Boolean variables is a substitution of these variables by a vector $v \in Bool^n$. A literal x_i or $\bar{x}_i$ is said to be satisfiable by an assignment if its variable is mapped to 1 or 0, respectively. A clause C is said to be satisfiable by an assignment v if at least one of its literals is satisfiable by v, in such case the value of the clause equals 1 ($C(v)=1$), otherwise it is said to be unsatisfiable ($C(v)=0$). A model for a SAT is an assignment where all clauses are satisfiable. The SAT problem asks to find a model for a given set of clauses.

SAT is a paradigmatic *NP*-complete problem in logic and computing theory [9]. It provides a natural mathematical formalism to encode information about many prob-

B. Buchberger and J.A. Campbell (Eds.): AISC 2004, LNAI 3249, pp. 143–156, 2004.
© Springer-Verlag Berlin Heidelberg 2004

lems in automated reasoning, planning, scheduling and several areas of engineering and design. In recent years, much effort has been spent to increase the efficiency of both complete and incomplete algorithms for SAT. Therefore, SAT-based methods have become an important complement to traditional specialized ones.

Real-world problems must be solved continually because real environment is always changing by adding or removing new constraints such as the evolution of the set of tasks to be performed in scheduling or planning applications. Problem solving is a never ending process [21]. Thus, one may have to deal with a series of problems that change over time where the current solution of an instance of a problem has to be adapted to handle the next one. In the CSP community, this concept is known as the dynamic CSP and was introduced by Dechter and Dechter [11] to maintain solutions in dynamic constraint networks. The incremental satisfiability problem (ISAT) is a generalisation of SAT which allows changes of a problem over time and can be considered as a prototypical dynamic CSP [13], [15].

ISAT was introduced by Hooker [14] as the problem of checking whether satisfiability is preserved when adding one clause at a time to an initial satisfiable set of clauses. It was solved using an implementation of the Davis-Putnam-Loveland procedure (DPL) [10], [17]. When adding a new clause, the procedure maintains building the search tree generated previously for the initial satisfiable set of clauses. Incremental DPL performs substantially faster than DPL for a large set of SAT problems [14].

A more general definition has been suggested by Kim et al. [16] to address practical situations that arise in various domains such as planning and electronic design automation (EDA). Given a set Φ of m Boolean functions in Conjunctive Normal Form $\varphi_i(X)$ over a set of variables X, $\Phi = \{\varphi_i(X) | \varphi_i(X) = \varphi_P(X) \wedge \varphi_{Si}(X)\}$ where each function has a common prefix function $\varphi_P(X)$ and a different suffix function $\varphi_{Si}(X)$. The problem is to determine the satisfiability of each function $\varphi_i(X)$. Kim et al. [16] applied ISAT to prove the untestability of non-robust delay fault in logic circuits. They formulated the encoding SAT instances as a sequence of closely related SAT instances sharing a common sub-sequence. They used a DPL-like approach to solve ISAT and showed that the results achieved using this methodology outperform those obtained when solving each SAT instance independently.

Hoos and O'Neill [15] introduced another formulation of ISAT as the dynamic SAT problem (DynSAT). It consists of adding or removing dynamically clauses from a given SAT instance. The problem is to determine for a given DynSAT instance whether it is satisfiable for each time. DynSAT can be solved simply as a set of independent SAT, but solving them together and considering that they share a common subset of clauses, may result in significant decrease in total run time. They presented an initial investigation of mainly two approaches for solving DynSAT. They used existing local search algorithms for SAT to solve the current SAT instance, while applying either random restart or trajectory continuation before solving the next SAT instance in the DynSAT sequence. Their empirical analysis on variants of WalkSAT algorithm [22] indicated that trajectory continuation approach is more efficient than random restart for hard unstructured problems derived from the SATLIB benchmark suite [24].

Gutierrez and Mali [13] presented a basic local search algorithm for ISAT integrating the ability of recycling model of the satisfiable set of clauses when adding new clauses. They presented experimental results on both random and structured ISAT instances generated using SAT instances from the SATLIB benchmark suite [24]. However, it is not clear how their algorithm could be compared to state-of-the-art local search-based algorithms for SAT.

Recently, a local search heuristic method called *Extremal Optimization* (EO) was proposed for solving hard optimization problems such as the graph partitioning [5], [7], the graph coloring, the spin glass [6] and the maximum satisfiability [19], [20]. EO is characterized by large fluctuations of the search process and extremal selection against worst variables in a sub-optimal solution. This make it able to explore efficiently the search space without loosing well-adapted parts of a solution. ISAT can be perceived as the problem of searching an equilibrium to a dynamic system perturbed by adding new clauses. By Extremal Optimization process the major parts of the system can be rearranged so that, a new equilibrium state of optimal adaptation can be reached. Therefore, it is worthwhile to use this heuristic to design an algorithm for ISAT.

The remainder of the paper is organized as follows. The next Section presents Extremal Optimization heuristic. Section 3 describes an implementation of EO to solve the SAT problem. In Section 4, we present a new EO-based algorithm for solving ISAT. In Section 5, we report on computational results on ISAT instances generated from SAT benchmark problems. In Section 6, we conclude and discuss directions for future work.

2 Extremal Optimization Heuristic

In 1987 Bak, Tang and Wiesenfeld [2] introduced the concept of *Self-Organized Criticality* (SOC) to describe the scale free behaviour exhibited by natural systems such as earthquakes, water flows in rivers, mass extinction and evolution to name a few. These systems are driven by their own dynamics to a SOC state characterized by power-law distribution of event sizes [1]. SOC is produced by self-organization in a long transitory period at the border of stability and chaos [1].

The Bak and Sneppen (BS) model of evolution [3] is the prototype of a wide class of SOC models related to various areas such as real evolution of bacteria populations [8] and macro-economical processes [23]. In BS model, species have an associated fitness value between 0 and 1 representing a time scale at which the species will mutate to a different species or become extinct. The species with higher fitness has more chance of surviving. All species are placed on the vertices of a graph and at each iteration, a selection process against the species with the poorest degree of adaptation is applied so that the smallest fitness value is replaced by a new random one which also impacts the fitness values of its neighbours. A state of optimal adaptation (SOC) above a certain threshold is attained after a sufficient number of steps. In this state almost all species have a *punctuated equilibrium* [3], [4] that makes them intimately connected. Any perturbation of this equilibrium involves large fluctuations, called *critical avalanches* [3], in the configuration of the fitness values, potentially making any configuration accessible when the system slides back to a self-organization state. The BS model demonstrated that the duration t of that avalanches follow a power-law distribution $P(t) \propto t^{-\tau}$, where the avalanche exponential τ is a small real number close to 1.

Extremal Optimization (EO) is a heuristic search method introduced by Boettcher and Percus [5] for solving combinatorial and physical optimization problems. It was motivated by the Bak-Sneppen model [3] which was converted into an incomplete optimization algorithm. EO heuristic search can be outlined as follows. Let consider a system described by a set of N species x_i with an associated fitness value λ_i also called individual cost [5]. The cost function $C(S)$ of a configuration S consists of the individual cost contributions λ_i. EO starts from an initial random state of the system and at each step performs search on a single configuration S. The variables are ranked from the worst to the best according to their fitness values and the variable with the smallest fitness value is randomly updated. The configuration with minimum cost is then maintained. The rank ordering allows EO to preserve well-adapted pieces of a solution, while updating a weak variable gives the system enough flexibility to explore various space configurations.

An improved variant of EO [6], [7], called τ-EO, consists to rank all variables from rank $n=1$ for the worst fitness to rank $n=N$ for the best fitness λ. According to the BS model, a power-law distribution over the rank order is considered. For a given value of τ,

$$P(n) \propto n^{-\tau}, \ (1 \leq n \leq N) \tag{1}$$

At each update, select a rank k according to $P(k)$ and change the state of the variable x_k. The variable with rank 1 (worst variable) will be chosen most frequently, while the higher ones will sometimes be updated. In this way, a bias against worst variables is maintained and no rank gets completely excluded from the selection process. However, the search performance depends on the value of the parameter τ. For $\tau=0$, the algorithm is simply a random walk through the search space. While for too large values of τ the process approaches a deterministic local search where only a small number of variables with particularly bad fitness would be chosen at each iteration. An optimal value of τ will allow even the variable with the highest rank to be selected during a given run time. Boettcher and Percus [7] have established a relation between τ, the run time t_{max} and the number N of the variables of the system, to estimate the optimal value of τ. Let $t=AN$ where A is a constant, then

$$\tau \sim 1 + \frac{\ln(A/\ln(N))}{\ln(N)} \ (N \to \infty, \ 1 \ll A \ll N) \tag{2}$$

At this optimal value, the best fitness variables are not completely excluded from the selection process and hence, more space configurations can be reached so that greatest performance can be obtained.

3   τ-EO for SAT

Given a SAT problem instance of n Boolean variables $X = \{x_1, x_2,..., x_n\}$ and m clauses $CF = (C_i)_{i \leq m}$. A variable fitness λ_i is defined as the negation of the total number of clauses violated by x_i:

$$\lambda_i = -\#unsat(x_i) = -\sum 1_{x_i \in C_j \text{ and } C_j(v)=0} \tag{3}$$

The best fitness is $\lambda_i = 0$ and the worst one is $\lambda_i = -m$. The cost function $C(S)$ consists of the individual cost contributions λ_i for each variable x_i. In order to maximize the total number of satisfied clauses, $C(S)$ has to be minimized. Thus,

$$C(S) = -\sum_{i=1}^{n} \lambda_i \tag{4}$$

Algorithm τ-EO-SAT
1.   Randomly generate a solution S. Set *Sbest* = S.
2.   If S satisfies *Clauses* then return (S).
3.   Evaluate λ_i for each x_i in *Variables* according to Equation 3.
4.   Rank x_i according to λ_i from the worst to the best.
5.   Select a rank j according to Equation 1.
6.   Set $S' = S$ in which the truth value of x_j is flipped
7.   If $C(S') < C(Sbest)$ then Set *Sbest* = S'.
8.   Set $S = S'$.
9.   If the number of steps does not exceed *MaxSteps* return to Step 2.
10. Return (No model can be found).
End τ-EO-SAT

Fig. 1. Generic τ-EO algorithm for SAT

Figure 1 provides the basic outlines of the algorithm τ-EO for SAT. Let *Variables*, *Clauses*, *MaxSteps* and τ be respectively the set of variables, the set of clauses, the given bound for iteration steps and the free parameter of EO.

(i)   Line 1. The search begins at a random truth assignment S and the current best assignment, *Sbest*, is set to S.
(ii) Line 2. S is returned as a model if it satisfies *Clauses*.
(iii) Lines 3-4. The variable individual fitnesses are evaluated and sorted using a Shell sort.

(iv) Lines 5-6. A variable is selected according to the power-law distribution with the parameter τ (Equation 1). Its value is then flipped.

(v) Lines 7-8. *Sbest* is related to the current assignment with the minimum cost (Equation 4). The current assignment is then updated.

(vi) Line 9. The optimization loop is executed for *MaxSteps* iterations.

(vii) Line 10. There is no guaranty that the algorithm finds a model to the problem if one exists. τ-EO-SAT is incomplete.

4 τ-EO for ISAT

Let $V = \{x_1, x_2, ..., x_n\}$ be a set of Boolean variables and φ_P a set of clauses over a subset of variables $X \subseteq V$. φ_P is found to be satisfiable and has a model S. Determine the satisfiability of $\varphi_P \cup \varphi_S$ for a given set of clauses φ_S over a subset of variables $X' \subseteq V$. We use the same classification of clauses of $\varphi_P \cup \varphi_S$ as suggested in [13] in order to perform minimum local changes on the model of φ_P. Let X_φ be the set of variables that appear in the clauses of φ. Let $\varphi_P = \varphi_{P1} \cup \varphi_{P2}$ and $\varphi_S = \varphi_{S1} \cup \varphi_{S2}$ where :

$$\varphi_{P1} = \left\{ C_i \in \varphi_P \middle| X_{C_i} \cap X_{\varphi'_{P1}} \neq \phi \right\} \text{ and } \varphi'_{P1} = \left\{ C_i \in \varphi_P \middle| X_{C_i} \cap X_{\varphi_S} \neq \phi \right\} \tag{5}$$

$$\varphi_{P2} = \varphi_P - \varphi_{P1} \tag{6}$$

$$\varphi_{S1} = \left\{ C_i \in \varphi_S \middle| X_{C_i} \cap X_{\varphi'_{S1}} \neq \phi \right\} \text{ and } \varphi'_{S1} = \left\{ C_i \in \varphi_S \middle| X_{C_i} \cap X_{\varphi_P} \neq \phi \right\} \tag{7}$$

$$\varphi_{S2} = \varphi_S - \varphi_{S1} \tag{8}$$

Let us make this point by presenting an example. Let $I = \varphi_P \cup \varphi_S$ be an ISAT instance, where

$$\varphi_P = \left\{ x_1 \vee x_2, \bar{x}_2 \vee x_3, x_3 \vee x_4 \vee x_{10}, x_6 \vee x_8, x_7 \vee x_8 \right\}$$

$$\varphi_S = \left\{ x_3 \vee x_5, x_5 \vee x_9, x_{10} \vee x_{11}, \bar{x}_{12} \vee x_{13} \vee x_{14} \right\}$$

From Equation 5, we have

$$\varphi'_{P1} = \left\{ \bar{x}_2 \vee x_3, x_3 \vee x_4 \vee x_{10} \right\}$$

$$\varphi_{P1} = \left\{ x_1 \vee x_2, \bar{x}_2 \vee x_3, x_3 \vee x_4 \vee x_{10} \right\}$$

and so from Equation 6, we have

$$\varphi_{P2} = \left\{ x_6 \vee x_8, x_7 \vee x_8 \right\}$$

In the same manner, from Equation 7 we have

$$\varphi'_{S1} = \{x_3 \vee x_5, x_{10} \vee x_{11}\}$$

$$\varphi_{S1} = \{x_3 \vee x_5, x_5 \vee x_9, x_{10} \vee x_{11}\}$$

and finally, from Equation 8 we have

$$\varphi_{S2} = \{\overline{x}_{12} \vee x_{13} \vee x_{14}\}$$

The clauses of φ_{P2} are discarded from the set of clauses of I since their model is not altered by any assignment to φ_{P1}. Although φ_{S2} is solved as an independent SAT instance. Subsequently, the objective turns into recycling the model of φ_P to $\varphi_{P1} \cup \varphi_{S1}$ rather than $\varphi_P \cup \varphi_S$.

The general structure of the algorithm τ-EO for ISAT is described in Figure 2.

- Clauses of φ_{P2} do not share any variable with those of φ_S and φ_{P1}, so they are removed from $\varphi_P \cup \varphi_S$.

- Clauses of φ_{S2} have no common variable with those of φ_P and φ_{S1}, so φ_{S2} is treated as an independent SAT problem. The algorithm τ-EO-SAT is then applied with $Variables = X_{\varphi_{S2}}$ and $Clauses = \varphi_{S2}$. The satisfiability of φ_{S2} is a necessary but no sufficient condition for the satisfiability of $\varphi_P \cup \varphi_S$.

- Clauses of $\varphi_{P1} \cup \varphi_{S1}$ have common variables in $X_{\varphi_{P1}} \cup X_{\varphi_{S1}}$. A particular SAT problem can then be addressed as follows. Given a set of clauses $\varphi_{P1} \cup \varphi_{S1}$ on variables $X_{\varphi_{P1}} \cup X_{\varphi_{S1}}$ where φ_{P1} is already satisfiable and has a model S. The problem is to determine the satisfiability of $\varphi_{P1} \cup \varphi_{S1}$ by reusing as much as possible the model of φ_{P1}. In a first step, a variant of τ-EO-SAT is used for solving this SAT instance. The algorithm starts with the assignment S to the variables $X_{\varphi_{P1}}$ and random Boolean values to the variables $X_{\varphi_{S1}}$. If no model can be found then a second step is performed. The algorithm starts with a completely randomly generated assignment. In this case, the variable fitness is defined as:

$$\lambda_i = -\#unsat(x_i) \times (1 + \#changes(x_i, \varphi_{P1})) \tag{9}$$

where $changes(x_i, \varphi_{P1})$ is the number of changes in the model of φ_{P1} obtained by flipping the truth value of x_i. Therefore, the best fitness is associated to the variable that both violates the least number of clauses and involves the least number of changes in the model S.

- Adding a single clause is considered as a particular case. Indeed, a clause is satisfiable if at least one of its literals is satisfiable. Moreover, a clause sharing only some variables with φ_P can be satisfied by simply assigning 1 or 0 to one positive or negative literal whose variable is not in X_{φ_P}. This can achieve significant gains in CPU time compared with that needed by τ-EO-SAT local search process.

Algorithm τ-EO-ISAT

If S satisfies φ_S then Return $\varphi_P \cup \varphi_S$ is satisfiable.

1. $\varphi_S = \{C\}$

   a. $X_{\varphi_P} \cap X_C = \phi$

   A model for C is found by assigning the truth value 1 or 0 to one positive or negative literal, respectively. Return that $\varphi_P \cup \varphi_S$ is satisfiable.

   b. $\left(X_{\varphi_P} \cap X_C \neq \phi\right)$ and $\left(X_C \not\subset X_{\varphi_P}\right)$

   Assign the truth value 1 or 0 to one positive or negative literal of $X_C - X_{\varphi_P}$, respectively. Return that $\varphi_P \cup \varphi_S$ is satisfiable.

   c. $X_C \subseteq X_{\varphi_P}$

   Determine φ_{P1}.

   Call τ-EO-SAT with *Variables* $= X_{\varphi_{P1}}$ and *Clauses* $= \varphi_{P1} \cup C$. The initial solution of the algorithm is the assignment S to the variables $X_{\varphi_{P1}}$ and the variable fitness λ_i is given by Equation 9.

   If $\varphi_{P1} \cup C$ is satisfiable then Return that $\varphi_P \cup \varphi_S$ is satisfiable Else Return that no model can be found.

2. $\varphi_S = \{C_1, C_2, ..., C_p\}$

   a. $X_{\varphi_P} \cap X_{\varphi_S} = \phi$

   Call τ-EO-SAT with *Variables* $= X_{\varphi_S}$ and *Clauses* $= \varphi_S$.

   If φ_S is satisfiable then Return that $\varphi_P \cup \varphi_S$ is satisfiable Else Return that no model can be found.

   b. $X_{\varphi_P} \cap X_{\varphi_S} \neq \phi$

   Determine $\varphi_{P1}, \varphi_{S1}, \varphi_{S2}$.

   (i) Call τ-EO-SAT with *Variables* $= X_{\varphi_{S2}}$ and *Clauses* $= \varphi_{S2}$.

   If no model can be found for φ_{S2} then Return that no model can be found for $\varphi_P \cup \varphi_S$.

   (ii) Call τ-EO-SAT with *Variables* $= X_{\varphi_{P1}} \cup X_{\varphi_{S1}}$ and *Clauses* $= \varphi_{P1} \cup \varphi_{S1}$.

   1. The algorithm starts with the assignment S to the variables $X_{\varphi_{P1}}$ and random Boolean values to $X_{\varphi_{S1}}$. If $\varphi_{P1} \cup \varphi_{S1}$ is satisfiable then Return that $\varphi_P \cup \varphi_S$ is satisfiable.

   2. The algorithm starts with completely randomly generated assignment where the variable fitness λ_i is given by Eqn. 9. If $\varphi_{P1} \cup \varphi_{S1}$ is satisfiable then Return that $\varphi_P \cup \varphi_S$ is satisfiable Else Return that no model can be found for $\varphi_P \cup \varphi_S$.

End τ-EO-ISAT

Fig. 2. Generic τ-EO algorithm for ISAT

5 Experimental Evaluation

In this section, we present an empirical evaluation of τ-EO-ISAT on ISAT instances obtained from both random and structured SAT instances. In order to compare the ISAT formulation considered in this work and that introduced by Hooker [14], two different types of ISAT instances were generated from each SAT instance. The first one denoted ISAT1, was obtained as follows. Let ISAT1 be an empty set of clauses. At each time one randomly chosen clause from the SAT instance was added to the ISAT1 instance and the resulting problem was solved. The series of problems was solved incrementally until all clauses had been added or no model for the current problem was found. The second ISAT instance, denoted ISAT2, was generated by randomly splitting the related SAT instance into two subsets φ_P and φ_S of m_P and $m_S = m - m_P$ clauses, respectively. Finding a model for φ_P is a necessary condition before solving the ISAT2 instance.

The SAT benchmarks were taken from the DIMACS [25] and SATLIB [24] archives. Six groups of satisfiable SAT instances were considered : ais* (4 instances of all interval series problem), ii8* (14 instances of Boolean circuit synthesis problem), par8-*-c (5 instances of learning the parity function problem), flat100-239 (100 instances of SAT-encoded graph colouring problem), uf125-538 (100 random 3-SAT instances) and g* (4 large instances of hard graph coloring problem). To have a reference point, we tested the performance of WalkSAT algorithm with the heuristic R-Novelty [18], which is one of the best-performing variant of the WalkSAT algorithm for SAT problem. The program code of WalkSAT was taken from the SATLIB archive [24]. The algorithm τ-EO-ISAT was coded in C and all experiments were performed on a 2.9 GHz Pentium 4 with 2 GB RAM running Linux. We ran τ-EO-ISAT on ISAT1 and ISAT2 instances with the parameter τ set to 1.4 according to approximately optimal parameter setting [20]. ISAT2 instances were generated setting m_P to $0.5m$. R-Novelty (WalkSAT with –rnovelty parameter) was run with a noise parameter p set to 0.6 as suggested in [18]. The programs were run 20 times on each instance with a maximum of 100000 flips allowed to solve each one except for the large problem instances g*, where a time limit of 300 seconds was allowed at each run.

The results achieved by τ-EO-ISAT on ISAT1 and ISAT2 instances are presented in Tables 1 and 2, respectively. Table 3 presents the results obtained with R-Novelty on related SAT instances. n and m denote the number of variables and clauses, respectively. For each instance, we give the success rate over 20 runs, the mean CPU time and the mean number of flips to find a solution and their respective standard deviations. Table 4 summarizes our results in terms of average success rate, average CPU time and average number of flips to solve each problem class. τ-EO-ISAT1 and τ-EO-ISAT2 refer to the algorithm τ-EO-ISAT on ISAT1 and ISAT2 instances, in that order. Mean CPU times related to τ-EO-ISAT2 include mean CPU times needed to find a model for φ_P .

In terms of average success rate, average CPU time and average number of flips at finding a solution τ-EO-ISAT2 outperformed τ-EO-ISAT1 and R-Novelty on the problems ais*, par8-*-c and uf125-538. R-Novelty was the best-performing algorithm on the problems flat100-239 and g*, however the results of τ-EO-ISAT2 on these problem instances were also quite competitive. Intuitively, this could be explained by the high connectivity inherent to the structure of these instances that involves much more difficulties in recycling an old model when new clauses are added. On the problems ii8*, the best average success rate was achieved by τ-EO-ISAT2, but with a significantly higher cost than R-Novelty.

Table 1. Results of τ-EO-ISAT on ISAT1 instances

Problem id	n	m	Success rate/ 20	CPU time		# Flips	
				Mean	Std	Mean	Std
ais6	61	581	20	0.4400	0.0320	8800.4	1245.2
ais8	113	1520	18	6.8452	1.6505	39292.5	6564.0
ais10	181	3151	7	6.6520	2.1420	8115.0	1824.9
ais12	265	5666	2	9.3290	0.0562	42638.0	2562.8
ii8a1	66	186	20	0.0023	0.0005	95.0	0.0
ii8a2	180	800	20	0.0179	0.0012	155.8	0.7
ii8a3	264	1552	20	0.0820	0.0015	328.7	65.4
ii8a4	396	2798	20	1.2400	0.0054	1126.2	32.1
ii8b1	336	2068	20	0.0142	0.0006	328.6	15.1
ii8b2	576	4088	20	4.6325	0.3235	5340.2	525.6
ii8b3	816	6108	20	7.8312	0.5350	45653.4	3540.6
ii8b4	1068	8214	5	16.1210	2.6453	4800.3	2945.7
ii8c1	510	3065	20	0.9200	0.0030	624.3	2.4
ii8c2	950	6689	20	9.5780	1.4210	5800.6	234.8
ii8d1	530	3207	20	1.3650	0.0450	712.5	26.5
ii8d2	930	6547	20	8.4500	1.4620	1912.5	185.1
ii8e1	520	3136	20	2.4520	0.3700	1415.2	164.0
ii8e2	870	6121	20	6.5642	1.1260	1846.5	138.8
par8-1-c	64	254	20	0.1762	0.0560	3623.5	325.8
par8-2-c	68	270	20	0.3320	0.0065	8264.0	237.5
par8-3-c	75	298	20	1.7328	0.1045	25605.2	1968.4
par8-4-c	67	266	20	0.4210	0.1540	9725.4	2154.7
par8-5-c	75	298	8	1.7620	0.2030	37640.4	7682.0
flat100-239 (100 inst.)	300	1117	3.60	8.2540	0.1054	54762.0	1542.5
uf125-538 (100 inst.)	125	538	19.25	0.6875	0.3260	9665.3	792.0
g125-17	2125	66272	14	142.9450	12.7620	2051374.0	158549.1
g125-18	2250	70163	20	1.4585	0.9500	24056.9	18669.4
g250-15	3750	233965	18	0.6326	0.9455	11615.2	15857.6
g250-20	7250	454622	16	186.5020	20.5490	1959751.8	232450.8

τ-EO-ISAT1 performed worse than the other two algorithms on all the instances taking significantly longer time. The intuitive explanation for that low performance is that an ISAT1 instance is a series of problems with incremental ratios of clauses to variables, where the first ones are relatively easy to satisfy (few clauses and many variables). If the series contains problems at the phase transition (values of ratio near the threshold), it is generally difficult to find a satisfying assignment to such problems and hence, more frequent local search repairs are performed (step 1.c. of the algorithm). It seems that this approach may not be suitable for stochastic local search algorithms, whereas the incremental systematic algorithm proposed by Hooker [14] was substantially faster than DPL [10], [17]. However, more complete tests are required to further investigate this hypothesis.

Table 2. Results of τ-EO-ISAT on ISAT2 instances

Problem id	n	m	Success rate/ 20	CPU time		# Flips	
				Mean	Std	Mean	Std
ais6	61	581	20	0.0540	0.0115	1080.0	320.1
ais8	113	1520	20	2.2310	1.1543	15403.5	1356.8
ais10	181	3151	8	3.0270	2.5230	6045.9	1582.2
ais12	265	5666	8	4.1985	1.7620	18458.1	2351.0
ii8a1	66	186	20	0.0008	0.0005	38.0	0.0
ii8a2	180	800	20	0.0045	0.0012	67.2	0.1
ii8a3	264	1552	20	0.0455	0.0125	205.4	54.2
ii8a4	396	2798	20	0.2480	0.0165	223.2	22.5
ii8b1	336	2068	20	0.0457	0.0054	104.7	24.1
ii8b2	576	4088	20	2.6800	0.2754	3216.4	412.5
ii8b3	816	6108	19	3.2670	2.2850	20255.4	654.1
ii8b4	1068	8214	20	4.3428	2.2490	1208.5	2543.9
ii8c1	510	3065	20	0.0542	0.0012	137.9	65.4
ii8c2	950	6689	20	3.6502	1.2760	2702.2	25.3
ii8d1	530	3207	20	1.3720	0.0086	823.9	85.2
ii8d2	930	6547	20	4.2365	0.0450	959.4	21.3
ii8e1	520	3136	20	0.0350	0.0014	123.1	8.4
ii8e2	870	6121	20	1.0395	0.0745	310.5	12.6
par8-1-c	64	254	20	0.0825	0.0150	2321.6	54.8
par8-2-c	68	270	20	0.0945	0.0172	2551.6	545.6
par8-3-c	75	298	20	0.0250	0.0120	550.6	185.4
par8-4-c	67	266	20	0.1278	0.0850	3195.0	1255.6
par8-5-c	75	298	15	0.9540	0.6535	21465.0	11520.1
flat100-239 (100 inst.)	300	1117	5.30	5.2050	0.1442	42435.7	1280.8
uf125-538 (100 inst.)	125	538	19.55	0.4382	0.0321	4359.6	458.0
g125-17	2125	66276	18	64.5205	8.3510	626452.0	95734.0
g125-18	2250	70163	20	1.2651	0.2090	11245.8	1562.6
g250-15	3750	233965	20	0.9510	0.0650	9756.1	455.3
g250-20	7250	454622	19	134.8600	12.2945	1295286.5	981005.0

As an initial result, τ-EO-ISAT2 solved all the problems in less computation time and provided better solution quality than τ-EO-ISAT1, but it did not dominate R-Novelty on all the problems. However, it should be noted that the mean times re-

Table 3. Results of R-Novelty on SAT benchmark instances

Problem id	n	m	Success rate/ 20	CPU time		# Flips	
				Mean	Std	Mean	Std
ais6	61	581	20	0.0984	0.0062	2185.8	84.4
ais8	113	1520	19	3.3941	0.8450	22065.4	5490.2
ais10	181	3151	6	4.9402	0.6376	8590.0	1262.5
ais12	265	5666	6	5.6430	1.2640	20652.8	1850.1
ii8a1	66	186	20	0.0015	0.0001	36.0	0.0
ii8a2	180	800	20	0.0097	0.0001	90.4	0.0
ii8a3	264	1552	20	0.0260	0.0075	270.0	10.2
ii8a4	396	2798	20	0.4724	0.0710	400.5	85.1
ii8b1	336	2068	20	0.0522	0.0150	115.9	52.5
ii8b2	576	4088	20	0.9451	0.0548	815.4	152.0
ii8b3	816	6108	18	1.8500	0.9640	2366.5	1175.3
ii8b4	1068	8214	12	2.1050	0.7571	2725.0	901.5
ii8c1	510	3065	20	0.2378	0.0001	174.0	0.0
ii8c2	950	6689	20	1.2482	0.8440	229.1	1.0
ii8d1	530	3207	20	0.6200	0.0118	210.2	5.0
ii8d2	930	6547	20	1.8550	0.0330	564.5	78.5
ii8e1	520	3136	20	0.6850	0.0235	355.0	12.0
ii8e2	870	6121	20	0.9850	0.1482	510.2	15.0
par8-1-c	64	254	20	0.0760	0.0050	2540.5	536.4
par8-2-c	68	270	20	0.1405	0.0095	2115.0	450.2
par8-3-c	75	298	20	0.9550	0.2050	16630.0	1285.0
par8-4-c	67	266	20	1.5304	0.1225	22440.8	672.0
par8-5-c	75	298	10	0.7710	0.0860	6856.0	1448.4
flat100-239 (100 inst.)	300	1117	7.10	4.2600	0.9520	37574.0	3715.0
uf125-538 (100 inst.)	125	538	19.40	0.5500	0.1060	4850.20	722.0
g125-17	2125	66272	20	71.4670	5.6390	931675.2	98821.0
g125-18	2250	70163	20	1.6651	0.2405	12608.4	5732.1
g250-15	3750	233965	20	0.7520	0.9460	4632.0	28730.2
g250-20	7250	454622	20	134.1025	7.6200	453581.0	110836.5

ported include times needed to find a model for φ_P. Hence, the results presented suggest that ISAT2 is a promising approach for incremental solving.

6 Conclusion

In this paper, we have presented a stochastic local search algorithm for solving the incremental satisfiability problem (ISAT). The satisfiability of the dynamic problem is maintained by recycling the current model of a sub-problem when new clauses are added. Local search repairs are performed by an Extremal Optimization algorithm based on the theory of self-organized criticality of many natural systems. We established experimentally its effectiveness by testing two types of ISAT instances derived from SAT benchmark problems. Indeed, one or more clauses can be added to the original problem either incrementally one at a time (ISAT1) or together at the same time (ISAT2). The experimental results suggest that the algorithm is more efficient on ISAT2 instances and at least as competitive as WalkSAT with R-Novelty heuristic for SAT. However, this work should be understood as an initial investigation and further experiments are necessary to both study the behaviour of the algorithm on large benchmark instances, and compare its performance with those of other methods. Additionally, we plan to investigate the effect of varying the ratio of the clauses being added as well as the free parameter of τ-EO on the performance of the algorithm. Finally, this work can be used as an algorithmic framework for incremental solving with other SAT solvers.

Table 4. Summary of average results for each problem group

Problem class	ais*	ii8*	par8-*-c	flat100-239	uf125-538	g*
			Average Success rate / 20			
τ-EO-ISAT1	11.75	18.92	17.60	3.60	19.25	17.00
τ-EO-ISAT2	14.00	19.92	19.00	5.30	19.55	19.25
R-Novelty	12.75	19.28	18	7.10	19.40	20.00
			Average CPU time secs			
τ-EO-ISAT1	5.8165	4.2335	0.8848	8.2540	0.6875	82.8845
τ-EO-ISAT2	2.3776	1.5015	0.2567	5.2050	0.4382	50.3991
R-Novelty	3.5189	0.7923	0.6945	4.2600	0.5500	51.9900
			Average flips			
τ-EO-ISAT1	24711.47	5009.98	16971.69	54762.00	9665.30	1011699.40
τ-EO-ISAT2	10246.87	2169.69	6016.76	42435.70	4359.60	485685.10
R-Novelty	13373.5	633.04	10116.45	37574.00	4850.20	350624.15

Acknowledgment

The author is grateful to the anonymous referees for their valuable comments and suggestions.

References

1. Adami, C.: Introduction to Artificial Life. Springer-Verlag Berlin Heidelberg New York (1999)
2. Bak, P., Tang, C., Wiesenfeld, K.: Self-Organized Criticality: An Explanation of $1/f$ noise. Physical Review Letters, V86 N23, (1987) 5211-5214
3. Bak, P., Sneppen, K.: Punctuated Equilibrium and Criticality in a Simple Model of Evolution. Physical Review Letters, 59. (1993) 381-384
4. Bak, P.: How Nature Works. Springer-Verlag, Berlin Heidelberg New York (1996)

5. Boettcher, S., Percus, A.G.: Nature's Way of Optimizing. Elsevier Science, Artificial Intelligence, 119. (2000) 275-286

6. Boettcher, S., Percus, A.G.: Optimization with Extremal Dynamics. Physical Review Letters, V86 N23. (2001) 5211-5214

7. Boettcher, S., Percus, A.G.: Extremal Optimization for Graph Partitioning. Physical Review E, V64 026114. (2001) 1-13

8. Bose, I., Chandhuri, I.: Bacteria Evolution and the Bak-Sneppen Model. International Journal of Modern Physics C, 12, N5, (2001) 675-683

9. Cook, S. A.: The Complexity of Theorem Proving Procedures. Proceedings of the 3rd Annual ACM Symposium of the Theory of Computation. (1971) 263-268

10. Davis, M., Putnam, H.: A computing Procedure for Quantification Theory. Journal of the Association for Computing Machinery, 7. (1960) 201-215

11. Dechter, R., Dechter, A.: Belief Maintenance in Dynamic Constraint Networks. in Proceedings of AAAI-88. MIT Press. (1988) 32-42

12. Eén, N., Sörensson, N.: Temporal Induction by Incremental SAT Solving. Electronic Notes in Theoretical Computer Science, 89, N°4. (2004)
http://www.elsevier.nl/locate/entcs/volume89.html

13. Gutierrez, J., Mali, A.D.: Local Search for Incremental Satisfiability. in Proceedings of the International Conference on Artificial Intelligence (IC-AI'02), Las Vegas, USA. (2002) 986-991

14. Hooker, J.N.: Solving the Incremental Satisfiability Problem. Journal of Logic Programming 15. (1993) 177-186

15. Hoos, H.H., O'Neill, K.: Stochastic Local Search Methods for Dynamic SAT – An Initial Investigation. In Proceedings of AAAI-2000. Workshop 'Leveraging Probability and Uncertainty in Computation'. (2000) 22-26

16. Kim, J., Whittemore, J., Marques-Silva, J.P., Sakallah, K.: Incremental Boolean Satisfiability and Its Application to Delay Fault Testing. IEEE/ACM International Workshop on Logic Synthesis. (1999)

17. Loveland, D.W.: Automated Theorem Proving: A Logical Basis. North-Holland (1978)

18. Mc Allester, D., Selman, B., Kautz, H.: Evidence for Invariants in Local Search. In Proceedings of IJCAI-97, (1997)

19. Menaï, M.B., Batouche, M.: EO for MAXSAT. in Proceedings of the International Conference on Artificial Intelligence (IC-AI'02), Las Vegas, USA. (2002) 954-958

20. Menaï, M.B., Batouche, M.: Efficient Initial Solution to Extremal Optimization Algorithm for Weighted MAXSAT Problem. in Proceedings of IEA/AIE 2003, Loughborough UK, LNAI 2718, (2003) 592-603

21. Michalewicz, Z., Fogel, D.B.: How to Solve It : Modern Heuristics. Springer-Verlag, Berlin Heidelberg New York (2000)

22. Selman, B., Kautz, H.A., Cohen B.: Noise Strategies for Improving Local Search. Proceedings of the 12th National Conference on Artificial Intelligence. (1994) 337-343

23. Yamano, T.: Regulation Effects on Market with Bak-Sneppen Model in High Dimensions. International Journal of Modern Physics C, 12, N9, (2001) 1329-1333

24. http://www.satlib.org/

25. http://dimacs.rutgers.edu/Challenges/

Solving Equations Involving Sequence Variables and Sequence Functions

Temur Kutsia[*]

Research Institute for Symbolic Computation
Johannes Kepler University Linz
A-4040 Linz, Austria
kutsia@risc.uni-linz.ac.at

Abstract. Term equations involving individual and sequence variables, and individual and sequence function symbols are studied. Function symbols can have either fixed or flexible arity. A new unification procedure for solving such equations is presented. Decidability of unification is proved. Completeness and almost minimality of the procedure is shown.

1 Introduction

We study term equations with sequence variables and sequence function symbols. A sequence variable can be instantiated by any finite sequence of terms, including the empty sequence. A sequence function abbreviates a finite sequence of functions all having the same argument lists[1]. An instance of such a function is $\mathsf{IntegerDivision}(x,y)$ that abbreviates the sequence $\mathsf{Quotient}(x,y), \mathsf{Remainder}(x,y)$.

Bringing sequence functions in the language naturally allows Skolemization over sequence variables: Let x, y be individual variables, $\overline{x}$ be a sequence variable, and p be a flexible arity predicate symbol. Then $\forall x \forall y \exists \overline{x}.p(x, y, \overline{x})$ Skolemizes to $\forall x \forall y.p(x, y, \overline{f}(x, y))$, where $\overline{f}$ is a binary Skolem sequence function symbol. Another example, $\forall \overline{y} \exists \overline{x}.p(\overline{y}, \overline{x})$, where $\overline{y}$ is a sequence variable, after Skolemization introduces a flexible arity sequence function symbol $\overline{g}$: $\forall \overline{y}.p(\overline{y}, \overline{g}(\overline{y}))$.

Equation solving with sequence variables plays an important role in various applications in automated reasoning, artificial intelligence, and programming. At the end of the paper we briefly review some of the works related to this topic.

We contribute to this area by introducing a new unification procedure for solving equations in the free theory with individual and sequence variables, and individual and sequence function symbols. Function symbols can have either fixed or flexible arity. We prove that solvability of an equation is decidable in such a theory, and provide a unification procedure that enumerates an almost minimal complete set of solutions. The procedure terminates if the set is finite. This work is an extension and refinement of our previous results [10].

We implemented the procedure (without the decision algorithm) in MATHE-MATICA [18] on the base of a rule-based programming system ρLOG[2] [13].

---

[*] Supported by the Austrian Science Foundation (FWF) under Project SFB F1302.
[1] Semantically, sequence functions can be interpreted as multi-valued functions.
[2] Available from http://www.ricam.oeaw.ac.at/people/page/marin/RhoLog/.

B. Buchberger and J.A. Campbell (Eds.): AISC 2004, LNAI 3249, pp. 157–170, 2004.

The paper is organized as follows: In Section 2 basic notions are introduced. In Section 3 decidability is proved. In Section 4 relation with order-sorted higher-order E-unification is discussed. In Section 5 the unification procedure is defined and its properties are studied. In Section 6 some of the related work is reviewed.

A longer version of this paper with full proofs is available on the web [12].

2   Preliminaries

We assume that the reader is familiar with the standard notions of unification theory [3]. We consider an alphabet consisting of the following pairwise disjoint sets of symbols: individual variables $\mathcal{V}_{\mathrm{Ind}}$, sequence variables $\mathcal{V}_{\mathrm{Seq}}$, fixed arity individual function symbols $\mathcal{F}_{\mathrm{Ind}}^{\mathrm{Fix}}$, flexible arity individual function symbols $\mathcal{F}_{\mathrm{Ind}}^{\mathrm{Flex}}$, fixed arity sequence function symbols $\mathcal{F}_{\mathrm{Seq}}^{\mathrm{Fix}}$, flexible arity sequence function symbols $\mathcal{F}_{\mathrm{Seq}}^{\mathrm{Flex}}$. Each set of variables is countable. Each set of function symbols is finite or countable. Besides, the alphabet contains the parenthesis '(', ')' and the comma ','. We will use the following denotations: $\mathcal{V} := \mathcal{V}_{\mathrm{Ind}} \cup \mathcal{V}_{\mathrm{Seq}}$; $\mathcal{F}_{\mathrm{Ind}} := \mathcal{F}_{\mathrm{Ind}}^{\mathrm{Fix}} \cup \mathcal{F}_{\mathrm{Ind}}^{\mathrm{Flex}}$; $\mathcal{F}_{\mathrm{Seq}} := \mathcal{F}_{\mathrm{Seq}}^{\mathrm{Fix}} \cup \mathcal{F}_{\mathrm{Seq}}^{\mathrm{Flex}}$; $\mathcal{F}^{\mathrm{Fix}} := \mathcal{F}_{\mathrm{Ind}}^{\mathrm{Fix}} \cup \mathcal{F}_{\mathrm{Seq}}^{\mathrm{Fix}}$, $\mathcal{F}^{\mathrm{Flex}} := \mathcal{F}_{\mathrm{Ind}}^{\mathrm{Flex}} \cup \mathcal{F}_{\mathrm{Seq}}^{\mathrm{Flex}}$; $\mathcal{F} := \mathcal{F}_{\mathrm{Ind}} \cup \mathcal{F}_{\mathrm{Seq}} = \mathcal{F}^{\mathrm{Fix}} \cup \mathcal{F}^{\mathrm{Flex}}$. The *arity* of $f \in \mathcal{F}^{\mathrm{Fix}}$ is denoted by $\mathcal{A}r(f)$. A function symbol $c \in \mathcal{F}^{\mathrm{Fix}}$ is called a *constant* if $\mathcal{A}r(c) = 0$.

Definition 1. *A term over $\mathcal{F}$ and $\mathcal{V}$ is either an individual or a sequence term defined as follows:*

1. *If $t \in \mathcal{V}_{\mathrm{Ind}}$ (resp. $t \in \mathcal{V}_{\mathrm{Seq}}$), then t is an individual (resp. sequence) term.*
2. *If $f \in \mathcal{F}_{\mathrm{Ind}}^{\mathrm{Fix}}$ (resp. $f \in \mathcal{F}_{\mathrm{Seq}}^{\mathrm{Fix}}$), $\mathcal{A}r(f) = n$, $n \geq 0$, and $t_1, \ldots, t_n$ are individual terms, then $f(t_1, \ldots, t_n)$ is an individual (resp. sequence) term.*
3. *If $f \in \mathcal{F}_{\mathrm{Ind}}^{\mathrm{Flex}}$ (resp. $f \in \mathcal{F}_{\mathrm{Seq}}^{\mathrm{Flex}}$) and $t_1, \ldots, t_n$ $(n \geq 0)$ are individual or sequence terms, then $f(t_1, \ldots, t_n)$ is an individual (resp. sequence) term.*

The *head* of a term $t = f(t_1, \ldots, t_n)$, denoted by $\mathcal{H}ead(t)$, is the function symbol f. We denote by $\mathcal{T}_{\mathrm{Ind}}(\mathcal{F}, \mathcal{V})$, $\mathcal{T}_{\mathrm{Seq}}(\mathcal{F}, \mathcal{V})$, and $\mathcal{T}(\mathcal{F}, \mathcal{V})$, respectively, the sets of all individual terms, all sequence terms, and all terms over $\mathcal{F}$ and $\mathcal{V}$. An *equation* over $\mathcal{F}$ and $\mathcal{V}$ is a pair $\langle s, t \rangle$, denoted by $s \approx t$, where $s, t \in \mathcal{T}_{\mathrm{Ind}}(\mathcal{F}, \mathcal{V})$.

Example 1. Let $x, y \in \mathcal{V}_{\mathrm{Ind}}$, $\overline{x} \in \mathcal{V}_{\mathrm{Seq}}$, $f \in \mathcal{F}_{\mathrm{Ind}}^{\mathrm{Flex}}$, $g \in \mathcal{F}_{\mathrm{Ind}}^{\mathrm{Fix}}$, $\overline{f} \in \mathcal{F}_{\mathrm{Seq}}^{\mathrm{Flex}}$, $\overline{g} \in \mathcal{F}_{\mathrm{Seq}}^{\mathrm{Fix}}$, $\mathcal{A}r(g) = 2$, and $\mathcal{A}r(\overline{g}) = 1$. Then $f(\overline{x}, g(x, y))$ and $f(\overline{x}, \overline{f}(x, \overline{x}, y))$ are individual terms; $\overline{f}(\overline{x}, \overline{f}(x, \overline{x}, y))$ and $\overline{g}(f(x, \overline{x}, y))$ are sequence terms; $f(\overline{x}, \overline{g}(\overline{x}))$, $f(\overline{x}, g(\overline{x}, y))$ and $f(\overline{x}, \overline{g}(x, y))$ are not terms; $f(\overline{x}, g(x, y)) \approx g(x, y)$ is an equation; $f(\overline{x}, g(\overline{x}, y)) \approx g(x, y)$, $\overline{x} \approx f(\overline{x})$ and $\overline{g}(x) \approx f(\overline{x})$ are not equations.

If not otherwise stated, the following symbols, maybe with indices, are used as metavariables: x and y – over individual variables; $\overline{x}$, $\overline{y}$, $\overline{z}$ – over sequence variables; v – over (individual or sequence) variables; f, g, h – over individual function symbols; $\overline{f}$, $\overline{g}$, $\overline{h}$ – over sequence function symbols; a, b, c – over individual constants; $\overline{a}$, $\overline{b}$, $\overline{c}$ – over sequence constants; s, t, r, q – over terms.

Let T be a term, a sequence of terms, or a set of terms. Then we denote by $\mathcal{I}\mathcal{V}ar(T)$ (resp. by $\mathcal{S}\mathcal{V}ar(T)$) the set of all individual (resp. sequence) variables in T; by $\mathcal{V}ar(T)$ the set $\mathcal{IV}ar(T) \cup \mathcal{SV}ar(T)$; by $\mathcal{IF}un(T)$ (resp. by $\mathcal{SF}un(T)$) the set of all individual (resp. sequence) function symbols in T; by $\mathcal{F}ix(T)$ (resp. by $\mathcal{F}lex(T)$) the set of all fixed (resp. flexible) arity function symbols in T.

Definition 2. *A* variable binding *is either a pair* $x \mapsto t$ *where* $t \in T_{\mathrm{Ind}}(\mathcal{F}, \mathcal{V})$ *and* $t \neq x$, *or an expression* $\overline{x} \mapsto \ulcorner t_1, \ldots, t_n \urcorner$ [3] *where* $n \geq 0$, *for all* $1 \leq i \leq n$ *we have* $t_i \in T(\mathcal{F}, \mathcal{V})$, *and if* $n = 1$ *then* $t_1 \neq \overline{x}$.

Definition 3. *A* sequence function symbol binding *is an expression of the form* $\overline{f} \mapsto \ulcorner \overline{g_1}, \ldots, \overline{g_m} \urcorner$, *where* $m \geq 1$, *if* $m = 1$ *then* $\overline{f} \neq \overline{g_1}$, *and either* $\overline{f}, \overline{g_1}, \ldots, \overline{g_m} \in \mathcal{F}^{\mathrm{Fix}}_{\mathrm{Seq}}$, *with* $Ar(\overline{f}) = Ar(\overline{g_1}) = \cdots = Ar(\overline{g_m})$, *or* $\overline{f}, \overline{g_1}, \ldots, \overline{g_m} \in \mathcal{F}^{\mathrm{Flex}}_{\mathrm{Seq}}$.

Definition 4. *A* substitution *is a finite set of bindings* $\{x_1 \mapsto t_1, \ldots, x_n \mapsto t_n, \overline{x}_1 \mapsto \ulcorner s^1_1, \ldots, s^1_{k_1} \urcorner, \ldots, \overline{x}_m \mapsto \ulcorner s^m_1, \ldots, s^m_{k_m} \urcorner, \overline{f}_1 \mapsto \ulcorner \overline{g^1_1}, \ldots, \overline{g^1_{l_1}} \urcorner, \ldots, \overline{f}_r \mapsto \ulcorner \overline{g^r_1}, \ldots, \overline{g^r_{l_r}} \urcorner\}$ *where* $n, m, r \geq 0$, $x_1, \ldots, x_n, \overline{x}_1, \ldots, \overline{x}_m$ *are distinct variables and* $\overline{f}_1, \ldots, \overline{f}_r$ *are distinct sequence function symbols.*

Lower case Greek letters are used to denote substitutions. The empty substitution is denoted by ε.

Definition 5. *The* instance *of a term* t *with respect to a substitution* σ, *denoted* $t\sigma$, *is defined recursively as follows:*

1. $x\sigma = \begin{cases} t, & \text{if } x \mapsto t \in \sigma, \\ x, & \text{otherwise.} \end{cases}$

2. $\overline{x}\sigma = \begin{cases} t_1, \ldots, t_n, & \text{if } \overline{x} \mapsto \ulcorner t_1, \ldots, t_n \urcorner \in \sigma, \ n \geq 0, \\ \overline{x}, & \text{otherwise.} \end{cases}$

3. $\underline{f}(t_1, \ldots, t_n)\sigma = f(t_1\sigma, \ldots, t_n\sigma)$.

4. $\overline{f}(t_1, \ldots, t_n)\sigma = \begin{cases} \overline{g_1}(t_1\sigma, \ldots, t_n\sigma), \ldots, \overline{g_m}(t_1\sigma, \ldots, t_n\sigma), & \text{if } \overline{f} \mapsto \ulcorner \overline{g_1}, \ldots, \overline{g_m} \urcorner \in \sigma, \\ \overline{f}(t_1\sigma, \ldots, t_n\sigma), & \text{otherwise.} \end{cases}$

Example 2. Let $\sigma = \{x \mapsto a, y \mapsto f(\overline{x}), \overline{x} \mapsto \ulcorner \urcorner, \overline{y} \mapsto \ulcorner a, \overline{x} \urcorner, \overline{g} \mapsto \ulcorner \overline{g_1}, \overline{g_2} \urcorner\}$. Then $f(x, \overline{x}, \overline{g}(y, \overline{g}()), \overline{y}))\sigma = f(a, \overline{g_1}(f(\overline{x}), \overline{g_1}(), \overline{g_2}()), \overline{g_2}(f(\overline{x}), \overline{g_1}(), \overline{g_2}()), a, \overline{x})$.

Definition 6. *The* application *of* σ *on* $\overline{f}$, *denoted* $\overline{f}\sigma$, *is a sequence of function symbols* $\overline{g_1}, \ldots, \overline{g_m}$ *if* $\overline{f} \mapsto \ulcorner \overline{g_1}, \ldots, \overline{g_m} \urcorner \in \sigma$. *Otherwise* $\overline{f}\sigma = \overline{f}$.

Applying a substitution θ on a sequence of terms $\ulcorner t_1, \ldots, t_n \urcorner$ gives a sequence of terms $\ulcorner t_1\theta, \ldots, t_n\theta \urcorner$.

Definition 7. *Let* σ *be a substitution.* (1) *The* domain *of* σ *is the set* $\mathcal{D}om(\sigma) = \{l \mid l\sigma \neq l\}$ *of variables and sequence function symbols.* (2) *The* codomain *of* σ *is the set* $\mathcal{C}od(\sigma) = \{l\sigma \mid l \in \mathcal{D}om(\sigma)\}$ *of terms and sequence function symbols* [4]. (3) *The* range *of* σ *is the set* $\mathcal{R}an(\sigma) = \mathcal{V}ar(\mathcal{C}od(\sigma))$ *of variables.*

---

[3] To improve readability, we write sequences that bind sequence variables between $\ulcorner$ and $\urcorner$.

[4] Note that the codomain of a substitution is a set of terms and sequence function symbols, not a set consisting of terms, sequences of terms, sequence function symbols, and sequences of sequence function symbols. For instance, $\mathcal{C}od(\{x \mapsto f(a), \overline{x} \mapsto \ulcorner a, a, b \urcorner, \overline{c} \mapsto \ulcorner \overline{c_1}, \overline{c_2} \urcorner\}) = \{f(a), a, b, \overline{c_1}, \overline{c_2}\}$.

Definition 8. *Let σ and ϑ be two substitutions:*

$$\sigma = \{\, x_1 \mapsto t_1, \ldots, x_n \mapsto t_n, \overline{x_1} \mapsto \ulcorner s_1^1, \ldots, s_{k_1}^1 \urcorner, \ldots, \overline{x_m} \mapsto \ulcorner s_1^m, \ldots, s_{k_m}^m \urcorner,$$
$$\overline{f_1} \mapsto \ulcorner \overline{f_1^1}, \ldots, \overline{f_{l_1}^1} \urcorner, \ldots, \overline{f_r} \mapsto \ulcorner \overline{f_1^r}, \ldots, \overline{f_{l_r}^r} \urcorner \},$$
$$\vartheta = \{\, y_1 \mapsto r_1, \ldots, y_{n'} \mapsto r_{n'}, \overline{y_1} \mapsto \ulcorner q_1^1, \ldots, q_{k'_1}^1 \urcorner, \ldots, \overline{y_{m'}} \mapsto \ulcorner q_1^{m'}, \ldots, q_{k'_{m'}}^{m'} \urcorner,$$
$$\overline{g_1} \mapsto \ulcorner \overline{g_1^1}, \ldots, \overline{g_{l'_1}^1} \urcorner, \ldots, \overline{g_{r'}} \mapsto \ulcorner \overline{g_1^{r'}}, \ldots, \overline{g_{l'_{r'}}^{r'}} \urcorner \}.$$

Then the composition *of σ and ϑ, $\sigma\vartheta$, is the substitution obtained from the set*

$$\{\, x_1 \mapsto t_1\vartheta, \ldots, x_n \mapsto t_n\vartheta, \overline{x_1} \mapsto \ulcorner s_1^1\vartheta, \ldots, s_{k_1}^1\vartheta \urcorner, \ldots, \overline{x_m} \mapsto \ulcorner s_1^m\vartheta, \ldots, s_{k_m}^m\vartheta \urcorner,$$
$$\overline{f_1} \mapsto \ulcorner \overline{f_1^1}\vartheta, \ldots, \overline{f_{l_1}^1}\vartheta \urcorner, \ldots, \overline{f_r} \mapsto \ulcorner \overline{f_1^r}\vartheta, \ldots, \overline{f_{l_r}^r}\vartheta \urcorner,$$
$$y_1 \mapsto r_1, \ldots, y_{n'} \mapsto r_{n'}, \overline{y_1} \mapsto \ulcorner q_1^1, \ldots, q_{k'_1}^1 \urcorner, \ldots, \overline{y_{m'}} \mapsto \ulcorner q_1^{m'}, \ldots, q_{k'_{m'}}^{m'} \urcorner,$$
$$\overline{g_1} \mapsto \ulcorner \overline{g_1^1}, \ldots, \overline{g_{l'_1}^1} \urcorner, \ldots, \overline{g_{r'}} \mapsto \ulcorner \overline{g_1^{r'}}, \ldots, \overline{g_{l'_{r'}}^{r'}} \urcorner \}$$

by deleting

1. *all the bindings $x_i \mapsto t_i\vartheta$ $(1 \leq i \leq n)$ for which $x_i = t_i\vartheta$,*
2. *all the bindings $\overline{x_i} \mapsto \ulcorner s_1^i\vartheta, \ldots, s_{k_i}^i\vartheta \urcorner$ $(1 \leq i \leq m)$ for which the sequence $s_1^i\vartheta, \ldots, s_{k_i}^i\vartheta$ consists of a single term $\overline{x_i}$,*
3. *all the sequence function symbol bindings $\overline{f_i} \mapsto \ulcorner \overline{f_1^i}\vartheta, \ldots, \overline{f_{l_i}^i}\vartheta \urcorner$ $(1 \leq i \leq r)$ such that the sequence $\overline{f_1^i}\vartheta, \ldots, \overline{f_{l_r}^i}\vartheta$ consists of a single function symbol $\overline{f_i}$,*
4. *all the bindings $y_i \mapsto r_i$ $(1 \leq i \leq n')$ such that $y_i \in \{x_1, \ldots, x_n\}$,*
5. *all the bindings $\overline{y_i} \mapsto \ulcorner q_1^i, \ldots, q_{k'_i}^i \urcorner$ $(1 \leq i \leq m')$ with $\overline{y_i} \in \{\overline{x_1}, \ldots, \overline{x_m}\}$,*
6. *all the sequence function symbol bindings $\overline{g_i} \mapsto \ulcorner \overline{g_1^i}, \ldots, \overline{g_{l'_i}^i} \urcorner$ $(1 \leq i \leq r')$ such that $\overline{g_i} \in \{\overline{f_1}, \ldots, \overline{f_r}\}$.*

Example 3. Let $\sigma = \{x \mapsto y, \overline{x} \mapsto \ulcorner \overline{y}, \overline{x} \urcorner, \overline{y} \mapsto \ulcorner f(a,b), y, \overline{g}(\overline{x}) \urcorner, \overline{f} \mapsto \ulcorner \overline{g}, \overline{h} \urcorner\}$ and $\vartheta = \{y \mapsto x, \overline{y} \mapsto \overline{x}, \overline{x} \mapsto \ulcorner \urcorner, \overline{g} \mapsto \ulcorner \overline{g_1}, \overline{g_2} \urcorner\}$ be two substitutions. Then $\sigma\vartheta = \{y \mapsto x, \overline{y} \mapsto \ulcorner f(a,b), x, \overline{g_1}(), \overline{g_2}() \urcorner, \overline{f} \mapsto \ulcorner \overline{g_1}, \overline{g_2}, \overline{h} \urcorner, \overline{g} \mapsto \ulcorner \overline{g_1}, \overline{g_2} \urcorner\}$.

Definition 9. *A substitution σ is called* linearizing away *from a finite set of sequence function symbols $\mathcal{Q}$ iff the following three conditions hold: (1) $Cod(\sigma) \cap \mathcal{Q} = \emptyset$. (2) For all $\overline{f}, \overline{g} \in Dom(\sigma) \cap \mathcal{Q}$, if $\overline{f} \neq \overline{g}$, then $\{\overline{f}\sigma\} \cap \{\overline{g}\sigma\} = \emptyset$. (3) If $\overline{f} \mapsto \ulcorner \overline{g_1} \ldots, \overline{g_n} \urcorner \in \sigma$ and $\overline{f} \in \mathcal{Q}$, then $\overline{g_i} \neq \overline{g_j}$ for all $1 \leq i < j \leq n$.*

Intuitively, a substitution linearizing away from $\mathcal{Q}$ either leaves a sequence function symbol in $\mathcal{Q}$ "unchanged", or "moves it away from" $\mathcal{Q}$, binding it with a sequence of distinct sequence function symbols that do not occur in $\mathcal{Q}$, and maps different sequence function symbols to disjoint sequences.

Let E be a set of equations over $\mathcal{F}$ and $\mathcal{V}$. By $\approx_E$ we denote the least congruence relation on $\mathcal{T}(\mathcal{F}, \mathcal{V})$ that is closed under substitution application and contains E. More precisely, $\approx_E$ contains E, satisfies reflexivity, symmetry, transitivity, congruence, and a special form of substitutivity: For all $s, t \in \mathcal{T}(\mathcal{F}, \mathcal{V})$, if $s \approx_E t$ and $s\sigma, t\sigma \in T(\mathcal{F}, \mathcal{V})$ for some σ, then $s\sigma \approx_E t\sigma$. Substitutivity in this form requires that $s\sigma$ and $t\sigma$ must be single terms, not arbitrary sequences of terms. The set $\approx_E$ is called an *equational theory* defined by E. In the sequel, we will also call the set E an equational theory, or E-theory. The *signature* of E is the set $\mathcal{S}ig(E) = \mathcal{IF}un(E) \cup \mathcal{SF}un(E)$. Solving equations in an E-theory is called *E-unification*. The fact that the equation $s \approx t$ has to be solved in an E-theory is written as $s\approx^?_E t$.

Definition 10. *Let E be an equational theory with $\mathcal{S}ig(E) \subseteq \mathcal{F}$. An E-unification problem over $\mathcal{F}$ is a finite multiset $\Gamma = \{s_1 \approx^?_E t_1, \ldots, s_n \approx^?_E t_n\}$ of equations over $\mathcal{F}$ and $\mathcal{V}$. An E-unifier of Γ is a substitution σ such that σ is linearizing away from $\mathcal{SF}un(\Gamma)$ and for all $1 \le i \le n$, $s_i\sigma \approx_E t_i\sigma$. The set of all E-unifiers of Γ is denoted by $\mathcal{U}_E(\Gamma)$, and Γ is E-unifiable iff $\mathcal{U}_E(\Gamma) \ne \emptyset$.*

If $\{s_1 \approx^?_E t_1, \ldots, s_n \approx^?_E t_n\}$ is a unification problem, then $s_i, t_i \in \mathcal{T}_{\text{Ind}}(\mathcal{F}, \mathcal{V})$ for all $1 \le i \le n$.

Example 4. Let $\Gamma = \{f(\overline{g}(\overline{x}, \overline{y}, a)) \approx^?_\emptyset f(\overline{g}(\overline{c}, b, x))\}$. Then $\{\overline{x} \mapsto \overline{c_1},\ \overline{y} \mapsto \ulcorner \overline{c_2}, b \urcorner$, $x \mapsto a, \overline{c} \mapsto \ulcorner \overline{c_1}, \overline{c_2} \urcorner\} \in \mathcal{U}_\emptyset(\Gamma)$.
  Let $\Gamma = \{f(\overline{g}(\overline{x}, \overline{y}, a)) \approx^?_\emptyset f(\overline{h}(\overline{c}, x))\}$. Then $\mathcal{U}_\emptyset(\Gamma) = \emptyset$. If we did not require the E-unifiers of a unification problem to be linearizing away from the sequence function symbol set of the problem, then Γ would have $\emptyset$-unifiers, e.g., $\{\overline{x} \mapsto \overline{c_1}, \overline{y} \mapsto \ulcorner \overline{c_2}, b \urcorner, x \mapsto a, \overline{g} \mapsto \overline{h}, \overline{c} \mapsto \ulcorner \overline{c_1}, \overline{c_2} \urcorner\}$ would be one of them.

In the sequel, if not otherwise stated, E stands for an equational theory, $\mathcal{X}$ for a finite set of variables, and $\mathcal{Q}$ for a finite set of sequence function symbols.

Definition 11. *A substitution σ is called* erasing *on $\mathcal{X}$ modulo E iff either $f(v)\sigma \approx_E f()$ for some $f \in \mathcal{S}ig(E)$ and $v \in \mathcal{X}$, or $\overline{x} \mapsto \ulcorner \urcorner \in \sigma$ for some $\overline{x} \in \mathcal{X}$. We call σ* non-erasing *on $\mathcal{X}$ modulo E iff σ is not erasing on $\mathcal{X}$ modulo E.*

Example 5. Any substitution containing $\overline{x} \mapsto \ulcorner \urcorner$ is erasing modulo $E = \emptyset$ on any $\mathcal{X}$ that contains $\overline{x}$.
  Let $E = \{f(\overline{x}, f(\overline{y}), \overline{z}) \approx f(\overline{x}, \overline{y}, \overline{z})\}$ and $\mathcal{X} = \{x, \overline{x}\}$. Then any substitution that contains $x \mapsto f()$, or $\overline{x} \mapsto \ulcorner \urcorner$, or $\overline{x} \mapsto \ulcorner t_1, \ldots, t_n \urcorner$ with $n \ge 1$ and $t_1 = \cdots = t_n = f()$, is erasing on $\mathcal{X}$ modulo E. For instance, the substitutions $\{x \mapsto f()\}$, $\{\overline{x} \mapsto f()\}$, $\{\overline{x} \mapsto \ulcorner \urcorner\}$, $\{\overline{x} \mapsto \ulcorner f(), f(), f(), f() \urcorner\}$ are erasing on $\mathcal{X}$ modulo E.

Definition 12. *A substitution σ* agrees *with a substitution ϑ on $\mathcal{X}$ and $\mathcal{Q}$ modulo E, denoted $\sigma =^{\mathcal{X},\mathcal{Q}}_E \vartheta$, iff (1) for all $x \in \mathcal{X}$, $x\sigma \approx_E x\vartheta$; (2) for all $\overline{f} \in \mathcal{Q}$, $\overline{f}\sigma = \overline{f}\vartheta$; (3) for all $\overline{x} \in \mathcal{X}$, there exist $t_1, \ldots, t_n, s_1, \ldots, s_n \in T(\mathcal{F}, \mathcal{V})$, $n \ge 0$, such that $\overline{x}\sigma = \ulcorner t_1, \ldots, t_n \urcorner$, $\overline{x}\vartheta = \ulcorner s_1, \ldots, s_n \urcorner$ and $t_i \approx_E s_i$ for each $1 \le i \le n$.*

Example 6. Let $\sigma = \{\overline{x} \mapsto \overline{a}\}$, $\vartheta = \{\overline{x} \mapsto \ulcorner \overline{b}, \overline{c} \urcorner, \overline{a} \mapsto \ulcorner \overline{b}, \overline{c} \urcorner\}$, and $\varphi = \{\overline{x} \mapsto \ulcorner \overline{b}, \overline{c} \urcorner, \overline{a} \mapsto \ulcorner \overline{b}, \overline{c} \urcorner\}$. Let also $\mathcal{X} = \{\overline{x}\}$, $\mathcal{Q} = \{\overline{a}\}$, and $E = \emptyset$. Then $\sigma\varphi =^{\mathcal{X},\mathcal{Q}}_E \vartheta$.

Definition 13. *A substitution σ is* more general *(resp. strongly more general) than ϑ on $\mathcal{X}$ and $\mathcal{Q}$ modulo E, denoted $\sigma \preceq^{\mathcal{X},\mathcal{Q}}_E \vartheta$ (resp. $\sigma \preceq^{\mathcal{X},\mathcal{Q}}_E \vartheta$), iff $\sigma\varphi =^{\mathcal{X},\mathcal{Q}}_E \vartheta$ for some substitution (resp. substitution non-erasing on $\mathcal{X}$ modulo E) φ.*

Example 7. Let $\sigma = \{\overline{x} \mapsto \overline{y}\}$, $\vartheta = \{\overline{x} \mapsto \ulcorner a, b \urcorner, \overline{y} \mapsto \ulcorner a, b \urcorner\}$, $\eta = \{\overline{x} \mapsto \ulcorner \urcorner, \overline{y} \mapsto \ulcorner \urcorner\}$, $\mathcal{X} = \{\overline{x}, \overline{y}\}$, $\mathcal{Q} = \emptyset$, $E = \emptyset$. Then $\sigma \trianglelefteq_E^{\mathcal{X},\mathcal{Q}} \vartheta$, $\sigma \preceq_E^{\mathcal{X},\mathcal{Q}} \vartheta$, $\sigma \trianglelefteq_E^{\mathcal{X},\mathcal{Q}} \eta$, $\sigma \not\preceq_E^{\mathcal{X},\mathcal{Q}} \eta$.

A substitution ϑ is an *E-instance* (resp. *strong E-instance*) of σ on $\mathcal{X}$ and $\mathcal{Q}$ iff $\sigma \trianglelefteq_E^{\mathcal{X},\mathcal{Q}} \vartheta$ (resp. $\sigma \preceq_E^{\mathcal{X},\mathcal{Q}} \vartheta$). The equivalence associated with $\trianglelefteq_E^{\mathcal{X},\mathcal{Q}}$ (resp. with $\preceq_E^{\mathcal{X},\mathcal{Q}}$) is denoted by $\doteq_E^{\mathcal{X},\mathcal{Q}}$ (resp. by $\approx_E^{\mathcal{X},\mathcal{Q}}$). The strict part of $\trianglelefteq_E^{\mathcal{X},\mathcal{Q}}$ (resp. $\preceq_E^{\mathcal{X},\mathcal{Q}}$) is denoted by $\lessdot_E^{\mathcal{X},\mathcal{Q}}$ (resp. $\prec_E^{\mathcal{X},\mathcal{Q}}$). Definition 13 implies $\prec_E^{\mathcal{X},\mathcal{Q}} \subseteq \trianglelefteq_E^{\mathcal{X},\mathcal{Q}}$.

Definition 14. *A set of substitutions S is called* minimal *(resp. almost minimal) with respect to $\mathcal{X}$ and $\mathcal{Q}$ modulo E iff two distinct elements of S are incomparable with respect to $\trianglelefteq_E^{\mathcal{X},\mathcal{Q}}$ (resp. $\preceq_E^{\mathcal{X},\mathcal{Q}}$).*

Minimality implies almost minimality, but not vice versa: A counterexample is the set $\{\sigma, \eta\}$ from Example 7.

Definition 15. *A* complete set of E-unifiers *of an E-unification problem Γ is a set S of substitutions such that (1) $S \subseteq \mathcal{U}_E(\Gamma)$, and (2) for each $\vartheta \in \mathcal{U}_E(\Gamma)$ there exists $\sigma \in S$ such that $\sigma \trianglelefteq_E^{\mathcal{X},\mathcal{Q}} \vartheta$, where $\mathcal{X} = Var(\Gamma)$ and $\mathcal{Q} = S\mathcal{F}un(\Gamma)$.*

The set S is a minimal *(resp. almost minimal) complete set of E-unifiers of Γ, denoted $mcu_E(\Gamma)$ (resp. $amcu_E(\Gamma)$) iff it is a complete set that is minimal (resp. almost minimal) with respect to $\mathcal{X}$ and $\mathcal{Q}$ modulo E.*

Proposition 1. *An E-unification problem Γ has an almost minimal complete set of E-unifiers iff it has a minimal complete set of E-unifiers.*

If Γ is not E-unifiable, then $mcu_E(\Gamma) = amcu_E(\Gamma) = \emptyset$. A minimal (resp. almost minimal) complete set of E-unifiers of Γ, if it exists, is unique up to the equivalence $\doteq_E^{\mathcal{X},\mathcal{Q}}$ (resp. $\approx_E^{\mathcal{X},\mathcal{Q}}$), where $\mathcal{X} = Var(\Gamma)$ and $\mathcal{Q} = S\mathcal{F}un(\Gamma)$.

Example 8. 1. $\Gamma = \{f(\overline{x}) \approx_\emptyset^? f(\overline{y})\}$. Then $mcu_\emptyset(\Gamma) = \{\{\overline{x} \mapsto \overline{y}\}\}$, $amcu_\emptyset(\Gamma) = \{\{\overline{x} \mapsto \overline{y}\}, \{\overline{x} \mapsto \ulcorner \urcorner, \overline{y} \mapsto \ulcorner \urcorner\}\}$.
2. $\Gamma = \{f(\overline{x}, x, \overline{y}) \approx_\emptyset^? f(f(\overline{x}), x, a, b)\}$. Then $mcu_\emptyset(\Gamma) = amcu_\emptyset(\Gamma) = \{\{x \mapsto f(), \overline{x} \mapsto \ulcorner \urcorner, \overline{y} \mapsto \ulcorner f(), a, b \urcorner\}\}$.
3. $\Gamma = \{f(a, \overline{x}) \approx_\emptyset^? f(\overline{x}, a)\}$. Then $mcu_\emptyset(\Gamma) = amcu_\emptyset(\Gamma) = \{\{\overline{x} \mapsto \ulcorner \urcorner\}, \{\overline{x} \mapsto a\}, \{\overline{x} \mapsto \ulcorner a, a \urcorner\}, \ldots\}$.
4. $\Gamma = \{f(\overline{x}, \overline{y}, x) \approx_\emptyset^? f(\overline{c}, a)\}$. Then $mcu_\emptyset(\Gamma) = amcu_\emptyset(\Gamma) = \{\{\overline{x} \mapsto \ulcorner \urcorner, \overline{y} \mapsto \overline{c}, x \mapsto a\}, \{\overline{x} \mapsto \overline{c}, \overline{y} \mapsto \ulcorner \urcorner, x \mapsto a\}, \{\overline{x} \mapsto \overline{c_1}, \overline{y} \mapsto \overline{c_2}, x \mapsto a, \overline{c} \mapsto \ulcorner \overline{c_1}, \overline{c_2} \urcorner\}\}$.

Definition 16. *A set of substitutions S is* disjoint *(resp. almost disjoint) wrt $\mathcal{X}$ and $\mathcal{Q}$ modulo E iff two distinct elements in S have no common E-instance (resp. strong E-instance) on $\mathcal{X}$ and $\mathcal{Q}$, i.e., for all $\sigma, \vartheta \in S$, if there exists φ such that $\sigma \trianglelefteq_E^{\mathcal{X},\mathcal{Q}} \varphi$ (resp. $\sigma \preceq_E^{\mathcal{X},\mathcal{Q}} \varphi$) and $\vartheta \trianglelefteq_E^{\mathcal{X},\mathcal{Q}} \varphi$ (resp. $\vartheta \preceq_E^{\mathcal{X},\mathcal{Q}} \varphi$), then $\sigma = \vartheta$.*

Disjointness implies almost disjointness, but not vice versa: Consider again the set $\{\sigma, \eta\}$ in Example 7.

Proposition 2. *If a set of substitutions S is disjoint (almost disjoint) wrt $\mathcal{X}$ and $\mathcal{Q}$ modulo E, then it is minimal (almost minimal) wrt $\mathcal{X}$ and $\mathcal{Q}$ modulo E.*

However, almost disjointness does not imply minimality: Again, take the set $\{\sigma, \eta\}$ in Example 7. On the other hand, minimality does not imply almost disjointness: Let $\sigma = \{x \mapsto f(a, y)\}$, $\vartheta = \{x \mapsto f(y, b)\}$, $\mathcal{X} = \{x\}$, $\mathcal{Q} = \emptyset$, and $E = \emptyset$. Then $\{\sigma, \vartheta\}$ is minimal but not almost disjoint with respect to $\mathcal{X}$ and $\mathcal{Q}$ modulo E, because $\sigma \preceq_E^{\mathcal{X}, \mathcal{Q}} \varphi$ and $\vartheta \preceq_E^{\mathcal{X}, \mathcal{Q}} \varphi$, with $\varphi = \{x \mapsto f(a, b)\}$, but $\sigma \neq \vartheta$. The same example can be used to show that almost minimality does not imply almost disjointness either. From these observations we can also conclude that neither minimality nor almost minimality imply disjointness.

The equational theory $E = \emptyset$ is called the *free theory with sequence variables and sequence function symbols*. Unification in the free theory is called the *syntactic sequence unification*. The theory $E = \{f(\overline{x}, f(\overline{y}), \overline{z}) \approx f(\overline{x}, \overline{y}, \overline{z})\}$ that we first encountered in Example 5 is called *the flat theory*, where f is the flat flexible arity individual function symbol. We call unification in the flat theory the *F-unification*. Certain properties of this theory will be used in proving decidability of the syntactic sequence unification.

3   Decidability and Unification Type

We show decidability of a syntactic sequence unification problem in three steps: First, we reduce the problem by unifiability preserving transformation to a unification problem containing no sequence function symbols. Second, applying yet another unifiability preserving transformation we get rid of all free flexible arity (individual) function symbols, obtaining a unification problem whose signature consists of fixed arity individual function symbols and one flat flexible arity individual function symbol. Finally, we show decidability of the reduced problem.

Let Γ be a general syntactic sequence unification problem and let $\mathcal{Q} = \mathcal{SFun}(\Gamma)$. Assume $\mathcal{Q} \neq \emptyset$. We transform Γ performing the following steps: (1) Introduce for each n-ary $\overline{f} \in \mathcal{Q}$ a new n-ary symbol $g_{\overline{f}} \in \mathcal{F}_{\mathrm{Ind}}^{\mathrm{Fix}}$. (2) Introduce for each flexible arity $\overline{f} \in \mathcal{Q}$ a new flexible arity symbol $g_{\overline{f}} \in \mathcal{F}_{\mathrm{Ind}}^{\mathrm{Flex}}$. (3) Replace each sequence function symbol $\overline{f}$ in Γ with the corresponding $g_{\overline{f}}$.

The transformation yields a new unification problem Λ that does not contain sequence function symbols. We impose the *first restriction on individual variables*, shortly **RIV1**, on Λ demanding that for any syntactic unifier λ of Λ and for any $x \in \mathcal{V}_{\mathrm{Ind}}$, $\mathcal{H}ead(x\lambda)$ must be different from any newly introduced individual function symbols.

Theorem 1. Γ *is syntactically unifiable iff* Λ *with the* **RIV1** *is syntactically unifiable.*

Remark 1. Unifiability of Λ without the **RIV1** does not imply unifiability of Γ: Let Γ be $\{f(x) \approx_\emptyset^? f(\overline{c})\}$. Then $\Lambda = \{f(x) \approx_\emptyset^? f(c_{\overline{c}})\}$. Γ is not unifiable, while $\{x \mapsto c_{\overline{c}}\}$ is a unifier of Λ, because $x \in \mathcal{V}_{\mathrm{Ind}}$ can be bound with $c_{\overline{c}} \in \mathcal{T}_{\mathrm{Ind}}(\mathcal{F}, \mathcal{V})$.

Next, our goal is to construct a general syntactic sequence unification problem without sequence function symbols that is unifiable (without restrictions) iff Λ with the **RIV1** is syntactically unifiable. We construct a finite set of individual terms $\mathcal{I}$ consisting of a new individual constant c, exactly one term of the form $h(y_1, \ldots, y_n)$ for each fixed arity $h \in \mathcal{IFun}(\Gamma)$ such that $n = \mathcal{Ar}(h)$ and $y_1, \ldots, y_n$ are distinct individual variables new for $\mathcal{I}$ and Λ, and exactly one term of the form $h(\overline{x})$ for each flexible arity $h \in \mathcal{IFun}(\Gamma)$ such that $\overline{x}$ is a sequence variable new for $\mathcal{I}$ and Λ.

Theorem 2. *Let Λ have the form $\{s \approx^?_{\emptyset} t\}$ with $\mathcal{IVar}(\Lambda) = \{x_1, \ldots, x_n\}$ and $g \in \mathcal{F}^{\mathrm{Fix}}$ be a new symbol with $\mathcal{Ar}(g) = n + 1$. Then Λ with the **RIV1** is syntactically unifiable iff there exist $r_1, \ldots, r_n \in \mathcal{I}$ such that the general syntactic unification problem (without sequence function symbols) $\{g(s, x_1, \ldots, x_n) \approx^?_{\emptyset} g(t, r_1, \ldots, r_n)\}$ is unifiable.*

Thus, we have to show that unifiability of a general syntactic unification problem Δ without sequence function symbols is decidable. We assume that $\mathcal{Flex}(\Delta) \neq \emptyset$, otherwise Δ would be a Robinson unification problem. We transform Δ performing the following steps: (1) Introduce a new flat symbol $seq \in \mathcal{F}^{\mathrm{Flex}}_{\mathrm{Ind}}$. (2) Introduce a new unary symbol $g_f \in \mathcal{F}^{\mathrm{Fix}}_{\mathrm{Ind}}$ for each $f \in \mathcal{Flex}(\Delta)$. (3) Replace each term $f(r_1, \ldots, r_m)$, $m \geq 0$, in Δ by $g_f(seq(r_1, \ldots, r_m))$.

The transformation yields a new general flat unification problem Θ. Sequence variables occur in Θ only as arguments of terms with the head seq. We impose the *second restriction on individual variables*, **RIV2**, on Θ demanding that, for any F-unifier ϑ of Θ and for any $x \in \mathcal{V}_{\mathrm{Ind}}$, $\mathcal{Head}(x\vartheta) \neq seq$.

Theorem 3. *Δ is syntactically unifiable iff Θ with the **RIV2** is F-unifiable.*

Remark 2. F-unifiability of Θ without the **RIV2** does not imply syntactic unifiability of Δ: Let Δ be $\{f(x) \approx^?_{\emptyset} f(a, b)\}$, $f \in \mathcal{F}^{\mathrm{Flex}}$. Then $\Theta = \{g_f(seq(x)) \approx^?_F g_f(seq(a, b))\}$. Obviously Δ is not unifiable, while $\{x \mapsto seq(a, b)\}$ is an F-unifier of Θ, because $seq(seq(a, b)) \approx_F seq(a, b)$.

Next, our goal is to construct a general F-unification problem that is F-unifiable (without restrictions) iff Θ with the **RIV2** is F-unifiable. First, we construct a finite set $\mathcal{J}$ of individual terms consisting of a new individual constant d and exactly one term of the form $h(y_1, \ldots, y_n)$ for each $h \in \mathcal{Fix}(\Theta)$ such that $n = \mathcal{Ar}(h)$ and $y_1, \ldots, y_n$ are distinct individual variables new for $\mathcal{J}$ and Θ.

Theorem 4. *Let Θ be $\{s \approx^?_F t\}$ with $\mathcal{IVar}(\Theta) = \{x_1, \ldots, x_n\}$ and $h \in \mathcal{F}^{\mathrm{Fix}}$ be a new symbol with $\mathcal{Ar}(h) = n + 1$. Then Θ with the **RIV2** is F-unifiable iff for some $r_1, \ldots, r_n \in \mathcal{J}$ the general F-unification problem $\{h(s, x_1, \ldots, x_n) \approx^?_F h(t, r_1, \ldots, r_n)\}$ is F-unifiable.*

Thus, we are left with proving that unifiability of an F-unification problem Φ, whose signature consists of fixed arity individual function symbols and the only flexible arity flat individual function symbol seq, is decidable.

Let Ψ be an F-unification problem obtained from Φ by replacing each $\overline{x} \in \mathcal{SVar}(\Phi)$ with a new individual variable x_ψ. It is easy to see that Φ is unifiable iff Ψ is. Indeed, replacing each variable binding $\overline{x} \mapsto \ulcorner s_1, \ldots, s_n \urcorner$ in a unifier of Φ with $x_\Psi \mapsto seq(s_1, \ldots, s_n)$ yields a unifier of Ψ, and vice versa.

We can consider Ψ as an elementary unification problem in the combined theory $E_1 \cup E_2$, where E_1 is a flat equational theory over $\{seq\}$ and $\mathcal{V}_{\text{Ind}}$, and E_2 is a free equational theory over $\mathcal{F}ix(\Psi)$ and $\mathcal{V}_{\text{Ind}}$. E_1-unification problems are, in fact, word equations, while E_2-unification is Robinson unification. Using the Baader-Schulz combination method [2], we can prove the following theorem:

Theorem 5. *F-unifiability of Ψ is decidable.*

Hence, unifiability of general syntactic sequence unification problem is decidable.

As for the unification type, in Example 8 we have seen that $mcu_\emptyset(\Gamma)$ is infinite for $\Gamma = \{f(a, \overline{x}) \approx_\emptyset^? f(\overline{x}, a)\}$. It implies that the syntactic sequence unification is at least infinitary. To show that it is not nullary, by Proposition 1, it is sufficient to prove existence of an almost minimal set of unifiers for every syntactic sequence unification problem. We do it in the standard way, by proving that for any Γ, every strictly decreasing chain $\sigma_1 \succ_\emptyset^{\mathcal{X}, \mathcal{Q}} \sigma_2 \succ_\emptyset^{\mathcal{X}, \mathcal{Q}} \cdots$ of substitutions in $\mathcal{U}_\emptyset(\Gamma)$ is finite, where $\mathcal{X} = \mathcal{V}ar(\Gamma)$ and $\mathcal{Q} = \mathcal{SF}un(\Gamma)$. Hence, syntactic sequence unification is infinitary.

4   Relation with Order-Sorted Higher-Order Unification

Syntactic sequence unification can be considered as a special case of order-sorted higher-order E-unification. Here we show the corresponding encoding in the framework described in [9]. We consider simply typed λ-calculus with the types i and o. The set of base sorts consists of $\text{ind}, \text{seq}, \text{seqc}, \text{o}$ such that the type of o is o and the type of the other sorts is i. We will treat individual and sequence variables as first order variables, sequence functions as second order variables and define a context Γ such that $\Gamma(x) = \text{ind}$ for all $x \in \mathcal{V}_{\text{Ind}}$, $\Gamma(\overline{x}) = \text{seq}$ for all $\overline{x} \in \mathcal{V}_{\text{Seq}}$, $\Gamma(\overline{f}) = \text{seq} \to \text{seqc}$ for each $\overline{f} \in \mathcal{F}_{\text{Seq}}^{\text{Flex}}$, and $\Gamma(\overline{f}) = \underbrace{\text{ind} \to \cdots \to \text{ind}}_{n \text{ times}} \to$ seqc for each $\overline{f} \in \mathcal{F}_{\text{Seq}}^{\text{Fix}}$ with $\mathcal{A}r(f) = n$. Individual function symbols are treated as constants. We assign to each $f \in \mathcal{F}_{\text{Ind}}^{\text{Flex}}$ a functional sort $\text{seq} \to \text{ind}$ and to each $f \in \mathcal{F}_{\text{Ind}}^{\text{Fix}}$ with $\mathcal{A}r(f) = n$ a functional sort $\underbrace{\text{ind} \to \cdots \to \text{ind}}_{n \text{ times}} \to \text{ind}$.

We assume equality constants $\approx_s$ for every sort s. In addition, we have two function symbols: binary $\ulcorner \urcorner$ of the sort $\text{seq} \to \text{seq} \to \text{seq}$ and a constant $[]$ of the sort seq. Sorts are partially ordered as $[\text{ind} \le \text{seqc}]$ and $[\text{seqc} \le \text{seq}]$. The equational theory is an AU-theory, asserting associativity of $\ulcorner \urcorner$ with $[]$ as left and right unit. We consider unification problems for terms of the sort ind where terms are in $\beta\eta$-normal form containing no bound variables, and terms whose head is $\ulcorner \urcorner$ are flattened. For a given unification problem Γ in this theory, we are looking for unifiers that obey the following restrictions: If a unifier σ binds a second order variable $\overline{f}$ of the sort $\text{seq} \to \text{seqc}$, then $\overline{f}\sigma = \lambda\overline{x}.\ulcorner \overline{g_1}(\overline{x}), \ldots, \overline{g_m}(\overline{x}) \urcorner$ and if σ binds a second order variable $\overline{f}$ of the sort $\underbrace{\text{ind} \to \cdots \to \text{ind}}_{n \text{ times}} \to \text{seqc}$, then $\overline{f}\sigma = \lambda x_1 \ldots x_n.\ulcorner \overline{g_1}(x_1, \ldots, x_n), \ldots, \overline{g_m}(x_1, \ldots, x_n)\urcorner$, where $m > 1$ and $\overline{g_1}, \ldots, \overline{g_m}$ are fresh variables of the same sort as f.

Hence, syntactic sequence unification can be considered as order-sorted second-order AU-unification with additional restrictions. Order-sorted higher-order syntactic unification was investigated in [9], but we are not aware of any work done on order-sorted higher-order equational unification.

5   Unification Procedure

In the sequel we assume that Γ, maybe with indices, and Γ' denote syntactic sequence unification problems. A *system* is either the symbol $\bot$ (representing failure), or a pair $\langle \Gamma; \sigma \rangle$. The inference system $\mathfrak{U}$ consists of the transformation rules on systems listed below. The function symbol $g \in \mathcal{F}_{\mathrm{Ind}}^{\mathrm{Flex}}$ in the rule PD2 is new. In the Splitting rule $\overline{f_1}$ and $\overline{f_2}$ are new sequence function symbols of the same arity as $\overline{f}$ in the same rule. We assume that the indices $n, m, k, l \geq 0$.

Projection (P):

$$\langle \Gamma;\ \sigma \rangle \Longrightarrow \langle \Gamma\vartheta;\ \sigma\vartheta \rangle, \quad \text{where } \vartheta \neq \varepsilon,\ \mathcal{D}om(\vartheta) \subseteq \mathcal{SV}ar(\Gamma) \text{ and } \mathcal{C}od(\vartheta) = \emptyset.$$

Trivial (T):

$$\langle \{s \approx_\emptyset^? s\} \cup \Gamma';\ \sigma \rangle \Longrightarrow \langle \Gamma';\ \sigma \rangle.$$

Orient 1 (O1):

$$\langle \{s \approx_\emptyset^? x\} \cup \Gamma';\ \sigma \rangle \Longrightarrow \langle \{x \approx_\emptyset^? s\} \cup \Gamma';\ \sigma \rangle, \text{ if } s \notin \mathcal{V}_{\mathrm{Ind}}.$$

Orient 2 (O2):

$$\langle \{f(s, s_1, \ldots, s_n) \approx_\emptyset^? f(\overline{x}, t_1, \ldots, t_m)\} \cup \Gamma';\ \sigma \rangle \Longrightarrow$$
$$\langle \{f(\overline{x}, t_1, \ldots, t_m) \approx_\emptyset^? f(s, s_1, \ldots, s_n)\} \cup \Gamma';\ \sigma \rangle, \qquad \text{if } s \notin \mathcal{V}_{\mathrm{Seq}}.$$

Solve (S):

$$\langle \{x \approx_\emptyset^? t\} \cup \Gamma';\ \sigma \rangle \Longrightarrow \langle \Gamma'\vartheta;\ \sigma\vartheta \rangle, \text{ if } x \notin \mathcal{IV}ar(t) \text{ and } \vartheta = \{x \mapsto t\}.$$

Total Decomposition (TD):

$$\langle \{f(s_1, \ldots, s_n) \approx_\emptyset^? f(t_1, \ldots, t_n)\} \cup \Gamma';\ \sigma \rangle \Longrightarrow$$
$$\langle \{s_1 \approx_\emptyset^? t_1, \ldots, s_n \approx_\emptyset^? t_n\} \cup \Gamma';\ \sigma \rangle$$

if $f(s_1, \ldots, s_n) \neq f(t_1, \ldots, t_n)$, and $s_i, t_i \in \mathcal{T}_{\mathrm{Ind}}(\mathcal{F}, \mathcal{V})$ for all $1 \leq i \leq n$.

Partial Decomposition 1 (PD1):

$$\langle \{f(s_1, \ldots, s_n) \approx_\emptyset^? f(t_1, \ldots, t_m)\} \cup \Gamma';\ \sigma \rangle \qquad\qquad\qquad \Longrightarrow$$
$$\langle \{s_1 \approx_\emptyset^? t_1, \ldots, s_{k-1} \approx_\emptyset^? t_{k-1}, f(s_k, \ldots, s_n) \approx_\emptyset^? f(t_k, \ldots, t_m)\} \cup \Gamma';\ \sigma \rangle$$

if $f(s_1, \ldots, s_n) \neq f(t_1, \ldots, t_m)$, for some $1 < k \leq min(n, m)$,
$s_k \in \mathcal{T}_{\mathrm{Seq}}(\mathcal{F}, \mathcal{V})$ or $t_k \in \mathcal{T}_{\mathrm{Seq}}(\mathcal{F}, \mathcal{V})$, and $s_i, t_i \in \mathcal{T}_{\mathrm{Ind}}(\mathcal{F}, \mathcal{V})$ for all $1 \leq i < k$.

Partial Decomposition 2 (PD2):

$$\langle \{f(\overline{f}(r_1, \ldots, r_k), s_1, \ldots, s_n) \approx_\emptyset^? f(\overline{f}(q_1, \ldots, q_l), t_1, \ldots, t_m)\} \cup \Gamma';\ \sigma \rangle \qquad \Longrightarrow$$
$$\langle \{g(r_1, \ldots, r_k) \approx_\emptyset^? g(q_1, \ldots, q_l), f(s_1, \ldots, s_n) \approx_\emptyset^? f(t_1, \ldots, t_m)\} \cup \Gamma';\ \sigma \rangle.$$

if $f(\overline{f}(r_1, \ldots, r_k), s_1, \ldots, s_n) \neq f(\overline{f}(q_1, \ldots, q_l), t_1, \ldots, t_m)$.

Sequence Variable Elimination 1 (SVE1):

$$\langle \{f(\overline{x}, s_1, \ldots, s_n) \approx_\emptyset^? f(\overline{x}, t_1, \ldots, t_m)\} \cup \Gamma';\ \sigma \rangle \Longrightarrow$$
$$\langle \{f(s_1, \ldots, s_n) \approx_\emptyset^? f(t_1, \ldots, t_m)\} \cup \Gamma';\ \sigma \rangle$$

if $f(\overline{x}, s_1, \ldots, s_n) \neq f(\overline{x}, t_1, \ldots, t_m)$.

Sequence Variable Elimination 2 (SVE2):

$$\langle\{f(\overline{x},s_1,\ldots,s_n)\approx^?_\emptyset f(t,t_1,\ldots,t_m)\}\cup\Gamma';\ \sigma\rangle\ \Longrightarrow$$
$$\langle\{f(s_1,\ldots,s_n)\vartheta\approx^?_\emptyset f(t_1,\ldots,t_m)\vartheta\}\cup\Gamma'\vartheta;\ \sigma\vartheta\rangle$$

if $\overline{x}\notin\mathcal{SV}ar(t)$ and $\vartheta=\{\overline{x}\mapsto t\}$.

Widening 1 (W1):

$$\langle\{f(\overline{x},s_1,\ldots,s_n)\approx^?_\emptyset f(t,t_1,\ldots,t_m)\}\cup\Gamma';\ \sigma\rangle\ \Longrightarrow$$
$$\langle\{f(\overline{x},s_1\vartheta,\ldots,s_n\vartheta)\approx^?_\emptyset f(t_1\vartheta,\ldots,t_m\vartheta)\}\cup\Gamma'\vartheta;\ \sigma\vartheta\rangle$$

if $\overline{x}\notin\mathcal{SV}ar(t)$ and $\vartheta=\{\overline{x}\mapsto\ulcorner t,\overline{x}\urcorner\}$.

Widening 2 (W2):

$$\langle\{f(\overline{x},s_1,\ldots,s_n)\approx^?_\emptyset f(\overline{y},t_1,\ldots,t_m)\}\cup\Gamma';\ \sigma\rangle\ \Longrightarrow$$
$$\langle\{f(s_1\vartheta,\ldots,s_n\vartheta)\approx^?_\emptyset f(\overline{y},t_1\vartheta,\ldots,t_m\vartheta)\}\cup\Gamma'\vartheta;\ \sigma\vartheta\rangle$$

where $\vartheta=\{\overline{y}\mapsto\ulcorner\overline{x},\overline{y}\urcorner\}$.

Splitting (Sp):

$$\langle\{f(\overline{x},s_1,\ldots,s_n)\approx^?_\emptyset f(\overline{f}(r_1,\ldots,r_k),t_1,\ldots,t_m)\}\cup\Gamma';\ \sigma\rangle\ \Longrightarrow$$
$$\langle\{f(s_1,\ldots,s_n)\vartheta\approx^?_\emptyset f(\overline{f_2}(r_1,\ldots,r_k),t_1,\ldots,t_m)\vartheta\}\cup\Gamma'\vartheta;\ \sigma\vartheta\rangle$$

if $\overline{x}\notin\mathcal{SV}ar(\overline{f}(r_1,\ldots,r_k))$ and $\vartheta=\{\overline{x}\mapsto\overline{f_1}(r_1,\ldots,r_k)\}\{\overline{f}\mapsto\ulcorner\overline{f_1},\overline{f_2}\urcorner\}$.

We may use the rule name abbreviations as subscripts, e.g., $\langle\Gamma_1;\ \sigma_1\rangle\Longrightarrow_{\mathsf{P}}$ $\langle\Gamma_2;\ \sigma_2\rangle$ for Projection. We may also write $\langle\Gamma_1;\ \sigma_1\rangle\Longrightarrow_{\mathsf{BT}}\langle\Gamma_2;\ \sigma_2\rangle$ to indicate that $\langle\Gamma_1;\ \sigma_1\rangle$ was transformed to $\langle\Gamma_2;\ \sigma_2\rangle$ by some *basic transformation* (i.e., non-projection) rule. **P**, **SVE2**, **W1**, **W2**, and **Sp** are non-deterministic rules.

A *derivation* is a sequence $\langle\Gamma_1;\sigma_1\rangle\Longrightarrow\langle\Gamma_2;\sigma_2\rangle\Longrightarrow\cdots$ of system transformations. A derivation is *fair* if any transformation rule which is continuously enabled is eventually applied. Any finite fair derivation $S_1\Longrightarrow S_2\Longrightarrow\cdots\Longrightarrow S_n$ is maximal, i.e., no further transformation rule can be applied on S_n.

Definition 17. *A* syntactic sequence unification procedure *is any program that takes a system* $\langle\Gamma;\ \varepsilon\rangle$ *as an input and uses the rules in* $\mathfrak{U}$ *to generate a tree of fair derivations, called the* unification tree for Γ, $\mathcal{UT}(\Gamma)$, *in the following way:*

1. *The root of the tree is labeled with* $\langle\Gamma;\ \varepsilon\rangle$;
2. *Each branch of the tree is a fair derivation either of the form* $\langle\Gamma;\ \varepsilon\rangle\Longrightarrow_{\mathsf{P}}$ $\langle\Gamma_1;\ \sigma_1\rangle\Longrightarrow_{\mathsf{BT}}\langle\Gamma_2;\ \sigma_2\rangle\Longrightarrow_{\mathsf{BT}}\cdots$ *or* $\langle\Gamma;\ \varepsilon\rangle\Longrightarrow_{\mathsf{BT}}\langle\Gamma_1;\ \sigma_1\rangle\Longrightarrow_{\mathsf{BT}}$ $\langle\Gamma_2;\ \sigma_2\rangle\Longrightarrow_{\mathsf{BT}}\cdots$. *The nodes in the tree are systems.*
3. *If several transformation rules, or different instances of the same transformation rule are applicable to a node in the tree, they are applied concurrently.*
4. *The decision procedure is applied to the root and to each node generated by a non-deterministic transformation rule, to decide whether the node contains a solvable unification problem. If the unification problem* Δ *in a node* $\langle\Delta;\ \delta\rangle$ *is unsolvable, then the branch is extended by* $\langle\Delta;\ \delta\rangle\Longrightarrow_{\mathsf{DP}}\bot$.

The leaves of $\mathcal{UT}(\Gamma)$ are labeled either with the systems of the form $\langle\emptyset;\ \sigma\rangle$ or with $\bot$. The branches of $\mathcal{UT}(\Gamma)$ that end with $\langle\emptyset;\ \sigma\rangle$ are called *successful branches*, and those with the leaves $\bot$ are *failed branches*. We denote by $Sol_\emptyset(\Gamma)$ the solution set of Γ, i.e., the set of all σ-s such that $\langle\emptyset;\ \sigma\rangle$ is a leaf of $\mathcal{UT}(\Gamma)$.

5.1   Soundness, Completeness and Almost Minimality

In this section we assume that $\mathcal{X} = \mathcal{V}ar(\Gamma)$ and $\mathcal{Q} = \mathcal{SF}un(\Gamma)$ for a syntactic sequence unification problem Γ. The soundness theorem is not hard to prove:

Theorem 6 (Soundness). *If* $\langle \Gamma; \varepsilon \rangle \Longrightarrow^{+} \langle \emptyset; \vartheta \rangle$, *then* $\vartheta \in \mathcal{U}_{\emptyset}(\Gamma)$.

Completeness can be proved by showing that for any unifier ϑ of Γ there exists a derivation from $\langle \Gamma; \varepsilon \rangle$ that terminates with success and the substitution in the last system of the derivation is strongly more general than ϑ:

Lemma 1. *For any* $\vartheta \in \mathcal{U}_{\emptyset}(\Gamma)$ *there exists a derivation of the form* $\langle \Gamma_0; \sigma_0 \rangle \Longrightarrow_{\mathsf{X}}$ $\langle \Gamma_1; \sigma_1 \rangle \Longrightarrow_{\mathsf{BT}} \langle \Gamma_2; \sigma_2 \rangle \Longrightarrow_{\mathsf{BT}} \cdots \Longrightarrow_{\mathsf{BT}} \langle \emptyset; \sigma_n \rangle$ *with* $\Gamma_1 = \Gamma$ *and* $\sigma_1 = \varepsilon$ *such that if* ϑ *is erasing on* $\mathcal{X}$ *then* $\mathsf{X} = \mathsf{P}$, *otherwise* $\mathsf{X} = \mathsf{BT}$, *and* $\sigma_n \preceq_{\emptyset}^{\mathcal{X},\mathcal{Q}} \vartheta$.

From Theorem 6, Lemma 1, and the fact that $\preceq_{E}^{\mathcal{X},\mathcal{Q}} \subseteq \preccurlyeq_{E}^{\mathcal{X},\mathcal{Q}}$, by Definition 17 and Definition 15 we get the completeness theorem:

Theorem 7 (Completeness). $\mathcal{S}ol_{\emptyset}(\Gamma)$ *is a complete set of unifiers of* Γ.

The set $\mathcal{S}ol_{\emptyset}(\Gamma)$, in general, is not minimal with respect to $\mathcal{V}ar(\Gamma)$ and $\mathcal{SF}un(\Gamma)$ modulo the free theory. Just consider $\Gamma = \{f(\overline{x}) \approx_{\emptyset}^{?} f(\overline{y})\}$, then $\mathcal{S}ol_{\emptyset}(\Gamma) = \{\{\overline{x} \mapsto \overline{y}\}, \{\overline{x} \mapsto \ulcorner\urcorner, \overline{y} \mapsto \ulcorner\urcorner\}\}$. However, it can be shown that $\mathcal{S}ol_{\emptyset}(\Gamma)$ is almost minimal. In fact, the following stronger statement holds:

Theorem 8 (Almost Disjointness). $\mathcal{S}ol_{\emptyset}(\Gamma)$ *is almost disjoint wrt* $\mathcal{X}$ *and* $\mathcal{Q}$.

Theorem 7, Theorem 8 and Proposition 2 imply the main result of this section:

Theorem 9 (Main Theorem). $\mathcal{S}ol_{\emptyset}(\Gamma) = amcu_{\emptyset}(\Gamma)$.

6   Conclusions and Related Work

We showed that general syntactic unification with sequence variables and sequence functions is decidable and has the infinitary type. We developed a unification procedure and showed its soundness, completeness and almost minimality.

Historically, probably the first attempt to implement unification with sequence variables (without sequence functions) was made in the system MVL [7]. It was incomplete because of restricted use of widening technique. The restriction was imposed for the efficiency reasons. No theoretical study of the unification algorithm of MVL, to the best of our knowledge, was undertaken.

Richardson and Fuchs [16] describe another unification algorithm with sequence variables that they call vector variables. Vector variables come with their length attached, that makes unification finitary. The algorithm was implemented but its properties have never been investigated.

Implementation of first-order logic in ISABELLE [14] is based on sequent calculus formulated using sequence variables (on the meta level). Sequence meta-variables are used to denote sequences of formulae, and individual meta-variables denote single formulae. Since in every such unification problem no sequence meta-variable occurs more that once, and all of them occur only on the top level, ISABELLE, in fact, deals with a finitary case of sequence unification.

Word equations [1, 8] and associative unification [15] can be modelled by syntactic sequence unification using constants, sequence variables and one flexible arity function symbol. In the similar way we can imitate the unification algorithm for path logics closed under right identity and associativity [17].

The SET-VAR prover [4] has a construct called vector of (Skolem) functions that resembles our sequence functions. However, unification does not allow to split vectors of functions between variables: such a vector of functions either entirely unifies with a variable, or with another vector of functions.

The programming language of MATHEMATICA uses pattern matching that supports sequence variables (represented as identifiers with "triple blanks", e.g., x___) and flexible arity function symbols. Our procedure (without sequence function symbols) can imitate the behavior of MATHEMATICA matching algorithm.

Buchberger introduced sequence functions in the THEOREMA system [6] to Skolemize quantified sequence variables. In the equational prover of THEOREMA [11] we implemented a special case of unification with sequence variables and sequence functions: sequence variables occurring only in the last argument positions in terms. It makes unification unitary. Similar restriction is imposed on sequence variables in the RELFUN system [5] that integrates extensions of logic and functional programming. RELFUN allows multiple-valued functions as well.

In [10] we described unification procedures for free, flat, restricted flat and orderless theories with sequence variables, but without sequence functions.

Under certain restrictions sequence unification problems have at most finitely many solutions: sequence variables in the last argument positions, unification problems with at least one ground side (matching as a particular instance), all sequence variables on the top level with maximum one occurrence. It would be interesting to identify more cases with finite or finitely representable solution sets.

Acknowledgements

I thank Bruno Buchberger and Mircea Marin for interesting discussions on the topic.

References

1. H. Abdulrab and J.-P. Pécuchet. Solving word equations. *J. Symbolic Computation*, 8(5):499–522, 1990.
2. F. Baader and K. U. Schulz. Unification in the union of disjoint equational theories: Combining decision procedures. *J. Symbolic Computation*, 21(2):211–244, 1996.
3. F. Baader and W. Snyder. Unification theory. In A. Robinson and A. Voronkov, editors, *Handbook of Automated Reasoning*, volume I, chapter 8, pages 445–532. Elsevier Science, 2001.

4. W. W. Bledsoe and Guohui Feng. SET-VAR. *J. Automated Reasoning*, 11(3):293–314, 1993.

5. H. Boley. *A Tight, Practical Integration of Relations and Functions*, volume 1712 of *LNAI*. Springer, 1999.

6. B. Buchberger, C. Dupré, T. Jebelean, F. Kriftner, K. Nakagawa, D. Vasaru, and W. Windsteiger. The THEOREMA project: A progress report. In M. Kerber and M. Kohlhase, editors, *Proc. of Calculemus'2000 Conference*, pages 98–113, St. Andrews, UK, 6–7 August 2000.

7. M. L. Ginsberg. User's guide to the MVL system. Technical report, Stanford University, Stanford, California, US, 1989.

8. J. Jaffar. Minimal and complete word unification. *J. ACM*, 37(1):47–85, 1990.

9. M. Kohlhase. A mechanization of sorted higher-order logic based on the resolution principle. PhD Thesis. Universität des Saarlandes. Saarbrücken, Germany, 1994.

10. T. Kutsia. Solving and proving in equational theories with sequence variables and flexible arity symbols. Technical Report 02-31, RISC-Linz, Austria, 2002.

11. T. Kutsia. Equational prover of THEOREMA. In R. Nieuwenhuis, editor, *Proc. of the 14th Int. Conference on Rewriting Techniques and Applications (RTA'03)*, volume 2706 of *LNCS*, pages 367–379, Valencia, Spain, 9–11 June 2003. Springer.

12. T. Kutsia. Solving equations involving sequence variables and sequence functions. Technical Report 04-01, RISC, Johannes Kepler University, Linz, Austria, 2004. http://www.risc.uni-linz.ac.at/people/tkutsia/papers/SeqUnif.ps.

13. M. Marin and T. Kutsia. On the implementation of a rule-based programming system and some of its applications. In B. Konev and R. Schmidt, editors, *Proc. of the 4th Int. Workshop on the Implementation of Logics (WIL'03)*, pages 55–68, Almaty, Kazakhstan, 2003.

14. L. Paulson. ISABELLE: the next 700 theorem provers. In P. Odifreddi, editor, *Logic and Computer Science*, pages 361–386. Academic Press, 1990.

15. G. Plotkin. Building in equational theories. In B. Meltzer and D. Michie, editors, *Machine Intelligence*, volume 7, pages 73–90. Edinburgh University Press, 1972.

16. J. Richardson and N. E. Fuchs. Development of correct transformation schemata for Prolog programs. In N. E. Fuchs, editor, *Proc. of the 7th Int. Workshop on Logic Program Synthesis and Transformation (LOPSTR'97)*, volume 1463 of *LNCS*, pages 263–281, Leuven, Belgium, 10–12 July 1997. Springer.

17. R. Schmidt. *E*-Unification for subsystems of S4. In T. Nipkow, editor, *Proc. of the 9th Int. Conference on Rewriting Techniques and Applications, RTA'98*, volume 1379 of *LNCS*, pages 106–120, Tsukuba, Japan, 1998. Springer.

18. S. Wolfram. *The Mathematica Book*. Cambridge University Press and Wolfram Research, Inc., fourth edition, 1999.

Verified Computer Algebra in ACL2
(Gröbner Bases Computation)

I. Medina-Bulo[1], F. Palomo-Lozano[1],
J.A. Alonso-Jiménez[2], and J.L. Ruiz-Reina[2]

[1] Depto. de Lenguajes y Sistemas Informáticos, Univ. de Cádiz
E.S. de Ingeniería de Cádiz, C/ Chile, s/n, 11003 Cádiz, España
{inmaculada.medina,francisco.palomo}@uca.es
[2] Depto. de Ciencias de la Computación e Inteligencia Artificial, Univ. de Sevilla
E.T.S. de Ingeniería Informática, Avda. Reina Mercedes, s/n, 41012 Sevilla, España
{jalonso,jruiz}@us.es

Abstract. In this paper, we present the formal verification of a COM-
MON LISP implementation of Buchberger's algorithm for computing
Gröbner bases of polynomial ideals. This work is carried out in the ACL2
system and shows how verified Computer Algebra can be achieved in an
executable logic.

1 Introduction

Computer Algebra has experienced a great development in the last decade, as
can be seen from the proliferation of Computer Algebra Systems (CAS). These
systems are the culmination of theoretical results obtained in the last half century. One of the main achievements is due to B. Buchberger. In 1965 he devised
an algorithm for computing Gröbner bases of multivariate polynomial ideals,
thus solving the ideal membership problem for polynomial rings. Currently, his
algorithm is available in most CAS and its theory, implementation and numerous
applications are widely documented in the literature, e.g. [2,4].

The aim of this paper is to describe the formal verification of a naive COM-
MON LISP implementation of Buchberger's algorithm. The implementation and
formal proofs have been carried out in the ACL2 system, which consists of a programming language, a logic for stating and proving properties of the programs,
and a theorem prover supporting mechanized reasoning in the logic.

The importance of Buchberger's algorithm in Computer Algebra justifies on
its own the effort of obtaining a formal correctness proof with a theorem prover,
and this is one of the motivations for this work. Nevertheless, this goal has
already been achieved by L. Théry in [13], where he gives a formal proof using
the COQ system and explains how an executable implementation in the OCAML
language is extracted from the algorithm defined in COQ. In contrast, in ACL2
we can reason directly about the LISP program implementing the algorithm, i.e.
about the very program which is executed by the underlying LISP system. There
is a price to pay: the logic of ACL2 is a quantifier-free fragment of first-order

B. Buchberger and J.A. Campbell (Eds.): AISC 2004, LNAI 3249, pp. 171–184, 2004.

logic, less expressive[1] than the logic of COQ, which is based on type theory. We show how it is possible to formalize all the needed theory within the ACL2 logic.

The formal proofs developed in ACL2 are mainly adapted from Chap. 8 of [1]. As the whole development consists of roughly one thousand ACL2 theorems and function definitions, we will only scratch its surface presenting the main results and a sketch of how the pieces fit together. We will necessarily omit many details that, we expect, can be inferred from the context.

2   The Acl2 System

ACL2 formalizes an applicative subset of COMMON LISP. In fact, the same language, based on prefix notation, is used for writing LISP code and stating theorems about it[2]. The logic is a quantifier-free fragment of first-order logic with equality. It includes axioms for propositional logic and for a number of LISP functions and data types. Inference rules include those for propositional calculus, equality and instantiation (variables in formulas are implicitly universally quantified). One important inference rule is the *principle of induction*, that permits proofs by well-founded induction on the ordinal ϵ_0 (the logic provides a constructive definition of the ordinals up to ϵ_0).

By the *principle of definition* new function definitions are admitted as axioms (using defun) only if its termination is proved by means of an ordinal measure in which the arguments of each recursive call, if any, decrease. In addition, the *encapsulation principle* allows the user to introduce new function symbols (using encapsulate) that are constrained to satisfy certain assumptions. To ensure that the constraints are satisfiable, the user must provide a witness function with the required properties. Within the scope of an encapsulate, properties stated as theorems need to be proved for the witnesses; outside, these theorems work as assumed axioms. Together, encapsulation and the derived inference rule, *functional instantiation*, provide a second-order aspect [5,6]: theorems about constrained functions can be instantiated with function symbols if they are proved to have the same properties.

The ACL2 theorem prover mechanizes the logic, being particularly well suited for obtaining automated proofs based on simplification and induction. Although the prover is automatic in the sense that once a proof attempt is started (with defthm) the user can no longer interact, nevertheless it is interactive in a deeper sense: usually, the role of the user is to lead the prover to a preconceived handproof, by proving a suitable collection of lemmas that are used as rewrite rules in subsequent proofs (these lemmas are usually discovered by the user after the inspection of failed proofs). We used this kind of interaction to obtain the formal proofs presented here. For a detailed description of ACL2, we refer the reader to the ACL2 book [5].

---

[1] Nevertheless, the degree of automation of the ACL2 theorem prover is higher than in other systems with more expressive logics.

[2] Although we are aware that prefix notation may be inconvenient for people not used to LISP, we will maintain it to emphasize the use of a *real* programming language.

3   Polynomial Rings and Ideals

Let $R = K[x_1, \ldots, x_k]$ be a polynomial ring on an arbitrary commutative field K, where $k \in \mathbb{N}$. The elements of R are polynomials in the indeterminates $x_1, \ldots, x_k$ with the coefficients in K. Polynomials are built from monomials of R, that is, power products like $c \cdot x_1^{a_1} \cdots x_k^{a_k}$, where $c \in K$ is the coefficient, $x_1^{a_1} \cdots x_k^{a_k}$ is the term, and $a_1, \ldots, a_k \in \mathbb{N}$.

Therefore, there are several algebraic structures that it is necessary to formalize prior to the notion of polynomial. A computational theory of multivariate polynomials on a coefficient field was developed in [8, 9]. This ACL2 formalization includes common operations and fundamental properties establishing a ring structure. The aim was to develop a reusable library on polynomials.

Regarding polynomial representation, we have used a sparse, normalized and uniform representation. That is, having fixed the number of variables, a canonical form can be associated to each polynomial. In this canonical representation all monomials are arranged in a strictly decreasing order, there are no null monomials and all of them have the same number of variables. The main advantage of this representation arises when deciding equality [9].

Monomial lists are used as the internal representation of polynomials. Monomials are also lists consisting of a coefficient and a term. Having selected a set of variables and an ordering on them, each term is uniquely represented by a list of natural numbers. Although most of the theory is done for an arbitrary field, via the encapsulation principle, we use polynomials over the field of rational numbers for our implementation of Buchberger's algorithm. This alleviates some proofs at the cost of some generality, as ACL2 can use its built-in linear arithmetic decision procedure. In any case, the general theory has to be eventually instantiated to obtain an executable algorithm.

The functions k-polynomialp and k-polynomialsp recognize polynomials and polynomial lists (with k variables and rational coefficients). Analogously, +, *, - and |0| stand for polynomial addition, multiplication, negation and the zero polynomial. Let us now introduce the notion of ideal, along with the formalization of polynomial ideals in ACL2.

Definition 1. *$I \subseteq R$ is an ideal of R if it is closed under addition and under the product by elements of R.*

Definition 2. *The ideal generated by $B \subseteq R$, denoted as $\langle B \rangle$, is the set of linear combinations of B with coefficients in R. We say that B is a basis of $I \subseteq R$ if $I = \langle B \rangle$. An ideal is finitely-generated if it has a finite basis.*

Hilbert's Basis Theorem implies that every ideal in $K[x_1, \ldots, x_k]$ is finitely-generated, if K is a field. Polynomial ideals can be expressed in ACL2 by taking this into account. Let C and F be lists of polynomials. The predicate $p \in \langle F \rangle$ can be restated as $\exists C\ p = lin\text{-}comb(C, F)$, where $lin\text{-}comb$ is a recursive function computing the linear combination of the elements in F with coefficients in C.

As ACL2 is a quantifier-free logic we use a common trick: we introduce a Skolem function assumed to return a list of coefficients witnessing the ideal membership. In ACL2 this can be expressed in the following way:

```
(defun-sk<k> in<> (p F)
  (exists (C) (and (k-polynomialsp C) (equal p (lin-comb C F)))))
```

The use of `exists` in this definition is just syntactic sugar. Roughly speaking, the above construction introduces a Skolem function `in<>-witness`, with arguments `p` and `F`, which is axiomatized to choose, if possible, a list `C` of polynomial coefficients such that when linearly combined with the polynomials in `F`, `p` is obtained. Thus, `C` is a witness of the membership of `p` to the ideal generated by `F`, and `in<>` is defined by means of `in<>-witness`. The following theorems establish that our definition of ideal in ACL2 meets the intended closure properties:

```
(defthm |p in <F> & q in <F> => p + q in <F>|
  (implies (and (k-polynomialp p) (k-polynomialp q) (k-polynomialsp F))
           (implies (and (in<> p F) (in<> q F)) (in<> (+ p q) F))))

(defthm |q in <F> => p * q in <F>|
  (implies (and (k-polynomialp p) (k-polynomialp q) (k-polynomialsp F))
           (implies (in<> q F) (in<> (* p q) F))))
```

Whenever a theorem about `in<>` is proved we have to provide ACL2 with a hint to construct the necessary witness. For example, to prove that polynomial ideals are closed under addition we straightforwardly built an intermediate function computing the witness of $p+q \in \langle F \rangle$ from those of $p \in \langle F \rangle$ and $q \in \langle F \rangle$.

Definition 3. *The congruence induced by an ideal I, written as $\equiv_I$, is defined by $p \equiv_I q \iff p - q \in I$.*

The definition of $\equiv_{\langle F \rangle}$ in ACL2 is immediate[3]:

```
(defun<k> =<> (p q F)
  (in<> (+ p (- q)) F))
```

Clearly, the ideal membership problem for an ideal I is solvable if, and only if, its induced congruence $\equiv_I$ is decidable. Polynomial reductions will help us to design decision procedures for that congruence.

4   Polynomial Reductions

Let $<_M$ be a well-founded ordering on monomials, $p \neq 0$ a polynomial and let $lm(p)$ denote the leader monomial of p with respect to $<_M$.

Definition 4. *Let $f \neq 0$ be a polynomial. The reduction relation on polynomials induced by f, denoted as $\rightarrow_f$, is defined such that $p \rightarrow_f q$ if p contains a monomial $m \neq 0$ such that there exists a monomial c such that $m = -c \cdot lm(f)$ and $q = p + c \cdot f$. If $F = \{f_1, \ldots, f_k\}$ is a finite set of polynomials, then the reduction relation induced by F is defined as $\rightarrow_F = \bigcup_{i=1}^{k} \rightarrow_{f_i}$.*

---

[3] For the sake of readability, we use `defun-sk<k>` and `defun<k>`, instead of `defun-sk` and `defun`. These are just macros which add an extra parameter `k` (the number of variables) to a function definition, so we do not have to specify it in each function application. When `k` is not involved, `defun` is used.

We have formalized polynomial reductions in the framework of abstract reductions developed in [11]. This approach will allow us to export, by functional instantiation, well-known properties of abstract reductions (for example, Newman's lemma) to the case of polynomial reductions, avoiding the need to prove them from scratch.

In [11], instead of defining reductions as binary relations, they are defined as the action of *operators* on elements, obtaining reduced elements. More precisely, the representation of a reduction relation requires defining three functions:

1. A unary predicate specifying the domain where the reduction is defined. In our case, polynomials, as defined by the function k-polynomialp.
2. A binary function, reduction, computing the application of an operator to a polynomial. In our case, operators are represented by structures $\langle m, c, f \rangle$ consisting of the monomials m and c, and the polynomial f appearing in the definition of the polynomial reduction relation (Def. 4).
3. A binary predicate checking whether the application of a given operator to a given object is *valid*. The application of an operator $\langle m, c, f \rangle$ to p is valid if p is a polynomial containing the monomial m, $f \neq 0$ is a polynomial in F and $c = -m/lm(f)$. Notice that the last requirement implies that $lm(f)$ must divide m. This validity predicate is implemented by a function validp.

These functions are just what we need to define in ACL2 all the concepts related to polynomial reductions. Let us begin defining $\leftrightarrow_F$ (the symmetric closure of $\rightarrow_F$). We need the notion of *proof step* to represent the connection of two polynomials by the reduction relation, in either direction (direct or inverse). Each proof step is a structure consisting of four fields: a boolean field marking the step direction, the operator applied, and the elements connected (elt1, elt2). A proof step is *valid* if one of its elements is obtained by a valid application of its operator to the other element in the specified direction. The function valid-proof-stepp (omitted here), checks the validity of a proof step.

The following function formalizes in ACL2 the relation $\leftrightarrow_F$. Note that due to the absence of existential quantification, the step argument is needed to explicitly introduce the proof step justifying that $p \leftrightarrow_F q$.

```
(defun <-> (p q step F)
  (and (valid-proof-stepp step F)
       (equal p (elt1 step)) (equal q (elt2 step)))))
```

Next, we define $\overset{*}{\leftrightarrow}_F$ (the equivalence closure of $\rightarrow_F$). This can be described by means of a sequence of concatenated proof steps, which we call a *proof*[4]. Note that again due to the absence of existential quantification, the proof argument explicitly introduces the proof steps justifying that $p \overset{*}{\leftrightarrow}_F q$.

```
(defun <->* (p q proof F)
  (if (endp proof)
```

---

[4] Notice that the meaning of the word "proof" here is different than in the expression "ACL2 proof". This proof is just a sequence of reduction steps. In fact, we are formalizing an algebraic proof system inside ACL2.

```
      (and (equal p q) (k-polynomialp p))
    (and (k-polynomialp p)
         (<-> p (elt2 (first proof)) (first proof) F)
         (<->* (elt2 (first proof)) q (rest proof) F))))
```

In the same way, we define the relation $\to_F^*$ (the transitive closure of $\to_F$), by a function called ->* (in this case, we also check that all proof steps are direct).

The following theorems establish that the congruence $\equiv_{\langle F\rangle}$ is equal to the equivalence closure $\overset{*}{\leftrightarrow}_F$. This result is crucial to connect the results about reduction relations to polynomial ideals.

```
(defthm |p =<F> q => p <->F* q|
  (let ((proof (|p =<F> q => p <->F* q|-proof p q F)))
    (implies (and (k-polynomialp p) (k-polynomialp q) (k-polynomialsp F)
                  (=<> p q F))
             (<->* p q proof F)))

(defthm |p <->F* q => p =<F> q|
  (implies (and (k-polynomialp p) (k-polynomialp q) (k-polynomialsp F)
                (<->* p q proof F))
           (=<> p q F)))
```

These two theorems establish that it is possible to obtain a sequence of proof steps justifying that $p \overset{*}{\leftrightarrow}_F q$ from a list of coefficients justifying that $p - q \in \langle F\rangle$, and vice versa. The expression (|p =<F> q => p <->F* q|-proof p q F) explicitly computes such proof, in a recursive way. This is typical in our development: in many subsequent ACL2 theorems, the proof argument in <->* or ->* will be locally-bound (through a let or let* form) to a function computing the necessary proof steps. As these functions are rather technical and it would take long to explain them, we will omit their definitions. But it is important to remark this constructive aspect of our formalization.

Next, we proceed to prove the Noetherianity of the reduction relation. In the sequel, < represents the polynomial ordering whose well-foundedness was proved in [9][5]. This ordering can be used to state the Noetherianity of the polynomial reduction. For this purpose, it suffices to prove that the application of a valid operator to a polynomial produces a smaller polynomial with respect to this well-founded relation:

```
(defthm |validp(p, o, F) => reduction(p, o) < p|
  (implies (and (k-polynomialp p) (k-polynomialsp F))
           (implies (validp p o F) (< (reduction p o) p))))
```

As a consequence of Noetherianity we can define the notion of normal form.

Definition 5. *A polynomial p is in normal form or is irreducible w.r.t. $\to_F$ if there is no q such that $p \to_F q$. Otherwise, p is said to be reducible. A polynomial q is a normal form of p w.r.t. $\to_F$ if $p \to_F^* q$ and q is irreducible w.r.t. $\to_F$.*

---

[5] As it is customary in ACL2, this is proved by means of an ordinal embedding into ϵ_0.

The notion of normal form of a polynomial can be easily defined in our framework. First, we define a function `reducible`, implementing a reducibility test: when applied to a polynomial p and to a list of polynomials F, it returns a valid operator, whenever it exists, or `nil` otherwise. The following theorems state the main properties of `reducible`:

```
(defthm |reducible(p, F) => validp(p, reducible(p, F), F)|
  (implies (reducible p F)
           (validp p (reducible p F) F)))
```

```
(defthm |~reducible(p, F) => ~validp(p, o, F)|
  (implies (not (reducible p F))
           (not (validp p o F))))
```

Now it is easy to define a function `nf` that computes a normal form of a given polynomial with respect to the reduction relation induced by a given list of polynomials. This function is simple: it iteratively tests reducibility and applies valid operators until an irreducible polynomial is found. Note that termination is guaranteed by the Noetherianity of the reduction relation and the well-foundedness of the polynomial ordering.

```
(defun<k> nf (p F)
  (if (and (k-polynomialp p) (k-polynomialsp F))
      (let ((red (reducible p F)))
        (if red (nf (reduction p red) F) p))
    p))
```

The following theorems establish that, in fact, `nf` computes normal forms. Again, in order to prove that $p \to_F^* nf_F(p)$, we have to explicitly define a function `|p ->F* nf(p, F)|-proof` which construct a proof justifying this. This function is easily defined by collecting the operators returned by `reducible`.

```
(defthm |p ->F* nf(p, F)|
  (let ((proof (|p ->F* nf(p, F)|-proof p F)))
    (implies (and (k-polynomialp p) (k-polynomialsp F))
             (->* p (nf p F) proof F))))
```

```
(defthm |nf(p, F) irreducible|
  (implies (and (k-polynomialp p) (k-polynomialsp F))
           (not (validp (nf p F) o F))))
```

Although `nf` is suitable for reasoning about normal form computation, it is not suitable for being used by an implementation of Buchberger's algorithm: for example, `nf` explicitly deals with operators, which are a concept of theoretical nature. At this point, we talk about the polynomial reduction function red_F^* used in Buchberger's algorithm. This function (whose definition we omit) do not make any use of operators but is modeled from the closure of the set extension of another function, red, which takes two polynomials as its input and returns the result of reducing the first polynomial with respect to the second one. The following theorem shows the equivalence between nf_F and red_F^*:

```
(defthm |nf(p, F) = red*(p, F)|
   (implies (and (k-polynomialp p) (k-polynomialsp F))
            (equal (nf p F) (red* p F))))
```

With this result, we can translate all the properties proved about nf to red*. This is typical in our formalization: we use some functions for reasoning, and other functions for computing, translating the properties from one to another by proving equivalence theorems. For example, we proved the stability of the ideal with respect to red* using this technique.

5   Gröbner Bases

The computation of normal forms with respect to a given ideal can be seen as a generalized polynomial division algorithm, and the normal form computed as the "remainder" of that division. The ideal membership problem can be solved taking this into account: compute the normal form and check for the zero polynomial. Unfortunately, it is possible that, for a given basis F, a polynomial in $\langle F \rangle$ cannot be reduced to the zero polynomial. This is where Gröbner bases come into play:

Definition 6. *G is a Gröbner basis of the ideal generated by F if $\langle G \rangle = \langle F \rangle$ and $p \in \langle G \rangle \iff p \to_G^* 0$.*

The key point in Buchberger's algorithm is that the property of being a Gröbner basis can be deduced by only checking that a finite number of polynomials (called *s-polynomials*) are reduced to zero:

Definition 7. *Let p and q be polynomials. Let m, m_1 and m_2 be monomials such that $m = \text{lcm}(lm(p), lm(q))$ and $m_1 \cdot lm(p) = m = m_2 \cdot lm(q)$. The s-polynomial induced by p and q is defined as $s\text{-}poly(p, q) = m_1 \cdot p - m_2 \cdot q$*

Theorem 1. *Let $\Phi(F) \equiv \forall p, q \in F \; s\text{-}poly(p, q) \to_F^* 0$. The reduction induced by F is locally confluent if $\Phi(F)$ is verified. That is:*

$$\Phi(F) \implies \forall p, q, r \; (r \to_F p \wedge r \to_F q \implies \exists s \; (p \to_F^* s \wedge q \to_F^* s))$$

This theorem was the most difficult to formalize and prove in our work. First, note that it cannot be stated as a single theorem in the quantifier-free ACL2 logic, due to the universal quantifier in its hypothesis, $\Phi(F)$. For this reason, we state its hypothesis by the following encapsulate (we omit the local witnesses and some nonessential technical details):

```
(encapsulate
 ((F () t) (s-polynomial-proof (p q) t))
   ...
 (defthm |Phi(F)|
   (let ((proof (s-polynomial-proof p q)))
     (and (k-polynomials (F))
          (implies (and (in<> p (F)) (in<> q (F)))
                   (->* (s-poly p q) (|0|) proof (F)))))))
```

The first line of this `encapsulate` presents the *signature* of the functions it introduces, and the theorem inside can be seen as an assumed property about these functions. In this case, we are assuming that we have a list of polynomials given by the 0-ary function F, with the property that every s-polynomial formed with pairs of elements of (F) is reduced to (|0|). This reduction is justified by a function `s-polynomial-proof` computing the corresponding sequence of proof steps representing the reduction to (|0|). We insist that F and `s-polynomial-proof` are not completely defined: we are only assuming |Phi(F)| about them.

Now, the conclusion of Th. 1 is established as follows:

```
(defthm |Phi(F) => local-confluence(->F)|
  (let ((proof2 (transform-local-peak-F proof1)))
    (implies (and (k-polynomial p) (k-polynomial q)
                  (<->* p q proof1 (F)) (local-peakp proof1))
             (and (<->* p q proof2 (F)) (valleyp proof2)))))
```

This theorem needs some explanation. Note that local confluence can be reformulated in terms of the "shape" of the involved proofs: a reduction is locally confluent if, and only if, for every local peak proof (that is, of the form $p \leftarrow r \rightarrow q$) there exists an equivalent valley proof (that is, of the form $p \xrightarrow{*} s \xleftarrow{*} q$). It is easy to define in ACL2 the functions `local-peakp` and `valleyp`, checking those shapes of proofs. Note that again due to the absence of existential quantification, the valley proof in the above theorem is given by a function `transform-local-peak-F`, such that from a given local peak proof, it computes an equivalent valley proof. The definition of this function is very long and follows the same case distinction as in the classical proof of this result; only in one of its cases (the one dealing with "overlaps"), `s-polynomial-proof` is used as an auxiliary function, reflecting in this way where the assumption about $\Phi(F)$ is necessary.

The last step in this section follows from general results of abstract reduction relations. In particular, if a reduction is locally confluent and Noetherian then its induced equivalence can be decided by checking if normal forms are equal. This has been proved in ACL2 [11] as a consequence of Newman's lemma, also proved there. We can reuse this general result by functional instantiation and obtain an ACL2 proof of the fact that, if $\Phi(F)$, $p \xleftrightarrow{*}_F q \iff nf_F(p) = nf_F(q)$.

With this result, and using the equality between nf_F and red_F^*, and the equality between $\equiv_{\langle F \rangle}$ and $\xleftrightarrow{*}_F$, it can be easily deduced that if $\Phi(F)$ then F is a Gröbner basis (of $\langle F \rangle$). This is established by the following theorem (notice that (F) is still the list of polynomials assumed to have property Φ by the above `encapsulate`):

```
(defthm |Phi(F) => (p in <F> <=> red*(p, F) = 0)|
    (implies (k-polynomial p)
             (iff (in<> p (F)) (equal (red* p (F)) (|0|)))))
```

6   Buchberger's Algorithm

Buchberger's algorithm obtains a Gröbner basis of a given finite set of polynomials F by the following procedure: if there is a s-polynomial of F such that its normal form is not zero, then this normal form can be added to the basis. This makes it reducible to zero (without changing the ideal), but new s-polynomials are introduced that have to be checked. This *completion* process is iterated until all the s-polynomials of the current basis are reducible to zero.

In order to formalize an executable implementation of Buchberger's algorithm in ACL2, several helper functions are needed. The function `initial-pairs` returns all the ordered pairs from the elements of a list. The main function computes the initial pairs from a basis and starts the real computation process.

```
(defun Buchberger (F)
  (Buchberger-aux F (initial-pairs F)))
```

Next, the function that computes a Gröbner basis from an initial basis is defined. This function takes the initial basis and a list of pairs as its input. The function `pairs` returns the ordered pairs built from its first argument and every element in its second argument. As all ACL2 functions must be total and we need to deal with polynomials with a fixed set of variables to ensure termination of the function, we have to explicitly check that the arguments remain in the correct domain. We will comment more about these "type conditions" in Sect. 7.

```
(defun<k> Buchberger-aux (F C)
  (if (and (naturalp k) (k-polynomialsp F) (k-polynomial-pairsp C))
      (if (endp C)
          F
        (let* ((p (first (first C))) (q (second (first C)))
               (h (red* (s-poly p q) F)))
          (if (equal h (|0|))
              (Buchberger-aux F (rest C))
            (Buchberger-aux (cons h F) (append (pairs h F) (rest C))))))
    F))
```

A measure has to be supplied to prove the termination of the above function, so that it can be admitted by the principle of definition. The following section explains this issue.

6.1 Termination

Termination of Buchberger's algorithm can be proved using a lexicographic measure on its arguments. This is justified by the following observations:

1. In the first recursive branch, the first argument keeps unmodified while the second argument structurally decreases since one of its elements is removed.
2. In the second recursive branch, the first argument decreases in a certain well-founded sense despite of the inclusion of a new polynomial. This is a consequence of Dickson's lemma.

Lemma 1 (Dickson). *Let $k \in \mathbb{N}$ and $m_1, m_2, \ldots$ an infinite sequence of monomials with k variables. Then, there exist indices $i < j$ such that m_i divides m_j.*

If we consider the sequence of terms consisting of the leader terms of the polynomials added to the first argument, Dickson's lemma implies termination of Buchberger's algorithm. This is because the polynomial added to the basis, h, is not 0 and it cannot be reduced by F. Consequently, its leader term is not divisible by *any* of the leader terms of the polynomials in F.

Dickson's lemma has been formalized in ACL2 in [7] and [12]. In both cases it has been proved by providing an ordinal measure on finite sequences of terms such that this measure decreases every time a new term not divisible by any of the previous terms in the sequence is added.

We have defined a measure along these lines to prove the termination of Buchberger's algorithm. In fact, our measure is defined on top of the measures used to prove Dickson's lemma in [7, 12], lexicographically combined with the length of the second argument. Although both proofs of Dickson's lemma are based on totally different ideas, the results obtained can be used interchangeably in our formalization.

6.2   Partial Correctness

In order to show that *Buchberger* computes a Gröbner basis, and taking into account the results of the previous section, we just have to prove that $p \in \langle F \rangle \iff p \in \langle Buchberger(F) \rangle$ and that $Buchberger(F)$ satisfies Φ. The following ACL2 theorems establish these two properties:

```
(defthm |<Buchberger(F)> = <F>|
  (implies (and (k-polynomialp p) (k-polynomialsp F))
           (iff (in<> p (Buchberger F)) (in<> p F))))

(defthm |Phi(Buchberger(F))|
  (let ((G (Buchberger F)) (proof (|Phi(Buchberger(F))|-proof p q F)))
    (implies (and (k-polynomialp p) (k-polynomialp q) (k-polynomialsp F)
                  (in<> p G) (in<> q G))
             (->* (s-poly p q) (|0|) proof G))))
```

The statement of this last theorem deserves some comments. Our ACL2 formulation of Th. 1 defines the property $\Phi(F)$ as the existence of a function such that for every s-polynomial of F, it computes a sequence of proof steps justifying its reduction to (|0|) (assumption |Phi(F)| in the encapsulate of the previous section). Thus, if we want to establish the property Φ for a particular basis (the basis returned by Buchberger in this case), we must explicitly define such function and prove that it returns the desired proofs for every s-polynomial of the basis. In this case the function is called |Phi(Buchberger(F))|-proof. For the sake of brevity, we omit the definition of this function, but it is very interesting to point out that it is based in a recursion scheme very similar to the recursive definition of Buchberger-aux. This function collects, every time a new s-polynomial is examined, the corresponding proof justifying its reduction to the zero polynomial.

6.3 Deciding Ideal Membership

Finally, we can compile the results above to define a decision procedure for ideal membership. This procedure just checks whether a given polynomial reduces to 0 with respect to the Gröbner basis returned by Buchberger's algorithm.

```
(defun<k> imdp (p F)
  (equal (red* p (Buchberger F)) (|0|)))
```

Theorem 2. $G = Buchberger(F) \implies (p \in \langle F \rangle \iff red_G^*(p) = 0)$.

The ACL2 theorem stating the soundness and completeness of the decision procedure follows, as an easy consequence of the correctness of Buchberger and the theorem |Phi(F) => (p in <F> <=> red*(p, F) = 0)| in Sect. 5.

```
(defthm |p en <F> <=> imdp(p, F)|
    (implies (and (k-polynomialp p) (k-polynomialsp F))
             (iff (in<> p F) (imdp p F))))
```

In this context, the theorem |Phi(F) => (p in <F> <=> red*(p, F) = 0)| is used by functional instantiation, replacing F by (lambda () (Buchberger F)) and s-polynomial-proof by |Phi(Buchberger(F))|-proof.

Note that all the functions used in the definition of the decision procedure are executable and therefore the procedure is also executable. Note also that we do not mention operators or proofs, neither when defining the decision procedure nor when stating its correctness. These are only intermediate concepts, which make reasoning more convenient.

7 Conclusions

We have shown how it is possible to use the ACL2 system in the formal development of Computer Algebra algorithms by presenting a verified implementation of Buchberger's algorithm and a verified decision procedure for the ideal membership problem. It is interesting to point out that all the theory needed to prove the correctness of the algorithm has been developed in the ACL2 logic, in spite of its (apparently) limited expressiveness.

We have benefited from work previously done in the system. In particular, all the results about abstract reductions were originally developed for a formalization of rewriting systems [11]. We believe that this is a good example of how seemingly unrelated formalizations can be reused in other projects, provided the system offers a minimal support for it. However, we feel that ACL2 could be improved to provide more comfortable mechanisms for functional instantiation and for abstraction in general. Encapsulation provides a good abstraction mechanism but functionally instantiating each encapsulated theorem is a tedious and error-prone task. Recently, several proposals have been formulated (e.g. polymorphism and abstract data types) to cope with this problem in ACL2 [6]. A graphical interface to visualize proof trees would be helpful too.

---

[6] A similar modularity issue has been reported in the CoQ system too [13].

Little work has been done on the machine verification of Buchberger's algorithm. As we mentioned in the introduction, the most relevant is the work of L. Théry [13] in COQ. T. Coquand and H. Persson [3] report an incomplete integrated development in Martin-Löf's type theory using AGDA. There is also a MIZAR project [10] to formalize Gröbner bases. The main difference between our approach and these works is the underlying logic. All these logics are very different from ACL2, which is more primitive, basically an untyped and quantifier-free logic of total recursive functions, and makes no distinction between the programming and the specification languages. In exchange, a high degree of automation can be achieved and executability is obtained for free.

We think that an advantage of our approach is that the implementation presented is compliant with COMMON LISP, a real programming language, and can be directly executed in ACL2 or in any compliant COMMON LISP. This is not the case of other systems, where the logic is not executable at all or the code has to be extracted by unverified means. Taking into account that LISP is the language of choice for the implementation of CAS, like MACSYMA and AXIOM, this is not just a matter of theoretical importance but also a practical one.

Our formal proof differs from Théry's. First, it is based on [1] instead of [4]. Second, we prove that Φ implies local-confluence instead of confluence: compare this with the proof of *SpolyImpConf* in [13]. Differences extend also to definitions, e.g. ideals and confluence, mainly motivated by the lack of existential quantification. Finally, [13] uses a non-constructive proof of Dickson's lemma by L. Pottier[7]. Our termination argument uses a proof of Dickson's lemma obtained by an ordinal embedding in ϵ_0, the only well-founded structure known to ACL2.

We would like to remark that although polynomial properties seem trivial to prove, this is not the case [8, 9]. It seems that this is not due to the simplicity of the ACL2 logic. In [10] the authors recognize that it was challenging to prove the associativity of polynomial multiplication in MIZAR, a system devoted to the formalization of mathematics. They were amazed by the fact that, in well-known Algebra treatises, these properties are usually reduced to the univariate case or their proofs are partially sketched and justified "by analogy". In some cases, the proofs are even left as an exercise. In the same way, the author of [13] had to devote a greater effort to polynomials due to problems arising during their formalization in COQ.

As for the user interaction required, we provided 169 definitions and 560 lemmas to develop a theory of polynomials (although this includes more than the strictly needed here) and 109 definitions and 346 lemmas for the theory of Gröbner bases and Buchberger's algorithm. All these lemmas are proved almost automatically. It is worth pointing out that of the 333 lemmas proved by induction, only 24 required a user-supplied induction scheme. Other lemmas needed a hint about the convenience of using a given instance of another lemma or keeping a function definition unexpanded. Only 9 functions required a hint for their termination proofs. Thus, the main role of the user is to provide the suitable sequence of definitions and lemmas to achieve the final correctness theorem.

---

[7] A new proof of Dickson's lemma in COQ by H. Persson has been proposed later.

Théry's implementation provides some standard optimizations that we do not include. Regarding future work, we are interested in studying how our verified implementation could be improved to incorporate some of the refinements built into the very specialized and optimized (and not formally verified) versions used in industrial-strength applications. It would be also interesting to use it to verify some application of Gröbner bases such as those described in [2].

An obvious improvement in the verified implementation is to avoid the "type conditions" in the body of `Buchberger-aux`, since these conditions are unnecessarily evaluated in every recursive call. But these conditions are needed to ensure termination. Until ACL2 version 2.7, there was no way to avoid this; but since the recent advent of ACL2 version 2.8 that is no longer true, since it is possible for a function to have two different bodies, one used for execution and another for its logical definition: this is done by previously proving that both bodies behave in the same way on the intended domain of the function. We plan to apply this new feature to our definition of Buchberger's algorithm.

References

1. Baader, F. & Nipkow, T.: *Term Rewriting and All That*. Cambridge University Press (1998)
2. Buchberger, B. & Winkler, F. (eds.): *Gröbner Bases and Applications*. London Mathematical Society Series **251** (1998)
3. Coquand, T. & Persson, H.: *Gröbner Bases in Type Theory*. Types for Proofs and Programs, International Workshop. LNCS **1657**:33–46 (1999)
4. Geddes, K. O.; Czapor, S. R. & Labahn, G.: *Algorithms for Computer Algebra*. Kluwer (1998)
5. Kaufmann, M.; Manolios, P. & Moore, J S.: *Computer-Aided Reasoning: An Approach*. Kluwer (2000)
6. Kaufmann, M. & Moore, J S.: *Structured Theory Development for a Mechanized Logic*. Journal of Automated Reasoning **26**(2):161–203 (2001)
7. Martín, F. J.; Alonso, J. A.; Hidalgo, M. J. & Ruiz, J. L.: *A Formal Proof of Dickson's Lemma in ACL2*. Logic for Programming, Artificial Intelligence and Reasoning. LNAI **2850**:49-58 (2003)
8. Medina, I.; Alonso, J. A. & Palomo, F.: *Automatic Verification of Polynomial Rings Fundamental Properties in ACL2*. ACL2 Workshop 2000. Department of Computer Sciences, University of Texas at Austin. TR–00–29 (2000)
9. Medina, I.; Palomo, F. & Alonso, J. A.: *Implementation in ACL2 of Well-Founded Polynomial Orderings*. ACL2 Workshop 2002 (2002)
10. Rudnicki, P.; Schwarzweller, C. & Trybulec, A.: *Commutative Algebra in the Mizar System*. Journal of Symbolic Computation **32**(1–2):143–169 (2001)
11. Ruiz, J. L.; Alonso, J. A.; Hidalgo, M. J. & Martín, F. J.: *Formal Proofs about Rewriting using ACL2*. Annals of Mathematics and Artificial Intelligence **36**(3): 239–262 (2002)
12. Sustyk, M.: *Proof of Dickson's Lemma Using the ACL2 Theorem Prover via an Explicit Ordinal Mapping*. Fourth International Workshop on the ACL2 Theorem Prover and Its Applications (2003)
13. Théry, L.: *A Machine-Checked Implementation of Buchberger's Algorithm*. Journal of Automated Reasoning **26**(2): 107–137 (2001)

Polynomial Interpretations
with Negative Coefficients

Nao Hirokawa and Aart Middeldorp

Institute of Computer Science, University of Innsbruck, 6020 Innsbruck, Austria
{nao.hirokawa,aart.middeldorp}@uibk.ac.at

Abstract. Polynomial interpretations are a useful technique for proving termination of term rewrite systems. We show how polynomial interpretations with negative coefficients, like $x - 1$ for a unary function symbol or $x - y$ for a binary function symbol, can be used to extend the class of rewrite systems that can be automatically proved terminating.

1 Introduction

This paper is concerned with automatically proving termination of first-order rewrite systems by means of polynomial interpretations. In the classical approach, which goes back to Lankford [16], one associates with every n-ary function symbol f a polynomial P_f over the natural numbers in n indeterminates, which induces a mapping from terms to polynomials in the obvious way. Then one shows that for every rewrite rule $l \rightarrow r$ the polynomial P_l associated with the left-hand side l is strictly greater than the polynomial P_r associated with the right-hand side r, i.e., $P_l - P_r > 0$ for all values of the indeterminates. In order to conclude termination, the polynomial P_f associated with an n-ary function symbol f must be strictly monotone in all n indeterminates. Techniques for finding appropriate polynomials as well as approximating (in general undecidable) polynomial inequalities $P > 0$ are described in several papers (e.g. [4, 6, 9, 15, 19]). As a simple example, consider the rewrite rules

$$x + 0 \rightarrow x \qquad\qquad x \times 0 \rightarrow 0$$
$$x + \mathsf{s}(y) \rightarrow \mathsf{s}(x + y) \qquad\qquad x \times \mathsf{s}(y) \rightarrow (x \times y) + x$$

Termination can be shown by the strictly monotone polynomial interpretations

$$\times_{\mathbb{N}}(x, y) = 2xy + y + 1 \qquad +_{\mathbb{N}}(x, y) = x + 2y \qquad \mathsf{s}_{\mathbb{N}}(x) = x + 1 \qquad 0_{\mathbb{N}} = 1$$

over the natural numbers:

$$x + 2 > x \qquad\qquad 2x + 2 > 1$$
$$x + 2y + 2 > x + 2y + 1 \qquad 2xy + 2x + y + 2 > 2xy + 2x + y + 1$$

Compared to other classical methods for proving termination of rewrite systems (like recursive path orders and Knuth-Bendix orders), polynomial interpretations are rather weak. Numerous natural examples cannot be handled because

B. Buchberger and J.A. Campbell (Eds.): AISC 2004, LNAI 3249, pp. 185–198, 2004.
© Springer-Verlag Berlin Heidelberg 2004

of the strict monotonicity requirement which precludes interpretations like $x + 1$ for binary function symbols. In connection with the dependency pair method of Arts and Giesl [1], polynomial interpretations become much more useful because strict monotonicity is no longer required; weak monotonicity is sufficient and hence $x + 1$ or even 0 as interpretation of a binary function symbol causes no problems. Monotonicity is typically guaranteed by demanding that all coefficients are positive.

In this paper we go a step further. We show that polynomial interpretations over the integers with negative coefficients like $x - 1$ and $x - y + 1$ can also be used for termination proofs. To make the discussion more concrete, let us consider a somewhat artificial example: the recursive definition

$$f(x) = \text{if } x > 0 \text{ then } f(f(x-1)) + 1 \text{ else } 0$$

from [8]. It computes the identity function over the natural numbers. Termination of the rewrite system

$$1: \ \mathsf{f}(\mathsf{s}(x)) \to \mathsf{s}(\mathsf{f}(\mathsf{f}(\mathsf{p}(\mathsf{s}(x))))) \qquad 2: \ \mathsf{f}(0) \to 0 \qquad 3: \ \mathsf{p}(\mathsf{s}(x)) \to x$$

obtained after the obvious translation is not easily proved. The (manual) proof in [8] relies on forward closures whereas powerful automatic tools like AProVE [11] and CiME [5] that incorporate both polynomial interpretations and the dependency pair method fail to prove termination. There are three dependency pairs (here $\mathsf{f}^\sharp$ and $\mathsf{p}^\sharp$ are new function symbols):

$$4: \ \mathsf{f}^\sharp(\mathsf{s}(x)) \to \mathsf{f}^\sharp(\mathsf{f}(\mathsf{p}(\mathsf{s}(x)))) \quad 5: \ \mathsf{f}^\sharp(\mathsf{s}(x)) \to \mathsf{f}^\sharp(\mathsf{p}(\mathsf{s}(x))) \quad 6: \ \mathsf{f}^\sharp(\mathsf{s}(x)) \to \mathsf{p}^\sharp(\mathsf{s}(x))$$

By taking the *natural* polynomial interpretation

$$\mathsf{f}_{\mathbb{Z}}(x) = \mathsf{f}_{\mathbb{Z}}^\sharp(x) = x \qquad \mathsf{s}_{\mathbb{Z}}(x) = x + 1 \qquad 0_{\mathbb{Z}} = 0 \qquad \mathsf{p}_{\mathbb{Z}}(x) = \mathsf{p}_{\mathbb{Z}}^\sharp(x) = x - 1$$

over the integers, the rule and dependency pair constraints reduce to the following inequalities:

$$1: \ x + 1 \geqslant x + 1 \qquad 3: \qquad x \geqslant x \qquad 5: \ x + 1 > x$$
$$2: \qquad 0 \geqslant 0 \qquad 4: \ x + 1 > x \qquad 6: \ x + 1 > x$$

These constraints are obviously satisfied. The question is whether we are allowed to conclude termination at this point. We will argue that the answer is affirmative and, moreover, that the search for appropriate natural polynomial interpretations can be efficiently implemented.

The approach described in this paper is inspired by the combination of the general path order and forward closures [8] as well as semantic labelling [24]. Concerning related work, Lucas [17, 18] considers polynomials with *real* coefficients for automatically proving termination of (context-sensitive) rewriting systems. He solves the problem of well-foundedness by replacing the standard order on $\mathbb{R}$ with $>_\delta$ for some fixed positive $\delta \in \mathbb{R}$: $x >_\delta y$ if and only if $x - y \geqslant \delta$. In addition, he demands that interpretations are uniformly bounded from below (i.e., there exists an $m \in \mathbb{R}$ such that $f_{\mathbb{R}}(x_1, \ldots, x_n) \geqslant m$ for all function symbols f and $x_1, \ldots, x_n \geqslant m$). The latter requirement entails that interpretations like $x - 1$ or $x - y + 1$ cannot be handled.

The remainder of the paper is organized as follows. In Section 3 we discuss polynomial interpretations with negative constants. Polynomial interpretations with negative coefficients require a different approach, which is detailed in Section 4. In Section 5 we discuss briefly how to find suitable polynomial interpretations automatically and we report on the many experiments that we performed.

2 Preliminaries

We assume familiarity with the basics of term rewriting [3, 21] and with the dependency pair method [1] for proving (innermost) termination. In the latter method a term rewrite system (TRS for short) is transformed into a collection of ordering constraints of the form $l \gtrsim r$ and $l > r$ that need to be solved in order to conclude termination. Solutions $(\gtrsim, >)$ must be *reduction pairs* which consist of a rewrite preorder $\gtrsim$ (i.e., a transitive and reflexive relation which is closed under contexts and substitutions) on terms and a compatible well-founded order $>$ which is closed under substitutions. Compatibility means that the inclusion $\gtrsim \cdot > \subseteq >$ or the inclusion $> \cdot \gtrsim \subseteq >$ holds. (Here $\cdot$ denotes relational composition.)

A general semantic construction of reduction pairs, which covers traditional polynomial interpretations, is based on the concept of algebra. If we equip the carrier A of an $\mathcal{F}$-algebra $\mathcal{A} = (A, \{f_A\}_{f \in \mathcal{F}})$ with a well-founded order $>$ such that every interpretation function is weakly monotone in all arguments (i.e., $f_A(x_1, \ldots, x_n) \geqslant f_A(y_1, \ldots, y_n)$ whenever $x_i \geqslant y_i$ for all $1 \leqslant i \leqslant n$, for every n-ary function symbol $f \in \mathcal{F}$) then $(\geqslant_A, >_A)$ is a reduction pair. Here the relations $\geqslant_A$ and $>_A$ are defined as follows: $s \geqslant_A t$ if $[\alpha]_A(s) \geqslant [\alpha]_A(t)$ and $s >_A t$ if $[\alpha]_A(s) > [\alpha]_A(t)$, for all assignments α of elements of A to the variables in s and t ($[\alpha]_A(\cdot)$ denotes the usual evaluation function associated with the algebra $\mathcal{A}$). In general, the relation $>_A$ is not closed under contexts, $\geqslant_A$ is a preorder but not a partial order, and $>_A$ is not the strict part of $\geqslant_A$. Compatibility holds because of the identity $\geqslant_A \cdot >_A = >_A$. We write $s =_A t$ if $[\alpha]_A(s) = [\alpha]_A(t)$ for all assignments α. We say that $\mathcal{A}$ is a model for a TRS $\mathcal{R}$ if $l =_A r$ for all rewrite rules in $\mathcal{R}$.

In this paper we use the following results from [10] concerning dependency pairs.

Theorem 1. *A TRS $\mathcal{R}$ is terminating if for every cycle $\mathcal{C}$ in its dependency graph there exists a reduction pair $(\gtrsim, >)$ such that $\mathcal{R} \subseteq \gtrsim$, $\mathcal{C} \subseteq \gtrsim \cup >$, and $\mathcal{C} \cap > \neq \varnothing$.*   □

Theorem 2. *A TRS $\mathcal{R}$ is innermost terminating if for every cycle $\mathcal{C}$ in its innermost dependency graph there exists a reduction pair $(\gtrsim, >)$ such that $\mathcal{U}(\mathcal{C}) \subseteq \gtrsim$, $\mathcal{C} \subseteq \gtrsim \cup >$, and $\mathcal{C} \cap > \neq \varnothing$.*   □

3   Negative Constants

3.1   Theoretical Framework

When using polynomial interpretations with negative constants like in the example of the introduction, the first challenge we face is that the standard order $>$ on $\mathbb{Z}$ is not well-founded. Restricting the domain to the set $\mathbb{N}$ of natural numbers makes an interpretation like $p_{\mathbb{Z}}(x) = x - 1$ ill-defined. Dershowitz and Hoot observe in [8] that if all (instantiated) subterms in the rules of the TRS are interpreted as non-negative integers, such interpretations can work correctly. Following their observation, we propose to modify the interpretation of p to $p_{\mathbb{N}}(x) = \max\{0, x - 1\}$.

Definition 3. *Let $\mathcal{F}$ be a signature and let $(\mathbb{Z}, \{f_{\mathbb{Z}}\}_{f \in \mathcal{F}})$ be an $\mathcal{F}$-algebra such that every interpretation function $f_{\mathbb{Z}}$ is weakly monotone in all its arguments. The interpretation functions of the induced algebra $(\mathbb{N}, \{f_{\mathbb{N}}\}_{f \in \mathcal{F}})$ are defined as follows: $f_{\mathbb{N}}(x_1, \ldots, x_n) = \max\{0, f_{\mathbb{Z}}(x_1, \ldots, x_n)\}$ for all $x_1, \ldots, x_n \in \mathbb{N}$.*

With respect to the interpretations in the introduction we obtain $s_{\mathbb{N}}(p_{\mathbb{N}}(x)) = \max\{0, \max\{0, x - 1\} + 1\} = \max\{0, x - 1\} + 1$, $p_{\mathbb{N}}(0_{\mathbb{N}}) = \max\{0, 0\} = 0$, and $p_{\mathbb{N}}(s_{\mathbb{N}}(x)) = \max\{0, \max\{0, x + 1\} - 1\} = x$.

Lemma 4. *If $(\mathbb{Z}, \{f_{\mathbb{Z}}\}_{f \in \mathcal{F}})$ is an $\mathcal{F}$-algebra with weakly monotone interpretations then $(\geqslant_{\mathbb{N}}, >_{\mathbb{N}})$ is a reduction pair.*

Proof. It is easy to show that the interpretation functions of the induced algebra are weakly monotone in all arguments. Routine arguments reveal that the relation $>_{\mathbb{N}}$ is a well-founded order which is closed under substitutions and that $\geqslant_{\mathbb{N}}$ is a preorder closed under contexts and substitutions. Moreover, the identity $>_{\mathbb{N}} \cdot \geqslant_{\mathbb{N}} \; = \; >_{\mathbb{N}}$ holds. Hence $(\geqslant_{\mathbb{N}}, >_{\mathbb{N}})$ is a reduction pair.   □

It is interesting to remark that unlike usual polynomial interpretations, the relation $>_{\mathbb{N}}$ does not have the (weak) subterm property. For instance, with respect to the interpretations in the example of the introduction, we have $s(0) >_{\mathbb{N}} p(s(0))$ and not $p(s(0)) >_{\mathbb{N}} p(0)$.

In recent modular refinements of the dependency pair method [23, 13, 22] suitable reduction pairs $(\gtrsim, >)$ have to satisfy the additional property of $\mathcal{C}_{\mathcal{E}}$-*compatibility*: $\gtrsim$ must orient the rules of the TRS $\mathcal{C}_{\mathcal{E}}$ consisting of the two rewrite rules $\mathsf{cons}(x, y) \rightarrow x$ and $\mathsf{cons}(x, y) \rightarrow y$, where cons is a fresh function symbol, from left to right. This is not a problem because we can simply define $\mathsf{cons}_{\mathbb{N}}(x, y) = \max\{x, y\}$. In this way we obtain a reduction pair $(\succsim, \succ)$ on terms over the original signature extended with cons such that $\gtrsim \cup \, \mathcal{C}_{\mathcal{E}} \subseteq \succsim$ and $> \; \subseteq \; \succ$.

Example 5. Consider the TRS consisting of the following rewrite rules:

1:	$\mathsf{half}(0) \rightarrow 0$	4:	$\mathsf{bits}(0) \rightarrow 0$
2:	$\mathsf{half}(s(0)) \rightarrow 0$	5:	$\mathsf{bits}(s(x)) \rightarrow s(\mathsf{bits}(\mathsf{half}(s(x))))$
3:	$\mathsf{half}(s(s(x))) \rightarrow s(\mathsf{half}(x))$		

The function $\mathsf{half}(x)$ computes $\lceil \frac{x}{2} \rceil$ and $\mathsf{bits}(x)$ computes the number of bits that are needed to represent all numbers less than or equal to x. Termination of this TRS is proved in [2] by using the dependency pair method together with the narrowing refinement. There are three dependency pairs:

$$6: \quad \mathsf{half}^\sharp(s(s(x))) \to \mathsf{half}^\sharp(x)$$
$$7: \quad \mathsf{bits}^\sharp(s(x)) \to \mathsf{bits}^\sharp(\mathsf{half}(s(x)))$$
$$8: \quad \mathsf{bits}^\sharp(s(x)) \to \mathsf{half}^\sharp(s(x))$$

By taking the interpretations $0_\mathbb{Z} = 0$, $\mathsf{half}_\mathbb{Z}(x) = x - 1$, $\mathsf{bits}_\mathbb{Z}(x) = \mathsf{half}_\mathbb{Z}^\sharp(x) = x$, and $s_\mathbb{Z}(x) = \mathsf{bits}_\mathbb{Z}^\sharp(x) = x + 1$, we obtain the following constraints over $\mathbb{N}$:

1:	$0 \geqslant 0$		5:	$x + 1 \geqslant x + 1$
2:	$0 \geqslant 0$		6:	$x + 2 > x$
3:	$x + 1 \geqslant \max\{0, x - 1\} + 1$		7:	$x + 2 > x + 1$
4:	$0 \geqslant 0$		8:	$x + 2 > x + 1$

These constraints are satisfied, so the TRS is terminating, but how can an inequality like $x + 1 \geqslant \max\{0, x - 1\} + 1$ be verified automatically?

3.2   Towards Automation

Because the inequalities resulting from interpretations with negative constants may contain the max operator, we cannot use standard techniques for comparing polynomial expressions. In order to avoid reasoning by case analysis ($x - 1 > 0$ or $x - 1 \leqslant 0$ for constraint 3 in Example 5), we approximate the evaluation function of the induced algebra.

Definition 6. *Given a polynomial P with coefficients in $\mathbb{Z}$, we denote the constant part by $c(P)$ and the non-constant part $P - c(P)$ by $n(P)$. Let $(\mathbb{Z}, \{f_\mathbb{Z}\}_{f \in \mathcal{F}})$ be an $\mathcal{F}$-algebra such that every $f_\mathbb{Z}$ is a weakly monotone polynomial. With every term t we associate polynomials $P_{left}(t)$ and $P_{right}(t)$ with coefficients in $\mathbb{Z}$ and variables in t as indeterminates:*

$$P_{left}(t) = \begin{cases} t & \text{if } t \text{ is a variable} \\ 0 & \text{if } t = f(t_1, \ldots, t_n),\ n(P_1) = 0,\ \text{and } c(P_1) < 0 \\ P_1 & \text{otherwise} \end{cases}$$

where $P_1 = f_\mathbb{Z}(P_{left}(t_1), \ldots, P_{left}(t_n))$ and

$$P_{right}(t) = \begin{cases} t & \text{if } t \text{ is a variable} \\ n(P_2) & \text{if } t = f(t_1, \ldots, t_n) \text{ and } c(P_2) < 0 \\ P_2 & \text{otherwise} \end{cases}$$

where $P_2 = f_\mathbb{Z}(P_{right}(t_1), \ldots, P_{right}(t_n))$. Let $\alpha \colon \mathcal{V} \to \mathbb{N}$ be an assignment. The result of evaluating $P_{left}(t)$ and $P_{right}(t)$ under α is denoted by $[\alpha]_\mathbb{Z}^l(t)$ and $[\alpha]_\mathbb{Z}^r(t)$. The result of evaluating a polynomial P under α is denoted by $\alpha(P)$.

According the following lemma, $P_{left}(t)$ is a lower bound and $P_{right}(t)$ is an upper bound of the interpretation of t in the induced algebra.

Lemma 7. *Let* $(\mathbb{Z}, \{f_{\mathbb{Z}}\}_{f \in \mathcal{F}})$ *be an $\mathcal{F}$-algebra such that every $f_{\mathbb{Z}}$ is a weakly monotone polynomial. Let t be a term. For every assignment $\alpha \colon \mathcal{V} \to \mathbb{N}$ we have*
$$[\alpha]_{\mathbb{Z}}^r(t) \geqslant [\alpha]_{\mathbb{N}}(t) \geqslant [\alpha]_{\mathbb{Z}}^l(t).$$

Proof. By induction on the structure of t. If $t \in \mathcal{V}$ then $[\alpha]_{\mathbb{Z}}^r(t) = [\alpha]_{\mathbb{Z}}^l(t) = \alpha(t) = [\alpha]_{\mathbb{N}}(t)$. Suppose $t = f(t_1, \ldots, t_n)$. According to the induction hypothesis, $[\alpha]_{\mathbb{Z}}^r(t_i) \geqslant [\alpha]_{\mathbb{N}}(t_i) \geqslant [\alpha]_{\mathbb{Z}}^l(t_i)$ for all i. Since $f_{\mathbb{Z}}$ is weakly monotone,
$$f_{\mathbb{Z}}([\alpha]_{\mathbb{Z}}^r(t_1), \ldots, [\alpha]_{\mathbb{Z}}^r(t_n)) \geqslant f_{\mathbb{Z}}([\alpha]_{\mathbb{N}}(t_1), \ldots, [\alpha]_{\mathbb{N}}(t_n)) \geqslant f_{\mathbb{Z}}([\alpha]_{\mathbb{Z}}^l(t_1), \ldots, [\alpha]_{\mathbb{Z}}^l(t_n))$$
By applying the weakly monotone function $\max\{0, \cdot\}$ we obtain $\max\{0, \alpha(P_2)\} \geqslant [\alpha]_{\mathbb{N}}(t) \geqslant \max\{0, \alpha(P_1)\}$ where $P_1 = f_{\mathbb{Z}}(P_{left}(t_1), \ldots, P_{left}(t_n))$ and $P_2 = f_{\mathbb{Z}}(P_{right}(t_1), \ldots, P_{right}(t_n))$. We have
$$[\alpha]_{\mathbb{Z}}^l(t) = \begin{cases} 0 & \text{if } n(P_1) = 0 \text{ and } c(P_1) < 0 \\ \alpha(P_1) & \text{otherwise} \end{cases}$$
and thus $[\alpha]_{\mathbb{Z}}^l(t) \leqslant \max\{0, \alpha(P_1)\}$. Likewise,
$$[\alpha]_{\mathbb{Z}}^r(t) = \begin{cases} \alpha(n(P_2)) & \text{if } c(P_2) < 0 \\ \alpha(P_2) & \text{otherwise} \end{cases}$$
In the former case, $\alpha(n(P_2)) = \alpha(P_2) - c(P_2) > \alpha(P_2)$ and $\alpha(n(P_2)) \geqslant 0$. In the latter case $\alpha(P_2) \geqslant 0$. So in both cases we have $[\alpha]_{\mathbb{Z}}^r(t) \geqslant \max\{0, \alpha(P_2)\}$. Hence we obtain the desired inequalities. □

Corollary 8. *Let* $(\mathbb{Z}, \{f_{\mathbb{Z}}\}_{f \in \mathcal{F}})$ *be an $\mathcal{F}$-algebra such that every $f_{\mathbb{Z}}$ is a weakly monotone polynomial. Let s and t be terms. If $P_{left}(s) - P_{right}(t) > 0$ then $s >_{\mathbb{N}} t$. If $P_{left}(s) - P_{right}(t) \geqslant 0$ then $s \geqslant_{\mathbb{N}} t$.* □

Example 9. Consider again the TRS of Example 5. By applying P_{left} to the left-hand sides and P_{right} to the right-hand sides of the rewrite rules and the dependency pairs, the following ordering constraints are obtained:

1: $0 \geqslant 0$	3: $x + 1 \geqslant x + 1$	5: $x + 1 \geqslant x + 1$	7: $x + 2 > x + 1$
2: $0 \geqslant 0$	4: $\quad\ 0 \geqslant 0$	6: $x + 2 > x$	8: $x + 2 > x + 1$

The only difference with the constraints in Example 5 is the interpretation of the term $\mathsf{s}(\mathsf{half}(x))$ on the right-hand side of rule 3. We have $P_{right}(\mathsf{half}(x)) = n(x - 1) = x$ and thus $P_{right}(\mathsf{s}(\mathsf{half}(x))) = x + 1$. Although $x + 1$ is less precise than $\max\{0, x - 1\} + 1$, it is accurate enough to solve the ordering constraint resulting from rule 3.

So once the interpretations $f_{\mathbb{Z}}$ are determined, we transform a rule $l \to r$ into the polynomial $P_{left}(l) - P_{right}(r)$. Standard techniques can then be used to test whether this polynomial is positive (or non-negative) for all values in $\mathbb{N}$ for the variables. The remaining question is how to find suitable interpretations for the function symbols. This problem will be discussed in Section 5.

4   Negative Coefficients

Let us start with an example which shows that negative coefficients in polynomial interpretations can be useful.

Example 10. Consider the following variation of a TRS in [2]:

$$
\begin{array}{llll}
1: & 0 \leqslant x \to \mathsf{true} & 7: & x - 0 \to x \\
2: & \mathsf{s}(x) \leqslant 0 \to \mathsf{false} & 8: & \mathsf{s}(x) - \mathsf{s}(y) \to x - y \\
3: & \mathsf{s}(x) \leqslant \mathsf{s}(y) \to x \leqslant y & 9: & \mathsf{if}(\mathsf{true}, x, y) \to x \\
4: & \mathsf{mod}(0, \mathsf{s}(y)) \to 0 & 10: & \mathsf{if}(\mathsf{false}, x, y) \to y \\
5: & \mathsf{mod}(\mathsf{s}(x), 0) \to 0 & & \\
6: & \mathsf{mod}(\mathsf{s}(x), \mathsf{s}(y)) \to \mathsf{if}(y \leqslant x, \mathsf{mod}(\mathsf{s}(x) - \mathsf{s}(y), \mathsf{s}(y)), \mathsf{s}(x)) & &
\end{array}
$$

There are 6 dependency pairs:

$$
\begin{array}{ll}
11: & \mathsf{s}(x) \leqslant^{\sharp} \mathsf{s}(y) \to x \leqslant^{\sharp} y \\
12: & \mathsf{s}(x) -^{\sharp} \mathsf{s}(y) \to x -^{\sharp} y \\
13: & \mathsf{mod}^{\sharp}(\mathsf{s}(x), \mathsf{s}(y)) \to \mathsf{if}^{\sharp}(y \leqslant x, \mathsf{mod}(\mathsf{s}(x) - \mathsf{s}(y), \mathsf{s}(y)), \mathsf{s}(x)) \\
14: & \mathsf{mod}^{\sharp}(\mathsf{s}(x), \mathsf{s}(y)) \to y \leqslant^{\sharp} x \\
15: & \mathsf{mod}^{\sharp}(\mathsf{s}(x), \mathsf{s}(y)) \to \mathsf{mod}^{\sharp}(\mathsf{s}(x) - \mathsf{s}(y), \mathsf{s}(y)) \\
16: & \mathsf{mod}^{\sharp}(\mathsf{s}(x), \mathsf{s}(y)) \to \mathsf{s}(x) -^{\sharp} \mathsf{s}(y)
\end{array}
$$

Since the TRS is non-overlapping, it is sufficient to prove innermost termination. The problematic cycle in the (innermost) dependency graph is $\mathcal{C} = \{15\}$. The usable rewrite rules for this cycle are $\mathcal{U}(\mathcal{C}) = \{7, 8\}$. We need to find a reduction pair $(\gtrsim, >)$ such that rules 4 and 5 are weakly decreasing (i.e., compatible with $\gtrsim$) and dependency pair 15 is strictly decreasing (with respect to $>$). The only way to achieve the latter is by using the observation that $\mathsf{s}(x)$ is semantically greater than the syntactically larger term $\mathsf{s}(x) - \mathsf{s}(y)$. If we take the natural interpretation $-_{\mathbb{Z}}(x, y) = x - y$, $\mathsf{s}_{\mathbb{Z}}(x) = x - 1$, and $0_{\mathbb{Z}} = 0$, together with $\mathsf{mod}^{\sharp}_{\mathbb{Z}}(x, y) = x$ then we obtain the following ordering constraints over the natural numbers:

$$
7: \ x \geqslant x \quad 8: \ \max\{0, x - y\} \geqslant \max\{0, x - y\} \quad 15: \ x + 1 > \max\{0, x - y\}
$$

4.1   Theoretical Framework

The constraints in the above example are obviously satisfied, but are we allowed to use an interpretation like $-_{\mathbb{Z}}(x, y) = x - y$ in (innermost) termination proofs? The answer appears to be negative because Lemma 4 no longer holds. Because the induced interpretation $-_{\mathbb{N}}(x, y) = \max\{0, x - y\}$ is not weakly monotone in its second argument, the order $\geqslant_{\mathbb{N}}$ of the induced algebra is not closed under contexts, so if $s \geqslant_{\mathbb{N}} t$ then it may happen that $C[s] \leqslant_{\mathbb{N}} C[t]$. Consequently, we do not obtain a reduction pair. However, if we have $s =_{\mathbb{N}} t$ rather than

$s \geqslant_\mathbb{N} t$, closure under contexts is obtained for free. So we could take $(=_\mathbb{N}, >_\mathbb{N})$ as reduction pair. This works fine in the above example because the induced algebra is a model of the set of usable rules $\{7, 8\}$ and $>_\mathbb{N}$ orients dependency pair 15. However, requiring that all dependency pairs in a cycle are compatible with $=_\mathbb{N} \cup >_\mathbb{N}$ is rather restrictive because dependency pairs that are transformed into a polynomial constraint of the form $x^2 \geqslant x$ or $x + 2y \geqslant x + y$ cannot be handled. So we will allow $\geqslant_\mathbb{N}$ for the orientation of dependency pairs in a cycle $\mathcal{C}$ but insist that at least one dependency pair in $\mathcal{C}$ is compatible with $>_\mathbb{N}$. (Note that the relation $=_\mathbb{N} \cup >_\mathbb{N}$ is properly contained in $\geqslant_\mathbb{N}$.) The theorems below state the soundness of this approach in a more abstract setting. The proofs are straightforward modifications from the ones in [13]. The phrase "there are no minimal $\mathcal{C}$-rewrite sequences" intuitively means that if a TRS $\mathcal{R}$ is non-terminating then this is due to a different cycle of the dependency graph.

Theorem 11. *Let $\mathcal{R}$ be a TRS and let $\mathcal{C}$ be a cycle in its dependency graph. If there exists an algebra $\mathcal{A}$ equipped with a well-founded order $>$ such that $\mathcal{R} \subseteq =_\mathcal{A}$, $\mathcal{C} \subseteq \geqslant_\mathcal{A}$, and $\mathcal{C} \cap >_\mathcal{A} \neq \varnothing$ then there are no minimal $\mathcal{C}$-rewrite sequences.*   □

In other words, when proving termination, a cycle $\mathcal{C}$ of the dependency graph can be ignored if the conditions of Theorem 11 are satisfied. A similar statement holds for innermost termination.

Theorem 12. *Let $\mathcal{R}$ be a TRS and let $\mathcal{C}$ be a cycle in its innermost dependency graph. If there exists an algebra $\mathcal{A}$ equipped with a well-founded order $>$ such that $\mathcal{U}(\mathcal{C}) \subseteq =_\mathcal{A}$, $\mathcal{C} \subseteq \geqslant_\mathcal{A}$, and $\mathcal{C} \cap >_\mathcal{A} \neq \varnothing$ then there are no minimal innermost $\mathcal{C}$-rewrite sequences.*   □

The difference with Theorem 11 is the use of the *innermost* dependency graph and, more importantly, the replacement of the set $\mathcal{R}$ of all rewrite rules by the set $\mathcal{U}(\mathcal{C})$ of *usable rules* for $\mathcal{C}$, which in general is a much smaller set. Very recently, it has been proved [13, 22] that the usable rules criterion can also be used for termination, provided the employed reduction pair is $\mathcal{C}_\mathcal{E}$-compatible. However, replacing $\mathcal{R}$ by $\mathcal{U}(\mathcal{C})$ in Theorem 11 would be unsound. The reason is that the TRS $\mathcal{C}_\mathcal{E}$ admits no non-trivial models.

Example 13. Consider the following non-terminating TRS $\mathcal{R}$:

$$1\colon\ \mathsf{h}(\mathsf{f}(\mathsf{a}, \mathsf{b}, x)) \to \mathsf{h}(\mathsf{f}(x, x, x)) \qquad 2\colon\ \mathsf{g}(x, y) \to x \qquad 3\colon\ \mathsf{g}(x, y) \to y$$

The only dependency pair $\mathsf{h}^\sharp(\mathsf{f}(\mathsf{a}, \mathsf{b}, x)) \to \mathsf{h}^\sharp(\mathsf{f}(x, x, x))$ forms a cycle in the dependency graph. There are no usable rules. If we take the polynomial interpretation $\mathsf{a}_\mathbb{Z} = 1$, $\mathsf{b}_\mathbb{Z} = 0$, $\mathsf{f}_\mathbb{Z}(x, y, z) = x - y$, and $\mathsf{h}^\sharp_\mathbb{Z}(x) = x$ then the dependency pair is transformed into $1 > 0$. Note that it is not possible to extend the interpretation to a model for $\mathcal{R}$. Choosing $\mathsf{h}_\mathbb{Z}(x) = 0$ will take care of rule 1, but there is no interpretation $\mathsf{g}_\mathbb{Z}$ such that $\max\{0, \mathsf{g}_\mathbb{Z}(x, y)\} = x$ and $\max\{0, \mathsf{g}_\mathbb{Z}(x, y)\} = y$ for all natural numbers x and y.

4.2   Towards Automation

How do we verify a constraint like $x + 1 > \max\{0, x - y\}$? The approach that we developed in Section 3.2 for dealing with negative constants is not applicable because Lemma 7 relies essentially on weak monotonicity of the polynomial interpretations.

Let $\mathcal{P}_{\geqslant 0}$ be a subset of the set of polynomials P with integer coefficients such that $\alpha(P) \geqslant 0$ for all $\alpha \colon \mathcal{V} \to \mathbb{N}$ for which membership is decidable. For instance, $\mathcal{P}_{\geqslant 0}$ could be the set of polynomials without negative coefficients. We define $\mathcal{P}_{<0}$ in the same way.

Definition 14. *Let $(\mathbb{Z}, \{f_{\mathbb{Z}}\}_{f \in \mathcal{F}})$ be an algebra. With every term t we associate a polynomial $Q(t)$ as follows:*

$$
Q(t) = \begin{cases}
t & \text{if } t \text{ is a variable} \\
P & \text{if } t = f(t_1, \ldots, t_n) \text{ and } P \in \mathcal{P}_{\geqslant 0} \\
0 & \text{if } t = f(t_1, \ldots, t_n) \text{ and } P \in \mathcal{P}_{<0} \\
v(P) & \text{otherwise}
\end{cases}
$$

where $P = f_{\mathbb{Z}}(Q(t_1), \ldots, Q(t_n))$. In the last clause $v(P)$ denotes a fresh abstract variable that we uniquely associate with P.

There are two kinds of indeterminates in $Q(t)$: ordinary variables occurring in t and abstract variables. The intuitive meaning of an abstract variable $v(P)$ is $\max\{0, P\}$. The latter quantity is always non-negative, like an ordinary variable ranging over the natural numbers, but from $v(P)$ we can extract the original polynomial P and this information may be crucial for a comparison between two polynomial expressions to succeed. Note that the polynomial P associated with an abstract variable $v(P)$ may contain other abstract variables. However, because $v(P)$ is different from previously selected abstract variables, there are no spurious loops like $P_1 = v(x - v(P_2))$ and $P_2 = v(x - v(P_1))$.

The reason for using $\mathcal{P}_{\geqslant 0}$ and $\mathcal{P}_{<0}$ in the above definition is to make our approach independent of the particular method that is used to test non-negativeness or negativeness of polynomials.

Definition 15. *With every assignment $\alpha \colon \mathcal{V} \to \mathbb{N}$ we associate an assignment $\alpha^* \colon \mathcal{V} \to \mathbb{N}$ defined as follows:*

$$
\alpha^*(x) = \begin{cases}
\max\{0, \alpha^*(P)\} & \text{if } x \text{ is an abstract variable } v(P) \\
\alpha(x) & \text{otherwise}
\end{cases}
$$

The above definition is recursive because P may contain abstract variables. However, since $v(P)$ is different from previously selected abstract variables, the recursion terminates and it follows that α^* is well-defined.

Theorem 16. *Let $(\mathbb{Z}, \{f_{\mathbb{Z}}\}_{f \in \mathcal{F}})$ be an algebra such that every $f_{\mathbb{Z}}$ is a polynomial. Let t be a term. For every assignment α we have $[\alpha]_{\mathbb{N}}(t) = \alpha^*(Q(t))$.*

Proof. We show that $[\alpha]_\mathbb{N}(t) = \alpha^*(Q(t))$ by induction on t. If t is a variable then $[\alpha]_\mathbb{N}(t) = \alpha(t) = \alpha^*(t) = \alpha^*(Q(t))$. Suppose $t = f(t_1, \ldots, t_n)$. Let $P = f_\mathbb{Z}(Q(t_1), \ldots, Q(t_n))$. The induction hypothesis yields $[\alpha]_\mathbb{N}(t_i) = \alpha^*(Q(t_i))$ for all i and thus

$$[\alpha]_\mathbb{N}(t) = f_\mathbb{N}(\alpha^*(Q(t_1)), \ldots, \alpha^*(Q(t_n)))$$
$$= \max\{0, f_\mathbb{Z}(\alpha^*(Q(t_1)), \ldots, \alpha^*(Q(t_n)))\} = \max\{0, \alpha^*(P)\}$$

We distinguish three cases, corresponding to the definition of $Q(t)$.

- First suppose that $P \in \mathcal{P}_{\geqslant 0}$. This implies that $\alpha^*(P) \geqslant 0$ and thus we have $\max\{0, \alpha^*(P)\} = \alpha^*(P)$. Hence $[\alpha]_\mathbb{N}(t) = \alpha^*(P) = \alpha^*(Q(t))$.
- Next suppose that $P \in \mathcal{P}_{<0}$. So $\alpha^*(P) < 0$ and thus $\max\{0, \alpha^*(P)\} = 0$. Hence $[\alpha]_\mathbb{N}(t) = 0 = \alpha^*(Q(t))$.
- In the remaining case we do not know the status of P. We have $Q(t) = v(P)$ and thus $\alpha^*(Q(t)) = \max\{0, \alpha^*(P)\}$ which immediately yields the desired identity $[\alpha]_\mathbb{N}(t) = \alpha^*(Q(t))$.

$\square$

Corollary 17. *Let $(\mathbb{Z}, \{f_\mathbb{Z}\}_{f \in \mathcal{F}})$ be an $\mathcal{F}$-algebra such that every $f_\mathbb{Z}$ is a polynomial. Let s and t be terms. If $Q(s) = Q(t)$ then $s =_\mathbb{N} t$. If $\alpha^*(Q(s) - Q(t)) > 0$ for all assignments $\alpha \colon \mathcal{V} \to \mathbb{N}$ then $s >_\mathbb{N} t$. If $\alpha^*(Q(s) - Q(t)) \geqslant 0$ for all assignments $\alpha \colon \mathcal{V} \to \mathbb{N}$ then $s \geqslant_\mathbb{N} t$.* $\square$

Example 18. Consider again dependency pair 15 from Example 10:

$$\mathsf{mod}^\sharp(\mathsf{s}(x), \mathsf{s}(y)) \to \mathsf{mod}^\sharp(\mathsf{s}(x) - \mathsf{s}(y), \mathsf{s}(y))$$

We have $Q(\mathsf{mod}^\sharp(\mathsf{s}(x), \mathsf{s}(y))) = x + 1$ and $Q(\mathsf{mod}^\sharp(\mathsf{s}(x) - \mathsf{s}(y), \mathsf{s}(y))) = v(x - y)$. Since $x + 1 - v(x - y)$ may be negative (when interpreting $v(x - y)$ as a variable), the above corollary cannot be used to conclude that 15 is strictly decreasing. However, if we estimate $v(x - y)$ by x, the non-negative part of $x - y$, then we obtain $x + 1 - x = 1$ which is clearly positive.

Given a polynomial P with coefficients in $\mathbb{Z}$, we denote the non-negative part of P by $N(P)$.

Lemma 19. *Let Q be a polynomial with integer coefficients. Suppose $v(P)$ is an abstract variable that occurs in Q but not in $N(Q)$. If Q' is the polynomial obtained from Q by replacing $v(P)$ with $N(P)$ then $\alpha^*(Q) \geqslant \alpha^*(Q')$ for all assignments $\alpha \colon \mathcal{V} \to \mathbb{N}$.*

Proof. Let $\alpha \colon \mathcal{V} \to \mathbb{N}$ be an arbitrary assignment. In $\alpha^*(Q)$ every occurrence of $v(P)$ is assigned the value $\alpha^*(v(P)) = \max\{0, \alpha^*(P)\}$. We have $\alpha^*(N(P)) \geqslant \alpha^*(P) \geqslant \alpha^*(v(P))$. By assumption, $v(P)$ occurs only in the negative part of Q. Hence Q is (strictly) anti-monotone in $v(P)$ and therefore $\alpha^*(Q) \geqslant \alpha^*(Q')$. $\square$

In order to determine whether $s \geqslant_\mathbb{N} t$ (or $s >_\mathbb{N} t$) holds, the idea now is to first use standard techniques to test the non-negativeness of $Q = Q(s) - Q(t)$ (i.e., we determine whether $\alpha(Q) \geqslant 0$ for all assignments α by checking whether $Q \in \mathcal{P}_{\geqslant 0}$). If Q is non-negative then we certainly have $\alpha^*(Q) \geqslant 0$ for all assignments α and thus $s \geqslant_{\mathbb{N}} t$ follows from Corollary 17. If non-negativeness cannot be shown then we apply the previous lemma to replace an abstract variable that occurs only in the negative part of Q. The resulting polynomial Q' is tested for non-negativeness. If the test succeeds then for all assignments α we have $\alpha^*(Q') \geqslant 0$ and thus also $\alpha^*(Q) \geqslant 0$ by the previous lemma. According to Corollary 17 this is sufficient to conclude $s \geqslant_{\mathbb{N}} t$. Otherwise we repeat the above process with Q'. The process terminates when there are no more abstract variables left that appear only in the negative part of the current polynomial.

5   Experimental Results

We implemented the techniques described in this paper in the Tyrolean Termination Tool [14]. We tested 219 terminating TRSs and 239 innermost terminating TRSs from three different sources:

- all 89 terminating and 109 innermost terminating TRSs from Arts and Giesl [2],
- all 23 TRSs from Dershowitz [7],
- all 116 terminating TRSs from Steinbach and Kühler [20, Sections 3 and 4].

Nine of these TRSs appear in more than one collection, so the total number is 219 for termination and 239 for innermost termination. In our experiments we use the dependency pair method with the recursive SCC algorithm of [12] for analyzing the dependency graph. The recent modular refinements mentioned after Theorem 12 are also used, except when we try to prove (full) termination with the approach of Section 4 (but when the TRS is non-overlapping we do use modularity since in that case innermost termination guarantees termination). All experiments were performed on a PC equipped with a 2.20 GHz Mobile Intel Pentium 4 Processor - M and 512 MB of memory.

Tables 1 and 2 show the effect of the negative constant method developed in Section 3. In Table 1 we prove termination whereas in Table 1 we prove innermost termination. In the columns labelled N we use the *natural* interpretation for certain function symbols that appear in many example TRSs:

$$0_{\mathbb{Z}} = 0 \qquad\qquad 1_{\mathbb{Z}} = 1 \qquad\qquad 2_{\mathbb{Z}} = 2 \qquad \cdots$$
$$\mathsf{s}_{\mathbb{Z}}(x) = x + 1 \qquad +_{\mathbb{Z}}(x, y) = x + y \qquad \times_{\mathbb{Z}}(x, y) = xy$$
$$\mathsf{p}_{\mathbb{Z}}(x) = x - 1 \ ^1 \qquad -_{\mathbb{Z}}(x, y) = x - y^1$$

For other function symbols we take *linear* interpretations

$$\mathsf{f}_{\mathbb{Z}}(x_1, \ldots, x_n) = a_1 x_1 + \cdots + a_n x_n + b$$

---

[1] In Tables 1 and 2 we do not fix the interpretation of p when -1 is not included in the indicated constant range and the natural interpretation of $-$ is not used.

Table 1. Negative constants: termination.

constant	0, 1			0, 1, 2			0, 1, −1		
coefficient	0, 1			0, 1, 2			0, 1		
interpretation	E	N	NE	E	N	NE	E	N	NE
success	179	158	182	180	159	183	188	164	191
	0.20	0.09	0.08	0.20	0.09	0.09	0.24	0.14	0.12
failure	40	61	37	39	60	36	31	55	28
	0.03	0.02	0.03	1.11	0.15	1.43	0.93	0.43	1.29
timeout	0	0	0	0	0	0	0	0	0
total time	36.41	15.72	16.56	78.48	23.88	67.29	73.01	46.14	59.45

Table 2. Negative constants: innermost termination.

constant	0, 1			0, 1, 2			0, 1, −1		
coefficient	0, 1			0, 1, 2			0, 1		
interpretation	E	N	NE	E	N	NE	E	N	NE
success	200	177	202	202	179	204	209	183	211
	0.19	0.10	0.09	0.19	0.10	0.09	0.23	0.14	0.12
failure	39	62	37	37	60	35	30	56	28
	0.04	0.03	0.05	1.19	0.16	1.47	0.92	0.43	1.26
timeout	0	0	0	0	0	0	0	0	0
total time	39.46	18.92	19.59	81.94	26.97	70.07	75.24	49.10	61.24

with $a_1, \ldots, a_n$ in the indicated coefficient range and b in the indicated constant range. In the columns labelled E we do not fix the interpretations in advance; for all function symbols we search for an appropriate linear interpretation. In the columns labelled NE we start with the default natural interpretations but allow other linear interpretations if (innermost) termination cannot be proved. Determining appropriate coefficients can be done by a straightforward but inefficient "generate and test" algorithm. We implemented a more involved algorithm in our termination tool, but we anticipate that the recent techniques described in [6] might be useful to optimize the search for coefficients.

We list the number of successful termination attempts, the number of failures (which means that no termination proof was found while fully exploring the search space implied by the options), and the number of timeouts, which we set to 30 seconds. The figures below the number of successes and failures indicate the average time in seconds.

Table 3. Negative coefficients.

interpretation	termination			innermost termination		
	E	N	NE	E	N	NE
success	109	102	114	181	161	185
	0.74	0.39	0.58	0.04	0.06	0.07
failure	71	96	66	30	62	28
	2.50	0.71	2.48	1.72	0.33	2.00
timeout	39	21	39	28	16	26
total time	1428.41	738.16	1400.60	899.01	511.14	848.25

By using coefficients from $\{0, 1, 2\}$ very few additional examples can be handled. For termination the only difference is [2, Example 3.18] which contains a unary function double that requires $2x$ as interpretation. The effect of allowing -1 as constant is more apparent. Taking default natural interpretations for certain common function symbols reduces the execution time considerably but also reduces the termination proving power. However, the NE columns clearly show that fixing natural interpretations initially but allowing different interpretations later is a very useful idea.

In Table 3 we use the negative coefficient method developed in Section 4. Not surprisingly, this method is more suited for proving innermost termination because the method is incompatible with the recent modular refinements of the dependency pair method when proving termination (for non-overlapping TRSs).

Comparing the first (last) three columns in Table 3 with the last three columns in Table 1 (Table 2) one might be tempted to conclude that the approach developed in Section 4 is useless in practice. We note however that several challenging examples can only be handled by polynomial interpretations with negative coefficients. We mention TRSs that are (innermost) terminating because of non-linearity (e.g. [2, Examples 3.46, 4.12, 4.25]) and TRSs that are terminating because the difference of certain arguments decreases (e.g. [2, Example 4.30]).

References

1. T. Arts and J. Giesl. Termination of term rewriting using dependency pairs. *Theoretical Computer Science*, 236:133–178, 2000.
2. T. Arts and J. Giesl. A collection of examples for termination of term rewriting using dependency pairs. Technical Report AIB-2001-09, RWTH Aachen, 2001.
3. F. Baader and T. Nipkow. *Term Rewriting and All That*. Cambridge University Press, 1998.
4. A. Ben Cherifa and P. Lescanne. Termination of rewriting systems by polynomial interpretations and its implementation. *Science of Computer Programming*, 9:137–159, 1987.

5. E. Contejean, C. Marché, B. Monate, and X. Urbain. C*i*ME version 2, 2000. Available at http://cime.lri.fr/.
6. E. Contejean, C. Marché, A.-P. Tomás, and X. Urbain. Mechanically proving termination using polynomial interpretations. Research Report 1382, LRI, 2004.
7. N. Dershowitz. 33 Examples of termination. In *French Spring School of Theoretical Computer Science*, volume 909 of *LNCS*, pages 16–26, 1995.
8. N. Dershowitz and C. Hoot. Natural termination. *Theoretical Computer Science*, 142(2):179–207, 1995.
9. J. Giesl. Generating polynomial orderings for termination proofs. In *Proc. 6th RTA*, volume 914 of *LNCS*, pages 426–431, 1995.
10. J. Giesl, T. Arts, and E. Ohlebusch. Modular termination proofs for rewriting using dependency pairs. *Journal of Symbolic Computation*, 34(1):21–58, 2002.
11. J. Giesl, R. Thiemann, P. Schneider-Kamp, and S. Falke. Automated termination proofs with AProVE. In *Proc. 15th RTA*, volume 3091 of *LNCS*, pages 210–220, 2004.
12. N. Hirokawa and A. Middeldorp. Automating the dependency pair method. In *Proc. 19th CADE*, volume 2741 of *LNAI*, pages 32–46, 2003.
13. N. Hirokawa and A. Middeldorp. Dependency pairs revisited. In *Proc. 16th RTA*, volume 3091 of *LNCS*, pages 249–268, 2004.
14. N. Hirokawa and A. Middeldorp. Tyrolean termination tool. In *Proc. 7th International Workshop on Termination*, Technical Report AIB-2004-07, RWTH Aachen, pages 59–62, 2004.
15. H. Hong and D. Jakuš. Testing positiveness of polynomials. *Journal of Automated Reasoning*, 21:23–28, 1998.
16. D. Lankford. On proving term rewriting systems are Noetherian. Technical Report MTP-3, Louisiana Technical University, Ruston, LA, USA, 1979.
17. S. Lucas. Polynomials for proving termination of context-sensitive rewriting. In *Proc. 7th FoSSaCS*, volume 2987 of *LNCS*, pages 318–332, 2004.
18. S. Lucas. Polynomials over the reals in proof of termination. In *Proc. 7th International Workshop on Termination*, Technical Report AIB-2004-07, RWTH Aachen, pages 39–42, 2004.
19. J. Steinbach. Generating polynomial orderings. *Information Processing Letters*, 49:85–93, 1994.
20. J. Steinbach and U. Kühler. Check your ordering – termination proofs and open problems. Technical Report SR-90-25, Universität Kaiserslautern, 1990.
21. Terese. *Term Rewriting Systems*, volume 55 of *Cambridge Tracts in Theoretical Computer Science*. Cambridge University Press, 2003.
22. R. Thiemann, J. Giesl, and P. Schneider-Kamp. Improved modular termination proofs using dependency pairs. In *Proc. 2nd IJCAR*, volume 3097 of *LNAI*, pages 75–90, 2004.
23. X. Urbain. Modular & incremental automated termination proofs. *Journal of Automated Reasoning*, 2004. To appear.
24. H. Zantema. Termination of term rewriting by semantic labelling. *Fundamenta Informaticae*, 24:89–105, 1995.

New Developments in Symmetry Breaking in Search Using Computational Group Theory

Tom Kelsey, Steve Linton, and Colva Roney-Dougal

School of Computer Science, University of St Andrews, Fife, Scotland
{tom,sal,colva}@dcs.st-and.ac.uk

Abstract. Symmetry-breaking in constraint satisfaction problems is a well-established area of AI research which has recently developed strong interactions with symbolic computation, in the form of computational group theory. GE-trees are a new conceptual abstraction, providing low-degree polynomial time methods for breaking value symmetries in constraint satisfaction problems. In this paper we analyse the structure of symmetry groups of constraint satisfaction problems, and implement several combinations of GE-trees and the classical SBDD method for breaking all symmetries. We prove the efficacy of our techniques, and present preliminary experimental evidence of their practical efficiency.

1 Introduction

Constraint systems are a generalization of the Boolean satisfiability problems that play a central role in theoretical computer science. Solving constraint satisfaction problems (CSPs) in general is thus NP-complete; but effective solving of constraint systems arising from real problems, such as airline scheduling, is of enormous industrial importance.

There has been a great deal of research interest in dealing with symmetries in CSPs in recent years. CSPs are often solved using AI search techniques involving backtrack search and propagation. Many approaches to dealing with symmetries (constraints and/or solutions that are interchangeable in terms of the structure of the problem) are themselves based on refinements of AI search techniques. These include imposing some sort of ordering (before search) on otherwise interchangeable elements, posting constraints at backtracks that rule out search in symmetric parts of the tree, and checking that nodes in the search tree are not symmetrically equivalent to an already-visited state. These approaches are collectively known as lexicographic ordering [1], symmetry breaking during search (SBDS) [2,3], and symmetry breaking by dominance detection (SBDD) [4,5]. Methods can be, and are, optimised for either finding the first solution or finding all solutions. In this paper we consider only the problem of finding all solutions.

A promising recent approach has been to consider the symmetries of a given CSP as a group of permutations. It then becomes possible to pose and answer questions about symmetry using computational group theory. In effect, the AI

B. Buchberger and J.A. Campbell (Eds.): AISC 2004, LNAI 3249, pp. 199–210, 2004.
© Springer-Verlag Berlin Heidelberg 2004

search for solutions proceeds as before, but with fast permutation group algorithms supplying information that restricts search to symmetrically inequivalent nodes. Both SBDS and SBDD have been successfully implemented in this way [6,7] using GAP–ECL^{i}PS^e, an interface between the ECL^{i}PS^e[8] constraint logic programming system and the GAP [9] computational group theory system.

An even more recent advance has been the theory of GE-trees [10].

The construction and traversal of a GE-tree breaks all symmetries in any CSP. In the special – but common – case that the CSP has only value symmetries (for example graph colouring problems, where the colours are distinct but interchangeable) a GE-tree can be constructed in low-degree polynomial time. GE-trees provide both a useful analytic framework for comparing symmetry breaking methods, and, for value symmetries, a practical method for efficient search.

In this paper we describe initial results, both theoretical and practical, concerning the integration of GE-tree construction with SBDD. We combine heuristic AI search with mathematical structures and algorithms in order to extend and enhance existing – mainly pure AI – techniques.

In the remainder of this paper we provide a formal framework for CSPs, variable and value symmetries and GE-trees. In the following section, we describe SBDD and make some observations about variations of this algorithm. In the next two sections we identify and discuss a mathematically special, but common, situation in which we can uncouple variable and value symmetries. We follow this with preliminary results for a range of symmetry breaking approaches involving various combinations of SBDD and GE-tree methods.

1.1   CSPs and Symmetries

Definition 1. *A CSP L is a set of constraints C acting on a finite set of variables $\Delta := \{A_1, A_2, \ldots, A_n\}$, each of which has finite domain of possible values $D_i := D(A_i) \subseteq \Lambda$. A solution to L is an instantiation of all of the variables in Δ such that all of the constraints in C are satisfied.*

Constraint logic programming systems such as ECL^{i}PS^e model CSPs using constraints over finite domains. The usual search method is depth-first, with values assigned to variables at choice points. After each assignment a partial consistency test is applied: domain values that are found to be inconsistent are deleted, so that a smaller search tree is produced. Backtrack search is itself a consistency technique, since any inconsistency in a current partial assignment (the current set of choice points) will induce a backtrack. Other techniques include forward-checking, conflict-directed backjumping and look-ahead.

Statements of the form ($var = val$) are called *literals*, so a partial assignment is a conjunction of literals. We denote the set of all literals by χ, and generally denote variables in Roman capitals and values by lower case Greek letters. We denote "constrained to be equal" by $\# =$.

Definition 2. *Given a CSP L, with a set of constraints C, and a set of literals χ, a symmetry of L is a bijection $f : \chi \to \chi$ such that a full assignment A of L satisfies all constraints in C if and only if $f(A)$ does.*

We denote the image of a literal $(X = \alpha)$ under a symmetry g by $(X = \alpha)g$. The set of all symmetries of a CSP form a *group*: that is, they are a collection of bijections from the set of all literals to itself that is closed under composition of mappings and under inversion. We denote the symmetry group of a CSP by G.

Note that under this definition of a symmetry, it is entirely possible to map a partial assignment that does not violate any constraints to a partial assignment which does. Suppose for instance that we had constraints $(X_1 \# = X_2)$, $(X_2 \# = X_3)$ where each X_i had domain $[\alpha, \beta]$. Then the symmetry group of the CSP would enable us to freely interchange X_1, X_2 and X_3. However, this means that we could map the partial assignment $(X_1 = \alpha) \wedge (X_3 = \beta)$ (which does not break either of the constraints) to $(X_1 = \alpha) \wedge (X_2 = \beta)$ (which clearly does). This shows that the interaction between symmetries and consistency can be complex.

There are various ways in which the symmetries of a CSP can act on the set of all literals, we now examine these in more detail.

Definition 3. *A* value symmetry *of a CSP is a symmetry $g \in G$ such that if $(X = \alpha)g = (Y = \beta)$ then $X = Y$.*

The collection of value symmetries form a *subgroup* of G: that is, the set of value symmetries is itself a group, which we denote by G^{Val}. We distinguish two types of value symmetries: a group G^{Val} acts via *pure value symmetries* if for all $g \in G^{\mathrm{Val}}$, whenever $(X = \alpha)g = (X = \beta)$, we have $(Y = \alpha)g = (Y = \beta)$. There are CSPs for which this does not hold. However, at a cost of making distinct, labelled copies of each domain we have the option to assume that G acts via pure value symmetries. If G^{Val} is pure then we may represent it as acting on the values themselves, rather than on the literals: we write αg to denote the image of α under $g \in G^{\mathrm{Val}}$.

We wish to define variable symmetries in an analogous fashion; however we must be a little more cautious at this point. We deal first with the standard case.

Definition 4. *Let L be a CSP for which all of the variables have the same domains, and let G be the full symmetry group of L. A* variable symmetry *of L is a symmetry $g \in G$ such that if $(X = \alpha)g = (Y = \beta)$, then $\alpha = \beta$.*

We need a slightly more general definition than this. Recall that if the value group does not act via pure value symmetries, we make labelled copies of the domains for each variable: afterwards we can no longer use Definition 4 to look for variable symmetries, as the variables no longer share a common domain. However, the domains of the variables do match up in a natural way. We describe $g \in G$ as being a variable symmetry if, whenever $(X = \alpha)g = (Y = \beta)$, the values α and β correspond to one another under this natural bijection. Formally, we have the following definition.

Definition 5. *Let L be a CSP with symmetry group G. Fix a total ordering on D_i for each i, and denote the elements of D_i by α_{ij}. A* variable symmetry *is a symmetry $g \in G$ such that if $(A_i = \alpha_{ij})g = (A_k = \alpha_{kl})$ then $l = j$.*

That is, the original value and the image value have the same position in the domain ordering. If all variables share a common domain then we recover

Definition 4 from Definition 5. Note that in the above definition, we may order at least one domain arbitrarily without affecting whether or not a symmetry is deemed to be a variable symmetry: it is the *relative ordering* of the domains that is crucial.

There may be several possible orderings of the domains, corresponding to different choices of variable group: the value group is often a normal subgroup of G (see section 3), and hence is uniquely determined, but there will usually be several (conjugate) copies of the variable group, corresponding to distinct split extensions. If there is a unique copy of the variable group, as well as a unique copy of the value group, then G is the direct product of these two normal subgroups. However, in the current context, ordering any one domain induces a natural order on all of the remaining domains. This is because our variables either have a common domain, or originally had a common domain which has now been relabelled into several distinct copies (one for each variable).

The collection of all variable symmetries (for a given ordering on each domain) is a subgroup of G, which we denote G^{Var}. We define a *pure variable symmetry* to be a variable symmetry such that if $(A_i = \alpha_{ij})g = (A_k = \alpha_{kj})$ then the value of k does not depend on j.

1.2   GE-Trees

In [10] we introduced the GE-tree, a search tree $\mathbf{T}$ for a CSP L with the property that searching $\mathbf{T}$ finds exactly one representative of each equivalence class of solutions under the symmetry group of L. Before we can define a GE-tree, we need a few more group-theoretic and search-based definitions.

We consider only search strategies in which all allowed options for one variable are considered before any values for other variables. This is a common, although not universal, pattern in constraint programming. Therefore, we consider search trees to consist of nodes which are labelled by variables (except for the leaves, which are unlabelled), and edges labelled by values. We think of this as meaning that the variable is set to that value as one traverses the path from the root of the tree toward the leaves. At a node $\mathcal{N}$, the partial assignment given by reading the labels on the path from the root to $\mathcal{N}$ (ignoring the label at $\mathcal{N}$ itself) is the *state* at $\mathcal{N}$. We will often identify nodes with their state, when the meaning is clear. By the *values in* $\mathcal{N}$ we mean the values that occur in literals in the state at $\mathcal{N}$, we denote this set by $\mathrm{Val}(\mathcal{N})$. We define $\mathrm{Var}(\mathcal{N})$ similarly for variables. We will often speak of a permutation as mapping a node $\mathcal{N}$ to a node $\mathcal{M}$, although strictly speaking the permutation maps the literals in the state at $\mathcal{N}$ to the literals in the state at $\mathcal{M}$ (in any order).

Definition 6. *Let G be a group of symmetries of a CSP. The stabiliser of a literal $(X = \alpha)$ is the set of all symmetries in G that map $(X = \alpha)$ to itself. This set is itself a group. The orbit of a literal $(X = \alpha)$, denoted $(X = \alpha)^G$, is the set of all literals that can be mapped to $(X = \alpha)$ by a symmetry in G. That is*

$$(X = \alpha)^G := \{(Y = \beta) : \exists g \in G \ s.t. \ (Y = \beta)g = (X = \alpha)\}.$$

The orbit of a node is defined similarly.

Given a collection S of literals, the *pointwise* stabiliser of S is the subgroup of G which stabilises each element of S individually. The *setwise* stabiliser of S is the subgroup of G that consists of symmetries mapping the set S to itself.

Definition 7. *A* GE-tree *(group equivalence tree) for a CSP with symmetry group G is any search tree $\mathbf{T}$ satisfying the following two axioms:*

1. *No node of $\mathbf{T}$ is isomorphic under G to any other node.*
2. *Given a full assignment $\mathcal{A}$, there is at least one leaf of $\mathbf{T}$ which lies in the orbit of $\mathcal{A}$ under G.*

Therefore, the nodes of $\mathbf{T}$ are representatives for entire orbits of partial assignments under the group, and the action of the group on the tree fixes every node. Of course, a GE-tree will be constructed dynamically, and the constraints of a CSP will generally prevent us from searching the whole of $\mathbf{T}$. We define a GE-tree to be *minimal* if the deletion of any node (and its descendants) will delete at least one full assignment.

One of the main results in [10] is a constructive proof of the following Theorem:

Theorem 1. *Let L be a CSP with only value symmetries. Then breaking all symmetries of L is tractable.*

The theorem is proved by giving a low-degree polynomial algorithm which constructs a minimal GE-tree for L. This algorithm is summarized as follows:

At each node $\mathcal{N}$ in the search tree do:
    Compute the pointwise stabiliser $G_{(\mathrm{Val}(\mathcal{N}))}$ of $\mathrm{Val}(\mathcal{N})$.
    Select a variable X which is not in $\mathrm{Var}(\mathcal{N})$.
    Compute the orbits of $G_{(\mathrm{Val}(\mathcal{N}))}$ on $\mathrm{Dom}(X)$.
    For each orbit $\mathcal{O}$ do:
        Construct a downedge from $\mathcal{N}$ labelled with an element from $\mathcal{O}$.
    End for.
End for.

It is shown in [10] that this can all be done in low-degree polynomial time. The result of only considering this reduced set of edges is that no two nodes in the resulting tree will be equivalent to each other under the subgroup of G that consists of value symmetries.

This theorem has the following immediate corollary:

Corollary 1. *Let L be any CSP with symmetry group G. Then breaking the subgroup of G that consists only of value symmetries is tractable (i.e. can be done in polynomial time).*

This begs the question: how may we break the remaining symmetries of a CSP? For certain groups of variable symmetries, such as the full symmetric group, the direct product of two symmetric groups, and the wreath product of two symmetric groups, it is relatively straightforward to develop tractable algorithms for constructing GE-trees. However at present we have no algorithm for arbitrary groups: the difficulty is that the action of the variable group does not, in general, map nodes of a search tree to nodes of a search tree. Therefore to solve this problem in general, we seek to develop hybrid approaches between our GE-tree construction and pre-existing symmetry breaking algorithms.

2 Symmetry Breaking by Dominance Detection

In this section we briefly describe symmetry breaking by dominance detection (SBDD). Let L be any CSP with symmetry group G. During backtrack search we maintain a record $\mathcal{F}$ of *fail sets* corresponding to the roots of completed subtrees. Each fail set consists of those $(var = val)$ assignments made during search to reach the root of the subtree. We also keep track of the set P of current ground variables: variables having unit domain, either by search decision or by propagation.

The next node in the search tree is *dominated* if there is a $g \in G$ and $S \in \mathcal{F}$ such that

$$Sg \subseteq P \quad .$$

In the event that we can find suitable g and S, it is safe to backtrack, since we are in a search state symmetrically equivalent to one considered previously. In practice, the cost of detecting dominance is often outweighed by the reduction in search.

SBDD works well in practice: empirical evidence suggests that SBDD can deal with larger symmetry groups than many other techniques. It is possible to detect dominance without using group theoretic techniques. However, this involves writing a bespoke detector for each problem, usually in the form of additional predicates in the constraint logic system. Using computational group theory enables generic SBDD, with the computational group theory system needing only a generating set for G to be able to detect dominance.

A symmetry breaking technique is called *complete* if it guarantees never to return two equivalent solutions. Both SBDD and GE-trees are complete, in fact, SBDD remains complete even when the dominance check is not performed at every node, provided that it is always performed at the leaves of the search tree (i.e. at solutions). This observation allows a trade-off between the cost of performing dominance checks and the cost of unnecessary search. Another possible approach is to use an incomplete, but presumably faster, symmetry breaking technique combined with a separate 'isomorph rejection' step to eliminate equivalent solutions.

The efficient algorithm for breaking value symmetries – GE-tree construction – can safely be combined with SBDD, as shown in Theorem 13 of [10]. The algorithm implied by this theorem performs a dominance check, using the full group G of the CSP L, at each node of a GE-tree for L under G^{Val}.

In the next two sections we will identify and discuss a mathematically special, but common, situation in which we can uncouple variable and value symmetries. We follow this with preliminary experimental results for a range of symmetry breaking approaches involving combinations of SBDD and GE-tree methods.

3   Complementary Variable Symmetries

We say that a group G is *generated* by a set X if every element of G can be written as product of elements of X and their inverses. For a great many CSPs, each element of the symmetry group can be uniquely written as a single value symmetry followed by a single variable symmetry. In this case we say that the variable symmetries form a *complement* to the value symmetries in the full symmetry group. Formally, a subgroup H_1 of a group H is a *complement* to a subgroup H_2 of H if the following conditions hold:

1. H_2 is a *normal* subgroup of H: this means that for all elements $h \in H$ the set $\{hh_2 \ : \ h_2 \in H_2\}$ is equal to the set $\{h_2h \ : \ h_2 \in H_2\}$.
2. $|H_1 \cap H_2| = 1$.
3. The set $\{h_1h_2 : h_1 \in H_1, \ h_2 \in H_2\}$ contains all elements of G.

If this is true for a CSP with symmetry group G when H_1 is G^{Var} and H_2 is G^{Val} then the CSP has *complementary variable symmetry*.

This holds, for instance, for any CSP where G is generated by pure value symmetries and pure variable symmetries. For example, in a graph colouring problem, the symmetries are generated by relabelling of the colours (pure value symmetries) and the automorphism group of the graph (pure variable symmetries).

Before going further, we collect a few facts describing the way in which the variable and value symmetries interact.

Lemma 1. *If G is generated by a collection of pure variable symmetries and some value symmetries, then G^{Val} is a normal subgroup of G.*

Proof. To show that G^{Val} is a normal subgroup of G, we show that for all $g \in G$ and all $h \in G^{\mathrm{Val}}$, the symmetry $g^{-1}hg \in G^{\mathrm{Val}}$.

Let $h \in G^{\mathrm{Val}}$ and let $g \in G$. Then g is a product of variable and value symmetries. Consider the literal $(X_i = \alpha_{ij})g^{-1}hg$. The image of a literal $(X_k = \alpha_{kl})$ under each variable symmetry in g depends only on k, not on l, and each value symmetry fixes the variables in each literal. Therefore as we move through the symmetries that make up g^{-1}, we will map X_i to various other variables, but each of these mappings will be inverted as we apply each of the symmetries that make up g. Thus $(X_i = \alpha_{ij})g^{-1}hg = (X_i = \alpha_{ij'})$ for some j', and so $g^{-1}hg$ is a value symmetry.

We note that if G contains any non-pure variable symmetries then G^{Val} is not, in general, a normal subgroup. Suppose that $g \in G^{\mathrm{Var}}$ maps $(X_i = \alpha_{ij}) \mapsto (X_k = \alpha_{kj})$, and also maps $(X_i = \alpha_{ij_1}) \mapsto (X_l = \alpha_{lj_1})$. Let $g \in G^{\mathrm{Val}}$ map $(X_i = \alpha_{ij}) \mapsto (X_i = \alpha_{ij_1})$. Then

$$
\begin{aligned}
(X_k = \alpha_{kj})g^{-1}hg &= (X_i = \alpha_{ij})hg \\
&= (X_i = \alpha_{ij_1})g \\
&= (X_l = \alpha_{lj_1}),
\end{aligned}
$$

but the map $(X_k = \alpha_{kj}) \mapsto (X_l = \alpha_{lj_1})$ is clearly *not* a value symmetry.

A symmetry group G for a CSP can quickly be tested for complementary value symmetry. There are several different ways of doing so, depending on how G has been constructed. If G has been input as a collection of pure value symmetries and a collection of pure variable symmetries then we always have complementary variable symmetry, and the construction of the subgroup of value symmetries and the group of variable symmetries is immediate.

So suppose that G has not been input in this form, and that we have G and a list of the domains for each variable. If the domains are not equal, and we have some value symmetries, then assume also that the bijections between each pair of domains are given.

We first check that the subgroup of value symmetries forms a normal subgroup of G, which can be done in time polynomial in $|\chi|$. We then form the *quotient group* of G by G^{Val}: this basically means that we divide out by the subgroup of all value symmetries, and can also be done in polynomial time. If the size of this quotient group is equal to the size of the group of variable symmetries, then the CSP has complementary variable symmetry, as it is clear that the only permutation of the set of all literals which lies in both the group of variable symmetries and the group of value symmetries is the identity map.

If the variables of the CSP share a common domain, the fastest way to find the group of variable symmetries (if this is not immediately clear from the way in which G has been described) is to compute the pointwise stabiliser in G of each of the values. This can be done in low-degree polynomial time [11].

In the next section we describe a new algorithm for symmetry breaking which is applicable to all CSPs with complementary variable symmetry.

4   SBDD on Complements

This approach can be summarized by saying that we construct a GE-tree for the set of value symmetries, and then search this using SBDD. However, we do not use SBDD on the full group, as this would involve also checking the value symmetries once more, but instead carry out SBDD only on the subgroup of variable symmetries.

In more detail, we proceed as follows. At each node $\mathcal{N}$ in the search tree we start by applying the GE-tree algorithm for G^{Val} to label $\mathcal{N}$ with a variable X and produce a short list of possible downedges from $\mathcal{N}$, say $\alpha_1, \alpha_2, \ldots, \alpha_k \in \mathrm{Dom}(X)$.

We now perform dominance detection on each of $\mathcal{N} \cup (X = \alpha_i)$, but in a 2-stage process separating out the variable mapping from the value mapping. That is, using only G^{Var} we check whether $\mathrm{Var}(\mathcal{N}) \cup \{X\}$ is dominated by the variables in any other node. Each time that we find dominance by a node $\mathcal{M}$, this implies that $\mathrm{Var}(\mathcal{M}g) \subseteq \mathrm{Var}(\mathcal{N} \cup \{X\})$. We therefore apply g to the literals in $\mathcal{M}$, and check whether there is an element of G^{Val} that can map the resulting collection of literals to $\mathcal{N} \cup (X = \alpha_i)$ for $1 \leq i \leq k$, bearing in mind that we now know precisely which literal in $\mathcal{M}g$ must be mapped to each literal in $\mathcal{N} \cap (X = \alpha_i)$, since the value group fixes the variables occurring in each literal. This latter query is therefore a low-degree polynomial time operation.

Only those α_i which are never dominated in this way are used to construct new nodes in **T**.

Theorem 2. *The tree* **T** *constructed as above is a GE-tree for the full symmetry group G.*

Proof. We start by showing that no two nodes of this tree are isomorphic under G. Let $\mathcal{M}$ and $\mathcal{N}$ be two distinct nodes, and suppose that there exists $g \in G$ such that $\mathcal{M}g = \mathcal{N}$. By assumption, we can write all elements of G as a product of a value symmetry and a variable symmetry, so write $g = lr$ where l is a value symmetry and r is a variable symmetry. Suppose (without loss of generality, since all group elements are invertible) that $\mathcal{M}$ is to the right of $\mathcal{N}$ in the search tree, so that we will have found $\mathcal{N}$ first during search. Then the partial assignment $\mathcal{N}' := \mathcal{N}r^{-1}$ is the image of $\mathcal{N}$ under an element of the variable group, and contains the same variables as $\mathcal{M}$. But this means that in our dominance check, we will discover that $\mathcal{M}$ is dominated by an ancestor of $\mathcal{N}$, contradicting the construction of **T**.

Next we show that any full assignment corresponds to at least one leaf of **T**. Let $\mathcal{A}$ be a full assignment, and let X_1 be the variable at the root of **T**. Then for some $\alpha_1 \in \mathrm{Dom}(X_1)$ the literal $(X_1 = \alpha_1) \in \mathcal{A}$. Since the downedges from the root are labelled with orbit representatives of G on $\mathrm{Dom}(X_1)$, there exists $\beta_1 \in \mathrm{Dom}(X_1)$ and $x_1 \in G^{\mathrm{Var}}$ such that $(X_1 = \alpha_1)x_1 = (X_1 = \beta_1)$ and β_1 is the label of a downedge from the root. Thus $\mathcal{A}x_1$ contains a node of the tree at depth 1.

Suppose that $\mathcal{A}x_i$ contains a node $\mathcal{N}$ of the tree at depth i, and suppose that the label at $\mathcal{N}$ is X_{i+1}. Note that since $\mathcal{A}$ is a full assignment, we must have $(X_{i+1} = \alpha_{i+1}) \in \mathcal{A}$, for some $\alpha_{i+1} \in \mathrm{Dom}(X_{i+1})$. We subdivide into two cases.

If there exists β_{i+1} in the orbit of α_{i+1} under $G^{\mathrm{Val}}_{(\mathcal{A})}$ such that β_{i+1} is the label of a downedge from $\mathcal{N}$, then letting $x_{i+1} \in G^{\mathrm{Val}}_{(\mathcal{A})}$ map $\alpha_{i+1} \mapsto \beta_{i+1}$ we see that $\mathcal{A}x_i x_{i+1}$ contains a node at depth $i + 1$.

Suppose instead that this is not the case. Then some orbit representative $(X_{i+1} = \beta_{i+1})$, in the orbit of $(X_{i+1} = \alpha_{i+1})$ under $G^{\mathrm{Val}}_{(\mathcal{N})}$, has been selected by the GE-tree technique but then rejected due to SBDD considerations. This means that there is a node $\mathcal{M}$ at depth $j \le i + 1$, which is to the left of $\mathcal{N}$ in **T**, and which dominates $\mathcal{N} \cup (X_{i+1} = \beta_{i+1})$. Hence there exists an element $g \in G$ such that $(\mathcal{N} \cup (X_{i+1} = \beta_{i+1}))g$ contains $\mathcal{M}$. Let $h \in G^{\mathrm{Var}}_{(\mathcal{N})}$ map $(X_{i+1} = \alpha_{i+1})$ to $(X_{i+1} = \beta_{i+1})$. Then $\mathcal{A}x_i h g$ contains $\mathcal{M}$. Either it is the case that there is a node at depth $i+1$ below $\mathcal{M}$ which can be reached from $\mathcal{A}x_i h g$ using only value symmetries, or there exists a node $\mathcal{M}'$ to the left of $\mathcal{M}$ which dominates some descendant of $\mathcal{M}$ that is contained in an image of $\mathcal{A}x_i h g$. Since this "mapping to left" operation can only be carried out a finite number of times, at some stage we must find a node of depth $i + 1$ which is contained in an image of $\mathcal{A}$.

We note at this point that the tree **T** described in this theorem is *not* necessarily minimal - the tree constructed by SBDD is not always minimal, so the same applies to any hybrid technique involving SBDD.

We finish this section with a brief discussion of the expected efficiency gain of this technique over plain SBDD. With SBDD, the cost at each node $\mathcal{N}$ of determining dominance is potentially exponential: for each completed subtree, we must determine whether or not it is possible to map the set of literals corresponding to the root of that subtree into the set of literals at $\mathcal{N}$. Best known algorithms for this run in moderately exponential time (that is, $O(e^{n^c})$ where $c < 1$ and n is the number of points that the group is acting on: namely $|\chi|$, the sum of the domain sizes). In our hybrid GE-tree and SBDD construction, the selection of a collection of orbit representatives for the down-edges for a node is done in low-degree polynomial time: there is a proof in [10] that the time is no worse than $O^{\sim}(n^4)$, but the actual bound is lower than this. We then perform SBDD on a reduced number of possible nodes, and with a smaller group.

5   Experiments

In [10] we showed that, for many CSPs, the cost of reformulating the problem into one with only value symmetry is clearly outweighed by the gains of the GE-tree construction, which is a polynomial-time method. In this paper we address the more standard question of CSPs with both value and variable symmetries. We take a highly symmetric CSP with known solutions, and compare symmetry breaking combinations.

The queens graph is a graph with n^2 nodes corresponding to squares of a chessboard. There is an edge between nodes iff they are on the same row, column, or diagonal, i.e. if two queens on those squares would attack each other in the absence of any intervening pieces. The colouring problem is to colour the queens graph with n colours. If possible, this corresponds to a set of n solutions to the n queens problem, forming a disjoint partition of the squares of the chessboard.

The problem is described by Martin Gardner. There is a construction for n congruent to 1 or 5 modulo 6, i.e. where n is not divisible by either 2 or 3. Other cases are settled on a case by case basis. Our CSP model has the cells of an $n \times n$ array as variables, each having domain $1 \ldots n$. The variable symmetries are those of a square (the dihedral group of order 8). The value symmetries are the $n!$ permutations of the domain values (the symmetric group of degree n).

Our symmetry breaking approaches (using GAP–ECL^{i}PS^e) are:

- SBDD only;
- GE-tree construction for the value symmetries, with SBDD only used to check the symmetric equivalence of solutions (GEtree+iso);
- GE-tree construction for the value symmetries, with SBDD – on the full symmetry group for the CSP – at each node (GEtree+SBDDfull);
- GE-tree construction for the value symmetries, with SBDD – on the symmetry group for the variables – at each node (GEtree+SBDDval).

Table 1 gives the GAP–ECL^{i}PS^{e}cpu times for a range of values for n. It seems clear that GE-tree combined with full SBDD is competitive with SBDD only. It also seems clear that only using SBDD (or any other isomorph rejection method)

Table 1. Experimental results

n		5	6	7	8
solutions		1	0	1	0
SBDD	GAP	0.39	0.48	1.01	112.44
	ECL	0.19	0.48	7.35	814.92
	Σ	0.68	0.96	8.36	927.36
GEtree+iso	GAP	0.42	0.37	1.49	127.15
	ECL	0.09	0.35	12.01	1677.26
	Σ	0.51	0.72	13.50	1804.73
GEtree+SBDDfull	GAP	0.42	0.52	1.51	195.15
	ECL	0.06	0.27	6.79	935.74
	Σ	0.48	0.79	8.30	1131.19
GEtree+SBDDval	GAP	0.77	0.93	3.96	930.77
	ECL	0.03	0.32	6.65	1146.46
	Σ	0.80	1.25	10.61	2077.23

is not competitive: the search tree is still large (since only value symmetries are being broken), with expensive variable symmetry breaking being postponed until the entire tree is searched. Interestingly, the results for GE-tree construction combined with SBDD on the complement of the value symmetries are not as good as expected. This approach combines polynomial time value symmetry breaking with dominance detection in a smaller algebraic structure than the full symmetry group. Therefore, on a heuristic level, we expect faster symmetry breaking than for GE-trees and full-group SBDD. Further experiments are required to pinpoint the reasons why the time gain is not as great as expected: the combination of symmetry breaking methods is, in general, an unexplored research area.

6 Conclusions

Symmetry breaking in constraint programming is an important area of interplay between artificial intelligence and symbolic computation. In this paper we have identified a number of important special structures and cases that can arise in the action of a symmetry group on the literals of a constraint problem. We have described a number of ways in which known symmetry breaking methods can be safely combined and described some new methods exploiting these newly identified structures.

Our initial experimental results demonstrate the applicability of our theoretical results, with more work to be done to overcome the apparent overheads of combining more than one symmetry breaking technique during the same search process. Other future work includes assessment of new heuristic approaches, including problem reformulation (to obtain a CSP with a more desirable symmetry group than that of a standard CSP model) and using dominance detection only at selected nodes in the tree (as opposed to every node, as currently implemented). We also aim to investigate both the theoretical and practical aspects of further useful ways of decomposing the symmetry group of a CSP.

Acknowledgements

The authors would like to thank Ian P. Gent for extremely helpful discussions. Our work is supported by EPSRC grants GR/R29666 and GR/S30580.

References

1. A.M. Frisch, B. Hnich, Z. Kiziltan, I. Miguel, and T. Walsh. Global constraints for lexicographic orderings. In P. van Hentenryck, editor, *Proceedings of the Eighth International Conference on Principles and Practice of Constraint Programming*, volume 2470 of *Lecture Notes in Computer Science*, pages 93–108. Springer, 2002.
2. R. Backofen and S. Will. Excluding symmetries in constraint-based search. In *Proceedings, CP-99*, pages 73–87. Springer, 1999.
3. I.P. Gent and B.M. Smith. Symmetry breaking in constraint programming. In W. Horn, editor, *Proc. ECAI 2000*, pages 599–603. IOS Press, 2000.
4. Torsten Fahle, Stefan Schamberger, and Meinolf Sellmann. Symmetry breaking. In T. Walsh, editor, *Proc. CP 2001*, pages 93–107, 2001.
5. Filippo Focacci and Michaela Milano. Global cut framework for removing symmetries. In T. Walsh, editor, *Proc. CP 2001*, pages 77–92, 2001.
6. Ian P. Gent, Warwick Harvey, and Tom Kelsey. Groups and constraints: Symmetry breaking during search. In Pascal Van Hentenryck, editor, *Proc. CP 2002*, pages 415–430. Springer-Verlag, 2002.
7. I.P. Gent, W. Harvey, T. Kelsey, and S.A. Linton. Generic SBDD using computational group theory. In Francesca Rossi, editor, *Proc. CP 2003*, pages 333–347. Springer-Verlag, 2003.
8. M. G. Wallace, S. Novello, and J. Schimpf. ECLiPSe : A platform for constraint logic programming. *ICL Systems Journal*, 12(1):159–200, May 1997.
9. The GAP Group. *GAP – Groups, Algorithms, and Programming, Version 4.2*, 2000. (http://www.gap-system.org).
10. Colva M. Roney-Dougal, Ian P. Gent, Tom Kelsey, and Steve A. Linton. Tractable symmetry breaking using restricted search trees. In *Proceedings, ECAI-04*, 2004. To appear.
11. Akos Seress. *Permutation group algorithms*. Number 152 in Cambridge tracts in mathematics. Cambridge University Press, 2002.

Recognition of Whitehead-Minimal Elements in Free Groups of Large Ranks

Alexei D. Miasnikov

The Graduate Center of CUNY
Computer Science
365 5th ave, New York, USA
amiasnikov@nyc.rr.com

Abstract. In this paper we introduce a pattern classification system to recognize words of minimal length in their automorphic orbits in free groups. This system is based on Support Vector Machines and does not use any particular results from group theory. The main advantage of the system is its stable performance in recognizing minimal elements in free groups with large ranks.

1 Introduction

This paper is a continuation of the work started in [5, 7]. In the previous papers we showed that pattern recognition techniques can be successfully used in abstract algebra and group theory in particular. The approach gives one an exploratory methods which could be helpful in revealing hidden mathematical structures and formulating rigorous mathematical hypotheses. Our philosophy here is that if irregular or non-random behavior has been observed during an experiment then there must be a pure mathematical reason behind this phenomenon, which can be uncovered by a proper statistical analysis.

In [7] we introduced a pattern recognition system that recognizes *minimal* (sometimes also called *Whitehead minimal*) words, i.e., words of minimal length in their automorphic orbits, in free groups. The corresponding probabilistic classification algorithm, a *classifier*, based on quadratic regression is very fast (linear time algorithm) and recognizes minimal words correctly with the high accuracy rate of more then 99%. However, the number of model parameters grows as a polynomial function of degree 4 on the rank of the free group. This limits applications of this system to free groups of small ranks (see Section 3.3).

In this paper we describe a probabilistic classification system to recognize Whitehead-minimal elements which is based on so-called *Support Vector Machines* [9, 10]. Experimental results described in the last section show that the system performs very well on different types of test data, including data generated in groups of large ranks.

The paper is structured as follows. In the next section we give a brief introduction to the Whitehead Minimization problem and discuss the limitations of the known deterministic procedure. In Section 3 we describe major components

B. Buchberger and J.A. Campbell (Eds.): AISC 2004, LNAI 3249, pp. 211–221, 2004.

of the classification system, including generation of training datasets and feature representation of elements in a free group. In the section we describe evaluation procedure and give empirical results on the performance of the system.

2   Whitehead's Minimization Problem

In this section we give a brief introduction to the Whitehead minimization problem.

Let X be a finite alphabet, $X^{-1} = \{x^{-1} \mid x \in X\}$ be the set of formal inverses of letters from X, and $X^{\pm 1} = X \cup X^{-1}$. For a word w in the alphabet $X^{\pm 1}$ by $|w|$ we denote the length of w. A word w is called *reduced* if it does not contain subwords of the type xx^{-1} or $x^{-1}x$ for $x \in X$. Applying reduction rules $xx^{-1} \to \varepsilon, x^{-1}x \to \varepsilon$ (where ε is the empty word) one can reduce each word w in the alphabet $X^{\pm 1}$ to a reduced word $\overline{w}$. The word $\overline{w}$ is uniquely defined and does not depend on the order in a particular sequence of reductions. The set $F = F(X)$ of all reduced words over $X^{\pm 1}$ forms a group with respect to multiplication defined by $u \cdot v = \overline{uv}$ (i.e., to compute the product of words $u, v \in F$ one has to concatenate them and then reduce). The group F with the multiplication defined as above is called a *free* group with *basis* X. The cardinality $|X|$ is called the *rank* of $F(X)$. Free groups play a central role in modern algebra and topology.

A bijection $\phi : F \to F$ is called an *automorphism* of F if $\phi(uv) = \phi(u)\phi(v)$ for every $u, v \in F$. The set $Aut(F)$ of all automorphisms of F forms a group with respect to composition of automorphisms. Every automorphism $\phi \in Aut(F)$ is completely determined by its images on elements from the basis X since $\phi(x_1 \dots x_n) = \phi(x_1) \dots \phi(x_n)$ and $\phi(x^{-1}) = \phi(x)^{-1}$ for any letters $x_i, x_i \in X^{\pm 1}$. An automorphism $t \in Aut(F(X))$ is called a *Whitehead's automorphism* if t satisfies one of the two conditions below:

1) t permutes elements in $X^{\pm 1}$;
2) t fixes a given element $a \in X^{\pm 1}$ and maps each element $x \in X^{\pm 1}, x \neq a^{\pm 1}$ to one of the elements $x, xa, a^{-1}x,$ or $a^{-1}xa$.

By $\Omega(X)$ we denote the set of all Whitehead's automorphisms of $F(X)$. It is known [8] that every automorphism from $Aut(F)$ is a product of finitely many Whitehead's automorphisms.

The automorphic orbit $Orb(w)$ of a word $w \in F$ is the set of all automorphic images of w in F:

$$Orb(w) = \{v \in F \mid \exists \varphi \in Aut(F) \text{ such that } \varphi(w) = v\}.$$

A word $w \in F$ is called *minimal* (or *automorphically minimal*) if $|w| \leq |\varphi(w)|$ for any $\varphi \in Aut(F)$. By w_{min} we denote a word of minimal length in $Orb(w)$. Notice that w_{min} is not unique. By $WC(w)$ (the *Whitehead's complexity* of w) we denote a minimal number of automorphisms $t_1, \dots, t_m \in \Omega(X)$ such that $t_m \dots t_1(w) = w_{min}$. The algorithmic problem which requires finding w_{min} for a given $w \in F$

is called the *Minimization Problem* for F, it is one of the principal problems in combinatorial group theory and topology. There is a famous Whitehead's decision algorithm for the Minimization Problem, it is based on the following result due to Whitehead ([11]): if a word $w \in F(X)$ is not minimal then there exists an automorphism $t \in \Omega(X)$ such that $|t(w)| < |w|$. Unfortunately, its complexity depends on cardinality of $\Omega(X)$ which is exponential in the rank of $F(X)$. We refer to [6] for a detailed discussion on complexity of Whitehead's algorithms.

In this paper we focus on the *Recognition Problem* for minimal elements in F. It follows immediately from the Whitehead's result that $w \in F$ is minimal if and only if $|t(w)| \geq |w|$ for every $t \in \Omega(X)$ (such elements sometimes are called *Whitehead's minimal*). This gives one a simple deterministic decision algorithm for the Recognition Problem, which is of exponential time complexity in the rank of F. Note, that the worst case in terms of the rank occur when the input word w is already minimal. In this situation all of the Whitehead automorphisms $\Omega(X)$ have to be applied.

Construction of a probabilistic classifier which recognizes words of minimal length allows one to solve the recognition problem quickly in expense of a small classification error. Such classifier can be used as a fast minimality check heuristic in a deterministic algorithm which solves the minimization problem.

It is convenient to consider the Minimization Problem only for cyclically reduced words in F. A word $w = x_1 \dots x_n \in F(X)$ $(x_i \in X^{\pm 1})$ is *cyclically reduced* if $x_1 \neq x_n^{-1}$. Clearly, every $w \in F$ can be presented in the form $w = u^{-1} \widetilde{w} u$ for some $u \in F(X)$ and a cyclically reduced element $\widetilde{w} \in F(X)$ such that $|w| = |\widetilde{w}| + 2|u|$. This $\widetilde{w}$ is unique and it is called a *cyclically reduced form* of w. Every minimal word in F is cyclically reduced, therefore, it suffices to construct a classifier only for cyclically reduced words in F.

3   Recognition of Minimal Words in Free Groups

One of the main applications of Pattern Recognition techniques is classification of a variety of given objects into categories. Usually classification algorithms or *classifiers* use a set of measurements (properties, characteristics) of objects, called *features*, which gives a descriptive representation for the objects. We refer to [2] for detailed introduction to pattern recognition techniques.

In this section we describe a particular pattern recognition system PR_{MIN} for recognizing minimal elements in free groups. The corresponding classifier is a supervised learning classifier which means that the decision algorithm is "trained" on a prearranged dataset, called *training* dataset in which each pattern is labelled with its true class label. The algorithm is based on Support Vector Machines (SVM) classification algorithm.

In Section 1 we have stressed that the number of parameters required to be estimated by the classification model based on quadratic regression is of order $O(n^4)$, where n is the rank of a free group F_n. This constitutes two main problems. First, in order to compute the parameters we have to multiply and decompose matrices of size equal to the number of the coefficients itself. For large n, the straightforward computation of such matrices might be impossible due to the memory size restrictions. Another problem, which is perhaps the major problem, is due to the fact that the number of observations in the training set needs to be about 100 times more than the number of the coefficients to be estimated. When n is large (for $n = 10$ the required number of observations is about 14,440,000) it is a significant practical limitation, especially when the data generation is time consuming.

One of the main attractive features of the Support Vector Machines is their ability to employ non-linear mapping without essential increase in the number of parameters to be estimated and, therefore, in computation time.

3.1   Data Generation: Training Datasets

A pseudo-random element w of $F(X)$ can be generated as a pseudo-random sequence $y_1, \ldots, y_l$ of elements $y_i \in X^{\pm 1}$ such that $y_i \neq y_{i+1}^{-1}$, where the length l is also chosen pseudo-randomly. However, it has been shown in [4] that randomly taken cyclically reduced words in F are already minimal with asymptotic probability 1. Therefore, a set of randomly generated cyclic words in F would be highly biased toward the class of minimal elements. To obtain fair training datasets we use the following procedure.

For each positive integer $l = 1, \ldots, L$ we generate pseudo-randomly and uniformly K cyclically reduced words from $F(X)$ of length l. Parameters L and K were chosen to be 1000 and 10 for pure practical reasons. Denote the resulting set by W. Then using the deterministic Whitehead algorithm we construct the corresponding set of minimal elements

$$W_{min} = \{w_{min} \mid w \in W\}.$$

With probability 0.5 we substitute each $v \in W_{min}$ with the word $\widetilde{t(v)}$, where t is a randomly and uniformly chosen automorphism from $\Omega(X)$ such that $|\widetilde{t(v)}| > |v|$ (if $|\widetilde{t(v)}| = |v|$ we chose another $t \in \Omega(X)$, and so on). Now, the resulting set L is a set of pseudo-randomly generated cyclically reduced words representing the classes of minimal and non-minimal elements in approximately equal proportions. It follows from the construction that our choice of non-minimal elements w is not quite representative, since all these elements have Whitehead's complexity one (which is not the case in general). One may try to replace the automorphism t above by a random finite sequence of automorphisms from Ω to get a more representative training set. However, we will see in Section 4 that the training dataset L is sufficiently good already, so we elected to keep it as it is.

From the construction we know for each element $v \in L$ whether it is minimal or not. Finally, we create a training set

$$D = \{< v, P(v) > \mid v \in L\},$$

where

$$P(v) = \begin{cases} 1, & v \text{ is minimal}; \\ 0, & \text{otherwise}. \end{cases}$$

3.2   Features

To describe the feature representation of elements from a free group $F(X)$ we need the following

Definition 1. *Labelled Whitehead Graph $WG(v) = (V, E)$ of an element $v \in F(X)$ is a weighted non-oriented graph, where the set of vertices V is equal to the set $X^{\pm 1}$, and for $x_i, x_j \in X^{\pm 1}$ there is an edge $(x_i, x_j) \in E$ if the subword $x_i x_j^{-1}$ (or $x_j x_i^{-1}$) occurs in the word v viewed as a cyclic word. Every edge (x_i, x_j) is assigned a weight l_{ij} which is the number of times the subwords $x_i x_j^{-1}$ and $x_j x_i^{-1}$ occur in v.*

Whitehead Graph is one of the main tools in exploring automorphic properties of elements in a free group [4, 8].

Now, let $w \in F(X)$ be a cyclically reduced word. We define features of element w as follows. Let $l(w)$ be a vector of edge weights in the Whitehead Graph $WG(w)$ with respect to a fixed order. We define a feature vector $f(w)$ by

$$f(w) = \frac{1}{|w|} l(w).$$

This is the basic feature vector in all our considerations.

3.3   Decision Rule

Below we give a brief description of the classification rule based on Support Vector Machine.

Let $D = \{w_1, \ldots, w_N\}$, $w \in F(X)$ be a training set and $D' = \{\mathbf{x}_1, \ldots, \mathbf{x}_N\}$, $\mathbf{x}_i = f(w_i)$ be the set of feature vectors with the corresponding labels $y_1, \ldots, y_N$, where

$$y_i = \begin{cases} +1, & \text{if } P(w_i) = 1; \\ -1, & \text{otherwise.} \end{cases}$$

Definition 2. *The margin of an example $(\mathbf{x}_i, y_i)$ with respect to a hyperplane $(\mathbf{w}, b)$ defined as the quantity*

$$\gamma_i = y_i(\mathbf{w}' \cdot \mathbf{x} + b).$$

Note that $\gamma_i > 0$ corresponds to the correct classification of $(\mathbf{x}_i, y_i)$.

Let $\gamma_+(\gamma_-)$ be the smallest margin among all positive (negative) points. Define the margin of separation

$$\gamma = \gamma_+ + \gamma_-.$$

A Support Vector Machine (SVM) is a statistical classifier that attempts to construct a decision hyperplane $(\mathbf{w}, b)$ in such a way that the margin of separation γ between positive and negative examples is maximized [9, 10].

We wish to find a hyper-plane which will separate the two classes such that all points on one side of the hyper-plane will be labelled +1, all points on the other side will be labelled -1. Define a discriminant function

$$g(\mathbf{x}) = \mathbf{w}^{*\prime} \cdot \mathbf{x} + b^*,$$

where $\mathbf{w}^*, b^*$ are the parameters of the optimal hyper-plane. Function $g(\mathbf{x})$ gives the distance from an arbitrary $\mathbf{x}$ to the optimal hyper-plane.

Parameters of the optimal hyperplane are obtained by maximizing the margin, which is equivalent to minimizing the cost function

$$\Phi(\mathbf{w}) = \|\mathbf{w}\|^2 = \mathbf{w}' \cdot \mathbf{w},$$

subject to the constraint that

$$y_i \left(\mathbf{w}' \cdot \mathbf{x}_i + b\right) - 1 \geq 0, \ i = 1, \dots, N.$$

This is an optimization problem with inequality constraints and can be solved by means of Lagrange multipliers. We form the Lagrangian

$$L(\mathbf{w}, b, \alpha) = \frac{1}{2}\mathbf{w}' \cdot \mathbf{w} - \sum_{i=1}^{N} \alpha_i \left[y_i \left(\mathbf{w}' \cdot \mathbf{x}_i + b\right) - 1\right],$$

where $\alpha_i \geq 0$ are the Lagrange multipliers. We need to minimize $L(\mathbf{w}, b, \alpha)$ with respect to $\mathbf{w}, b$ while requiring that derivatives of $L(\mathbf{w}, b, \alpha)$ with respect to all the α_i vanish, subject to the constraint that $\alpha_i \geq 0$. After solving the optimization problem the discriminant function

$$g(x) = \sum_{i=1}^{N} y_i \alpha_i^* \mathbf{x}_i' \cdot \mathbf{x} + b^*.$$

where α_i^*, b^* are the parameters of the optimal decision hyperplane. It shows that the distance can be computed as a weighted sum of the training data and the Lagrange multipliers, and that the training vectors $\mathbf{x}_i$ are only used in inner products.

One can extend linear case to non-linearly separable data by introducing a kernel function

$$K(\mathbf{x}_i, \mathbf{x}_j) = \varphi(\mathbf{x}_i) \cdot \varphi(\mathbf{x}_j),$$

where $\varphi(\mathbf{x})$ is some non-linear mapping into (possibly infinite) space H,

$$\varphi : \mathbb{R}^n \longmapsto H.$$

Since Support Vector Machines use only inner products to compute the discriminant function, given kernel $K(x_i, x_j)$, we can train a SVM without ever having to know $\varphi(x)$ [3]. The implication of this is that the number of parameters that has to be learned by the SVM does not depend on the choice of the kernel and, therefore, mapping φ. This gives an obvious computational advantage when mapping the original feature space into a higher dimensional space which is the main obstacle in the previous approach based on quadratic regression.

Examples of typical kernel functions are:

- Linear :
$$K(\mathbf{x}_i, \mathbf{x}_j) = \mathbf{x}_i' \cdot \mathbf{x}_j$$
- Polynomial:
$$K(\mathbf{x}_i, \mathbf{x}_j) = (1 + \mathbf{x}_i' \cdot \mathbf{x}_j)^d$$
- Exponential:
$$K(\mathbf{x}_i, \mathbf{x}_j) = e^{-\gamma \|\mathbf{x}_i - \mathbf{x}_j\|^2}$$
- Neural Networks:
$$K(\mathbf{x}_i, \mathbf{x}_j) = \tanh\left(\theta_1 \mathbf{x}_i' \cdot \mathbf{x}_j - \theta_2\right)$$

Now we can define the decision rule used by the system. The classification algorithm has to predict the value $P(w)$ of the predicate P for a given word w. The corresponding decision rule is

$$\text{Decide } P(w) = \begin{cases} 1, \text{ if } g(f(w)) \geq 0; \\ 0, \text{ otherwise.} \end{cases}$$

4 Evaluation of the System

4.1 Test Datasets

To test and evaluate our pattern recognition system we generate several test datasets of different types:

- A test set S_e which is generated by the same procedure as for the training set D, but independently of D.
- A test set S_R of pseudo-randomly generated cyclically reduced elements of $F(X)$, as described in Section 3.1.
- A test set S_P of pseudo-randomly generated cyclically reduced *primitive* elements in $F(X)$. Recall that $w \in F(X)$ is primitive if and only if there exists a sequence of Whitehead automorphisms $t_1 \ldots t_m \in \Omega(X)$ such that $t_m \ldots t_1(x) = w$ for some $x \in X^{\pm 1}$. Elements in S_P are generated by the procedure described in [6], which, roughly speaking, amounts to a random choice of $x \in X^{\pm 1}$ and a random choice of a sequence of automorphisms $t_1 \ldots t_m \in \Omega(X)$.
- A test set S_{10} which is generated in a way similar to the procedure used to generate the training set D. The only difference is that the non-minimal elements are obtained by applying not one, but several randomly chosen automorphisms from $\Omega(X)$. The number of such automorphisms is chosen uniformly randomly from the set $\{1, \ldots, 10\}$, hence the name.

For more details on the generating procedure see [6].

To show that performance of Support Vector Machines is acceptable for free groups, including groups of large ranks, we run experiments with groups of ranks 3,5,10,15,20. For each group we construct the training set D and test sets S_e, S_{10}, S_R, S_P using procedures described previously. Some statistics of the datasets are given in Table 1.

Table 1. Description of the training and test datasets in free groups F_{10}, F_{15} and F_{20}.

Dataset	size	% min	% non-min	(min,avg,max) word lengths
D	20000	49.1	50.9	(3,558.2,1306)
S_e	5000	48.9	51.1	(3,559,1292)
S_{10}	5000	49.1	50.9	(3,1016.5,13381)
S_R	5000	98.3	1.7	(3,501.2,999)
S_P	3850	0.0	100.0	(3,194.7,8719)

a) F_3;

Dataset	size	% min	% non-min	(min,avg,max) word lengths
D	20000	48.5	51.5	(5,581.3,1388)
S_e	5000	49.2	50.8	(8,583.7,1382)
S_{10}	5000	48.0	52.0	(7,1693.22,28278)
S_R	5000	97.2	2.8	(6,504.2,999)
S_P	2900	0.0	100.0	(5,656.9,22430)

c) F_5;

Dataset	size	% min	% non-min	(min,avg,max) word lengths
D	9660	48.9	51.1	(26,617.4,1461)
S_e	4811	49.2	50.8	(26,619.7,1443)
S_{10}	4837	49.5	50.5	(29,2589.8,65274)
S_R	4867	96.5	3.5	(18,512.7,999)
S_P	165	0.0	100.0	(12,150.8,1459)

a) F_{10};

Dataset	size	% min	% non-min	(min,avg,max) word lengths
D	9357	49.5	50.5	(41,635.3,1472)
S_e	4685	49.2	50.8	(40,642.5,1462)
S_{10}	4722	49.7	50.3	(46,3056.6,53422)
S_R	4755	95.3	4.7	(26,523.8,999)
S_P	870	0.0	100.0	(28,1109.3,4981)

b) F_{15};

Dataset	size	% min	% non-min	(min,avg,max) word lengths
D	9144	49.6	50.4	(47,658.3,1488)
S_e	4576	49.3	50.7	(48,659.8,1484)
S_{10}	4597	49.1	50.9	(64,3351.4,68316)
S_R	4643	94.0	6.0	(48,534.9,999)
S_P	182	0.0	100.0	(66,945.1,4762)

c) F_{20};

4.2   Accuracy Measure

To evaluate the performance of the classification system PR_{MIN} we define an accuracy measure A.

Let D_{eval} be a test data set and

$$K = |\{w \mid decide(w) = P(w), \ w \in D_{eval}\}|$$

be the number of correctly classified elements in D_{eval}. To evaluate the performance of a given pattern classification system we use a simple accuracy measure:

$$A = \frac{K}{|D_{eval}|},$$

which gives the fraction of the correctly classified elements from the test set D_{eval}.

Notice, that the numbers of correctly classified elements follow the Binomial distribution and A can be viewed as an estimate of probability p of a word being classified correctly.

We are interested in constructing a confidence interval for probability p. For binomial variates, exact confidence intervals do not exist in general. One can obtain an approximate $100(1 - \alpha)\%$ confidence interval $[p_S, p_L]$ by solving the following equations for p_S and p_L:

$$\sum_{i=0}^{K} \binom{|D_{eval}|}{i} p_S^i (1 - p_S)^{|D_{eval}|-i} = \alpha/2,$$

$$\sum_{i=K}^{|D_{eval}|} \binom{|D_{eval}|}{i} p_L^i (1 - p_L)^{|D_{eval}|-i} = \alpha/2$$

for a given α.

Exact solutions to the equations above can be obtained by re-expressing in terms of the incomplete beta function (see [1] for details).

4.3   Results of Experiments

Experiments were repeated with the following types of kernel functions:

K^1: linear;
K^2: quadratic $(b\mathbf{x}_i' \cdot \mathbf{x}_j + c)^2$;
K^3: polynomial of degree 3 $(b\mathbf{x_i}' \cdot \mathbf{x_j} + c)^3$;
K^4: polynomial of degree 4 $(b\mathbf{x_i}' \cdot \mathbf{x_j} + c)^4$;
K^e: Gaussian $e^{-\gamma||\mathbf{x_i} - \mathbf{x_j}||^2}$,

where $\mathbf{x_i}$, $\mathbf{x_j}$ are the feature vectors obtained with mapping f_{WG} and $\mathbf{x_i} \cdot \mathbf{x_j}$ is the inner product of $\mathbf{x_i}$ and $\mathbf{x_j}$.

The results of the experiments presented in Table 2. It shows that SVM with appropriate kernel perform well not only on free groups of small ranks but on groups of large ranks as well. The experiments confirmed observations, made previously, that classes of minimal and non-minimal words are not linearly separable. Moreover, once the rank and, therefore dimensionality of the feature space

Table 2. Performance of the Support Vector Machine classifier in free groups F_3, F_5, F_{10}, F_{15} and F_{20}.

Kernel	All elements					Elements with $|w| > 100$				
	F_3	F_5	F_{10}	F_{15}	F_{20}	F_3	F_5	F_{10}	F_{15}	F_{20}
K^1	.844	.805	.729	.676	.644	.859	.810	.738	.680	.648
K^2	.995	.978	.881	.782	.710	.999	.977	.880	.792	.711
K^3	.996	.988	.962	.888	.772	1.00	.996	.968	.894	.773
K^4	.996	.989	.984	.951	.832	1.00	.997	.991	.956	.834
K^e	.995	.986	.982	.988	.990	1.00	.999	.999	.998	.995

a) accuracy evaluated on the set S_e;

Kernel	All elements					Elements with $|w| > 100$				
	F_3	F_5	F_{10}	F_{15}	F_{20}	F_3	F_5	F_{10}	F_{15}	F_{20}
K^1	.806	.783	.760	.670	.676	.818	.798	.768	.717	.687
K^2	.993	.988	.885	.811	.750	.997	.971	.893	.814	.751
K^3	.994	.993	.969	.893	.809	1.00	.998	.976	.897	.810
K^4	.995	.993	.993	.954	.866	1.00	.998	.997	.957	.870
K^e	.995	.993	.985	.986	.989	1.00	1.00	.999	.999	.994

b) accuracy evaluated on the set S_{10};

Kernel	All elements					Elements with $|w| > 100$				
	F_3	F_5	F_{10}	F_{15}	F_{20}	F_3	F_5	F_{10}	F_{15}	F_{20}
K^1	.880	.833	.743	.710	.691	.915	.887	.768	.729	.727
K^2	.990	.986	.890	.812	.754	.999	.969	.903	.830	.790
K^3	.991	.991	.961	.893	.824	1.00	.996	.985	.911	.842
K^4	.992	.991	.973	.940	.867	1.00	.997	.999	.970	.883
K^e	.990	.986	.980	.973	.970	1.00	1.00	.999	.989	.973

c) accuracy evaluated on the set S_R;

Kernel	All elements					Elements with $|w| > 100$				
	F_3	F_5	F_{10}	F_{15}	F_{20}	F_3	F_5	F_{10}	F_{15}	F_{20}
K^1	.732	.798	.770	.694	.610	.674	.682	.769	.690	.612
K^2	.999	.997	.824	.722	.610	.993	.993	.785	.723	.626
K^3	1.00	1.00	.915	.777	.632	1.00	1.00	.877	.756	.648
K^4	1.00	1.00	.982	.821	.659	1.00	1.00	.985	.816	.665
K^e	1.00	1.00	1.00	1.00	1.00	1.00	1.00	1.00	1.00	1.00

d) accuracy evaluated on the set S_P.

grows, quadratic mapping does not guarantee the high classification accuracy. As one might expect, the accuracy increases when the degree of the polynomial mapping increases. Nevertheless, even with the polynomial kernel K^4 of the degree $d = 4$, Support Vector Machine is not able to perform accurate classification in groups F_{15} and F_{20}. However, Gaussian kernel produces stable and accurate results for all test datasets, including sets of elements in free groups of large ranks. This indicates that points in one of the classes (minimal or non-minimal) are compactly distributed in the feature space and can be accurately described as a Gaussian. We also can observe that Gaussian representation can be applied to only one of the classes. If the opposite was true, then the problem of separating the two classes would be much simpler and at least the quadratic mapping should have been as accurate as K^e.

We conclude this section with the following conclusions:

1. With appropriate kernel function Support Vector Machines approach performs very well in the task of classification of Whitehead-minimal words in free groups of various ranks, including groups of large ranks.
2. The best over all results are obtained with the Gaussian kernel K^e. This indicates that one of the classes is compact and can be bounded by a Gaussian function.
3. Regression approach is still would be preferable for groups of small ranks due to its simplicity and smaller resource requirements. However, the SMVs should be used for groups of larger ranks where the size of the training sets required to perform regression with non-linear preprocessing mapping becomes practically intractable.

References

1. Milton Abramowitz and Irene Stegun. *Handbook of Mathematical Functions with Formulas, Graphs, and Mathematical Table.* Dover Publications, Inc., 1972.
2. R. O. Duda, P. E. Hart, and D. G. Stork. *Pattern Classification.* Wiley-Interscience, 2nd edition, 2000.
3. S. Haykin. *Neural networks: A comprehensive foundation.* Prentice Hall, Upper Sadle River, NJ, 1999.
4. I. Kapovich, P. Schupp, and V. Shpilrain. Generic properties of whitehead's algorithm, stabilizers in $aut(f_k)$and one-relator groups. Preprint, 2003.
5. A.D. Miasnikov and R.M. Haralick. Regression analysis and automorphic orbits in free groups of Rank 2. 17th International Conference on Pattern Recognition. To appear, 2004.
6. A.D. Miasnikov and A.G. Myasnikov. Whitehead method and genetic algorithms. Contemporary Mathematics, 349:89-114, 2004.
7. R.M. Haralick, A.D. Miasnikov and A.G. Myasnikov. Pattern Recognition Approaches to Solving Combinatorial Problems in Free Groups. Contemporary Mathematics, 349:197-213, 2004.
8. L. R. and P. Schupp. *Combinatorial Group Theory,* volume 89 of *Series of Modern Studies in Math.* Springer-Verlag, 1977.
9. V.N. Vapnik. *Statistical Learning Theory.* John Wiley and Sons, New York, 1998.
10. V.N. Vapnik. *The nature of statistical learning theory.* Springer-Verlag, New York, 2000.
11. J. H. C. Whitehead. On equivalent sets of elements in a free group. *Annals of Mathematic,* 37(4):782-800, 1936.

Four Approaches to Automated Reasoning with Differential Algebraic Structures[*]

Jesús Aransay[1], Clemens Ballarin[2], and Julio Rubio[1]

[1] Dpto. de Matemáticas y Computación. Univ. de La Rioja. 26004 Logroño, Spain
{jesus-maria.aransay,julio.rubio}@dmc.unirioja.es
[2] Institut für Informatik, Technische Univ. München, D-85748 Garching, Germany
ballarin@in.tum.de

Abstract. While implementing a proof for the Basic Perturbation Lemma (a central result in Homological Algebra) in the theorem prover Isabelle one faces problems such as the implementation of algebraic structures, partial functions in a logic of total functions, or the level of abstraction in formal proofs. Different approaches aiming at solving these problems will be evaluated and classified according to features such as the degree of mechanization obtained or the direct correspondence to the mathematical proofs. From this study, an environment for further developments in Homological Algebra will be proposed.

1 Introduction

EAT [15] and Kenzo [5] are software systems written under Sergeraert's direction for symbolic computation in algebraic topology and homological algebra. These systems have been used to compute remarkable results (for instance, some homology groups of iterated loop spaces) previously unknown. Both of them are based on the intensive use of functional programming techniques, which enable in particular to encode and handle at runtime the infinite data structures appearing in algebraic topology algorithms. As pointed out in [4], algebraic topology is a field where challenging problems remain open for computer algebra systems and theorem provers.

In order to increase the reliability of the systems, a project to formally analyze fragments of the programs was undertaken. In the last years, several results have been found related to the algebraic specification of data structures, some of them presented in [9]. Following these results, the algorithms dealing with these data structures have to be studied. The long term goal of our research project is to get certified versions of these algorithms, which would ensure the correctness of the computer algebra systems to a degree much greater than current hand-coded programs. To this end, a tool for extracting code from mechanized proofs could be used. As a first step towards this general goal, our concrete objective in this paper is to explore several possibilities to implement proofs of theorems in algebraic topology by using a theorem prover. As theorem prover, the tactical

---

[*] Partially supported by MCyT, project TIC2002-01626 and by CAR ACPI-2002/06.

prover Isabelle [12] has been chosen, mainly, because it is the system the authors are most familiar with.

A first algorithm for which we intend to implement a proof is the Basic Perturbation Lemma (hereinafter, BPL), since its proof has an algorithm associated used in Kenzo as one of the central parts of the program. In Section 2, the statement of the BPL, as well as a collection of lemmas leading to its proof, will be given. In Section 3, a naïve attempt to implement the proofs of these lemmas is introduced. Section 4 describes an approach using the existing tools in Isabelle. In Section 5, we try to avoid the problems arising from partiality and provide a generic algebraic structure embedding most of the objects appearing in the problem. In Section 6, an approach based on a new Isabelle feature, instantiation of locales, will be commented on. The paper ends with a conclusions section.

2   Some Homological Algebra

In the following definitions, some notions of homological algebra are briefly introduced (for further details, see [10], for instance).

Definition 1. *A graded group C_* is a family of abelian groups indexed by the integers, $C_* = \{C_n\}_{n \in \mathbb{Z}}$, with each C_n an abelian group. A graded group homomorphism $f: A_* \to B_*$ of degree k ($\in \mathbb{Z}$) between two graded groups A_* and B_* is a family of group homomorphisms, $f = \{f_n\}_{n \in \mathbb{Z}}$, with $f_n: A_n \to B_{n+k}$ a group homomorphism $\forall n \in \mathbb{Z}$. A chain complex is a pair (C_*, d_{C_*}), where C_* is a graded group, and d_{C_*} (the differential map) is a graded group homomorphism $d_{C_*}: C_* \to C_*$ of degree -1 such that $d_{C_*} d_{C_*} = 0_{\hom D_* D_*}$. A chain complex homomorphism between two chain complexes (A_*, d_{A_*}) and (B_*, d_{B_*}) is a graded group homomorphism $f: A_* \to B_*$ (degree 0) such that $f d_{A_*} = d_{B_*} f$.*

Let us note that the same family of homomorphisms $f = \{f_n\}_{n \in \mathbb{Z}}$ can be considered as a graded group homomorphism or a chain complex homomorphism. If no confusion arises, C_* will represent both a graded group and a chain complex; in the case of a chain complex homomorphism, the differential associated to C_* will be denoted by d_{C_*}.

Definition 2. *A reduction $D_* \Rightarrow C_*$ between two chain complexes is a triple (f, g, h) where: (a) the components f and g are chain complex homomorphisms $f: D_* \to C_*$ and $g: C_* \to D_*$; (b) the component h is a homotopy operator on D_*, that is, a graded group homomorphism $h: D_* \to D_*$ of degree 1; (c) the following relations are satisfied: (1) $fg = \mathrm{id}_{C_*}$; (2) $gf + d_{D_*}h + hd_{D_*} = \mathrm{id}_{D_*}$; (3) $fh = 0_{\hom D_* C_*}$; (4) $hg = 0_{\hom C_* D_*}$; (5) $hh = 0_{\hom D_* D_*}$.*

Reductions are relevant since the homology of chain complexes is preserved by them and they allow to pass from a chain complex where the homology is unknown to a new one where homology is computable.

Definition 3. *Let D_* be a chain complex; a perturbation of the differential d_{D_*} is a homomorphism of graded groups $\delta_{D_*}: D_* \to D_*$ (degree -1) such that $d_{D_*} +$*

δ_{D_*} is a differential for the underlying graded group of D_*. A perturbation δ_{D_*} of d_{D_*} satisfies the nilpotency condition with respect to a reduction $(f, g, h) \colon D_* \Rightarrow C_*$ whenever the composition $\delta_{D_*} h$ is pointwise nilpotent, that is, $(\delta_{D_*} h)^n(x) = 0$ for an $n \in \mathbb{N}$ depending on each x in D_*.

Under certain conditions, reductions can be "perturbed" to obtain new reductions easier to work with. This is expressed in the BPL.

Theorem 1. Basic Perturbation Lemma – Let $(f, g, h) \colon D_* \Rightarrow C_*$ be a chain complex reduction and $\delta_{D_*} \colon D_* \to D_*$ a perturbation of the differential d_{D_*} satisfying the nilpotency condition with respect to the reduction (f, g, h). Then, a new reduction $(f', g', h') \colon D'_* \Rightarrow C'_*$ can be obtained where the underlying graded groups of D_* and D'_* (resp. C_* and C'_*) are the same, but the differentials are perturbed: $d_{D'_*} = d_{D_*} + \delta_{D_*}, d_{C'_*} = d_{C_*} + \delta_{C_*}$, and $\delta_{C_*} = f \phi \delta_{D_*} g$; $f' = f\phi$; $g' = (1 - h\phi\delta_{D_*})g$; $h' = h\phi$, where $\phi = \sum_{i=0}^{\infty} (-1)^i (\delta_{D_*} h)^i$.

The BPL is a key result in algorithmic homological algebra (in particular, it is crucial for EAT [15] and Kenzo [5]). It ensures that when a perturbation is discarded (usually in chain complexes of infinite nature), a new reduction between chain complexes can be algorithmically obtained and thus the process to obtain a chain complex with computable homology can be implemented in a symbolic computation system. The BPL first appeared in [16] and was rewritten in modern terms in [3]. Since then, plenty of proofs have been described in the literature (see, for instance, [7, 14]). We are interested in a proof due to Sergeraert [14]. This proof is divided into two parts:

Part 1. Let ψ be $\sum_{i=0}^{\infty} (-1)^i (h\delta_{D_*})^i$. From the BPL hypothesis, the following equalities are proved: $\psi h = h\phi$; $\delta_{D_*} \psi = \phi\delta_{D_*}$; $\psi = 1 - h\delta_{D_*}\psi = 1 - \psi h\delta_{D_*} = 1 - h\phi\delta_{D_*}$; $\phi = 1 - \delta_{D_*} h\phi = 1 - \phi\delta_{D_*} h = 1 - \delta_{D_*}\psi h$.

Part 2. With these equalities, it is possible to give a collection of lemmas providing the new reduction between the chain complexes (and therefore producing the algorithm associated to the BPL).

In the rest of the paper, we focus on the second part. The collection of lemmas will be now presented, and later the sketch of a proof, which combines these lemmas, will be given. The sketch of the proof shows the constructive nature of the proof, which would permit us to obtain an algorithm from it. In the following sections we will explain the different attempts we have studied to implement the proofs of these lemmas.

Lemma 1. Let $(f, g, h) \colon D_* \Rightarrow C_*$ be a chain complex reduction. There exists a canonical and explicit chain complex isomorphism between D_* and the direct sum $\ker(gf) \oplus C_*$. In particular, $F \colon \mathrm{im}(gf) \to C_*$ and $F^{-1} \colon C_* \to \mathrm{im}(gf)$, defined by $F(x) = f(x)$ and $F^{-1}(x) = g(x)$, are inverse isomorphisms of chain complexes and $\mathrm{im}(gf) = \ker(id_{D_*} - gf)$.

Let us denote by $\mathrm{inc}_{\ker(p)}$ the canonical inclusion homomorphism $\mathrm{inc}_{\ker(p)} \colon \ker(p) \to D_*$ given by $x \mapsto x$, with $p = d_{D_*} h + h d_{D_*}$. It is well defined since $\ker(p)$ is a chain subcomplex of D_*.

Lemma 2. *Let D_* be a chain complex, $h\colon D_* \to D_*$ (degree 1) a homomorphism of graded groups, satisfying $hh = 0$ and $hd_{D_*}h = h$. Let $p = d_{D_*}h + hd_{D_*}$ (from the reduction properties in* Definition 2 *follows that $id_{D_*} - p = gf$). Then $(id_{D_*} - p, \mathrm{inc}_{\ker(p)}, h)$ is a reduction from D_* to $\ker(p)$.*

Lemma 2 is used to give a (very easy) constructive proof of the following result.

Lemma 3. *Assuming the conditions of the BPL, and the equalities of Part 1, there exists a canonical and explicit reduction $D'_* \Rightarrow \ker(p')$, where $p' = d_{D'_*}h' + h'd_{D'_*}$.*

Lemma 4. *Assuming the conditions of the BPL, and the equalities of Part 1, there exists a canonical and explicit isomorphism of graded groups between $\ker(p)$ and $\ker(p')$, where $p = d_{D_*}h + hd_{D_*}$ and $p' = d_{D'_*}h' + h'd_{D'_*}$.*

Lemma 5. *Let A_* be a chain complex, B_* a graded group and $F\colon A_* \to B_*$, $F^{-1}\colon B_* \to A_*$ inverse isomorphisms between graded groups. Then, the graded group homomorphism (degree -1) $d_{B_*} := Fd_{A_*}F^{-1}$ is a differential on B_* such that F and F^{-1} become inverse isomorphisms between chain complexes.*

Lemma 6. *Let $(f, g, h)\colon A_* \Rightarrow B_*$ be a reduction and $F\colon B_* \to C_*$ a chain complex isomorphism. Then (Ff, gF^{-1}, h) is a reduction from A_* to C_*.*

Sketch of the BPL proof – By applying Lemma 3, a reduction $D'_* \Rightarrow \ker(p')$ is obtained. Then, by Lemma 4, a graded group isomorphism between $\ker(p')$ and $\ker(p)$ is built. Now, from Lemma 1, one can conclude that $\ker(p) = \ker(id_{D_*} - gf) = \mathrm{im}(gf) \cong C_*$, and an explicit isomorphism *of graded groups* between $\ker(p')$ and C_* is defined (by composition). The differential of $\ker(p')$ has to be transferred to C_* by applying Lemma 5, giving a new chain complex C'_*, with the property that $\ker(p') \cong C'_*$ as chain complexes. By applying Lemma 6 to $D'_* \Rightarrow \ker(p')$ and $\ker(p') \cong C'_*$, an explicit reduction from D'_* to C'_* is obtained. When the homomorphisms obtained in the different lemmas are consecutively composed, the equalities in the BPL statement are exactly produced.

3   A Symbolic Approach

Our first approach consists in considering the objects in the statements (the homomorphisms) as generic elements of an algebraic structure where equational reasoning can be carried out. The idea is to identify, for instance, the elements of a ring of homomorphisms with the elements of a generic ring. Then, calculations in this ring are identified with proof steps in the reasoning domain (homomorphisms in the example). We call this a *symbolic approach* since homomorphisms are represented simply by symbols (as elements of a generic ring) without reference to their nature as functions. In our case, one of the additional difficulties of the proofs we intend to implement is that most of them require the properties of the various domains (with also elements of different nature and type) involved in the proof; but when trying to implement mathematics in computers, an abstraction process is always needed to produce the translation of elements of the computer system employed into the elements of mathematics. This process is of great importance, since it may clarify what has been done and to which extent we can rely on the results obtained in the computer, and will be briefly described at the end of this section; for this symbolic approach, the abstraction process can lead to identify the different structures embedding homomorphisms with a simple structure embedding all of them; the idea will be again used in Section 5 and its importance will be observed also in Section 6.

The proof of Lemma 2 illustrates most of the problems that have to be solved: the implementation of complex algebraic structures and the implementation of homomorphisms. In addition, one must work with the homomorphisms in two different levels simultaneously: equationally, like elements of an algebraic structure, and also like functions over a concrete domain and codomain. These reasons made us choose this lemma to seek the more appropriate framework. Along this section this framework is precisely defined in the symbolic approach, then the proved lemmas are explained and finally some properties of this framework are enumerated[1].

The following abstractions can be carried out:

- The big chain complex (D_*, d_{D_*}) is by definition a graded group with a differential operator, and $(\ker p, d_{D_*})$ is a chain subcomplex of it. The endomorphisms of (D_*, d_{D_*}) are the elements of a generic ring R.
- Some special homomorphisms between (D_*, d_{D_*}) and $(\ker p, d_{D_*})$ and endomorphisms of $(\ker p, d_{D_*})$ are seen as elements of R (for instance, the identity, the null homomorphism or some contractions).

Some of the computations developed under this construction of a generic ring can be then identified with some parts of the proof of Lemma 2. On the other hand, some other properties can not be proved in this framework since some (relevant) information about the structures involved and their properties is lost. This framework, being too generic, permits to avoid the problems of the concrete implementation of homomorphisms.

We will now give an example of how some of the properties having to be proved in the lemma can be represented in this framework. According to Def. 2 we have to prove the 5 characteristic properties of the reduction given in the conclusion of the lemma. From the five equalities, two fit in this framework and can be derived equationally inside of it. The first one is property (5), i.e.

$$hh = 0_R$$

The proof is trivial since the statement follows directly from premises of Lemma 2 and its proof can be implemented as in the mathematical proof. The second example of a proof that can be given inside this framework is property (3), i.e.

$$\text{if } hdh = h \text{ and } hh = 0 \text{ and } p = dh + hd \text{ then } (1_R - p)h = 0_R,$$

whose implementation in Isabelle can be given as

---
[1] A similar pattern will be followed in the other approaches.

lemma *(in ring) property_three: assumes $d \in$ carrier R and $h \in$ carrier R
and $h * h = 0$ and $h * d * h = h$ and $p = d * h + h * d$
shows $(1 - p) * h = 0$*
proof - *from prems show ?thesis by algebra* **qed**

Some comments on this proof: firstly, regarding the accuracy of the statement
and, again, the abstraction process mentioned before, it should be pointed out
that from the comparison of the proposed lemma in Isabelle and the property stated in Lemma 2 one difference is observed; namely, Lemma 2 says
$(\mathrm{id}_{D_*} - p)h = 0_{\mathrm{hom}\, D_*\, \ker p}$ whereas the Isabelle proof corresponds exactly to
$(\mathrm{id}_{D_*} - p)h = 0_{\mathrm{hom}\, D_*\, D_*}$ since there is only one ring involved in the Isabelle context. This is not a problem in this concrete equality since the value of $0_{\mathrm{hom}\, D_*\, \ker p}$
and $0_{\mathrm{hom}\, D_*\, D_*}$ is equal at every point (because $\ker p$ is a graded subgroup of
D_*); so far, the proof implemented in Isabelle, although having a formal difference with the original proof, can be considered to be an exact implementation
of the mathematical proof. Nevertheless, this kind of situation is the one that
will lead us to seek a more appropriate environment where the different domains
and algebraic structures can be represented.

It is also worth to emphasize the tool support for this style of reasoning:

1. use of *locales* is of advantage since it clarifies the syntax, shortens the proof
   and creates contexts with local assumptions; in our case just by adding *(in
   ring)* in our lemma a specific context is built where R is a generic ring and
   all the theorems proved in the existing *ring* locale appear like facts.
2. the *algebra* tactic valid for rings automates proofs looking for a normal form
   of the given expression (0, in this case).

This approach has some advantages. First of all, it is quite simple and intuitive,
which has the consequence that proofs are quite close to the mathematical proofs
obtained and can be easily understood. As will be seen in Section 4, when more
elaborate approaches are discussed, it is also possible that the size of the proofs
turns them into something unfeasible. Moreover, Isabelle has among its standard libraries enough theories to produce proofs in the context of the symbolic
approach in a completely automatic way; these generic proofs can be used where
only equational reasoning is required. In addition to this, the basic idea of this
approach will be useful in Sections 5 and 6.

There are also some drawbacks of this method. Firstly, we cannot prove the
other properties needed to complete the implementation the proof of the lemma.
They can not be proved, since information about the domain of the homomorphisms or about the concrete definition of the homomorphisms in such domains is
required. For instance, it is not possible to derive with the tools of this framework
that $p\big|_{\ker p} x = 0$. A second disadvantage observed is the high level of abstraction
required to pass from the mathematical context to the given context in Isabelle.
The type assigned to homomorphisms in this framework, where they are considered ring elements, is just a generic type α, whereas something more similar to
the mathematical definition of homomorphism would be at least desirable (for
instance, $\alpha \to \beta$). In particular, neither the differential group (D_*, d_{D_*}) nor the

elements of the domain D_* appear. Therefore, the conceptual distance between Isabelle code and the mathematical content is too large. From a more practical point of view, the previous discussion on types shows that the computational content of homomorphisms (that is to say, their interpretation as functions) is lost, which prevents code extraction.

4   A Set Theoretic Approach

The previous approach offers a high level of abstraction but is insufficient to prove our theorems. In this section we present a framework where the algebraic structures are again represented by records, but homomorphisms are now represented as functions.

There are two main components in this framework. One is the algebraic structures involved and the other one is the homomorphisms between these structures. In Isabelle, declarations of constants (including algebraic structures or homomorphisms) consist of a type[2] and a defining axiom. Algebraic structures are implemented over extensible record types (see [11]); this permits record subtyping and parametric polymorphism, and thus algebraic structures can be derived using inheritance. As an example, a differential group can be defined in Isabelle as

> ***record*** α *diff_group* $= \alpha$ *monoid* $+$
>   *diff* $:: \alpha \Rightarrow \alpha$
> *diff_group* $C \equiv$ *ab_group* $C \wedge$ *diff* $\in$ hom $C\,C\,\wedge$
> $\forall\,x \in$ *carrier* C. *diff diff* $x = $ *one* C

For homomorphisms, again a type and a definition must be declared. In the Isabelle type system all functions are total, and therefore homomorphisms (which are partial functions over generic types) are implemented through total functions; the homomorphisms between two given algebraic structures will be the set of all total functions between two types, for which the homomorphism axioms hold[3]

> hom $:: [(\alpha,\,\theta)$ *monoid_scheme*, $(\beta,\,\gamma)$ *monoid_scheme*$] \Rightarrow (\alpha \Rightarrow \beta)$ *set*
> hom $A\,B \equiv \{f.\ f \in$ *carrier* $A \rightarrow$ *carrier* $B \wedge f(x *_A y) = (fx) *_B (fy)\}$

Since in Isabelle equality $(=)$ is total on types, problems arise when comparing homomorphisms (or partial functions, in general). A way of getting out of this situation is to use "arbitrary", which denotes (for every type) an arbitrary but unknown value. Doing so is sound, because types are inhabited. Partial functions can be simulated by assigning them to "arbitrary" outside of their domain.

With these definitions we have implemented the complete proof of Lemma 1. This proof takes advantage of the above mentioned inheritance among algebraic structures, starting from sets, with the definition of a bijection between sets, and then introducing the inherited structures step by step. Following this scheme,

---

[2] In Isabelle syntax, $f :: \alpha => \beta$ denotes a total function from type α to type β.
[3] In Isabelle syntax, $\{x.f(x)\}$ denotes set comprehension.

800 lines of Isabelle code were needed to both specify the required structures and implement the proofs, and a readable proof was obtained.

We would like to point out the following about this approach:

1. A higher degree of accuracy (in comparison to the previous approach) has been obtained; now the relationship between the objects implemented in the proof assistant and the mathematical concepts present in the proof is much clearer.

2. The representation of algebraic structures in Isabelle and Kenzo is performed in the same way, using records where each field represents an operator. On the other hand, homomorphisms have a simpler representation in Isabelle since they are just elements of a functional type whereas in Kenzo they are also records, allowing that way to keep explicit information over their domain and codomain.

3. Converting homomorphisms into total functions by the use of "arbitrary" makes things a bit more complex. From the formal point of view, it is difficult to identify a homomorphism containing an arbitrary element, i.e., a conditional statement "if $x \in G$ then fx else arbitrary", with any mathematical object; there is a gap in the abstraction process that cannot be directly filled. A solution to clarify this process by identifying the elements outside the domain with a special element of the homomorphism's codomain will be proposed in Section 5; with this idea are avoided the non defined values of the total function representing a concrete homomorphism. This solution has been proposed before and its disadvantages are well known, but making a careful use of it can make mechanization easier.

4. Homomorphisms are represented by conditional statements, and working with n homomorphisms at the same time one has to consider 2^n cases. There are two solutions to this problem. The first one consists in enhancing the homomorphisms with an algebraic structure allowing to reason with them like elements of this structure in an equational way (for instance, an abelian group or a ring). With this implementation, proofs can sometimes avoid to get into the concrete details of the representation of homomorphisms in the theorem prover. To some extent this can be understood as the development of a new level of abstraction; some steps (or computations) are developed at the level of the elements of the homomorphisms domains whereas some are developed at the level of the structure where homomorphisms are embedded. This forces to implement proof steps more carefully and also to make a correct choice of the homomorphisms that can be provided with an algebraic structure, and will be discussed in Section 5. A second possible solution not implemented yet using equivalence classes will be mentioned as future work in Section 7.

5   A Homomorphism Approach

If we capitalize the advantages of both the symbolic approach and the set theoretic approach and we combine them carefully, it is possible to obtain a new framework where the level of abstraction is as suitable as in the set theoretic approach and the degree of mechanization is as high as in the symbolic approach. The idea is to develop a structure where the domain of the homomorphisms and also the homomorphisms can be found at the same time, with homomorphisms represented by elements of functional type (or something similar), which would allow working with them at a concrete level, but with the same homomorphisms being part of an algebraic structure. Since various algebraic structures are involved in the problem (at least (D_*, d_{D_*}) and $(\ker p, d_{D_*})$), there is not a simple algebraic structure containing all the homomorphisms appearing in the problem. Clearly endomorphisms of (D_*, d_{D_*}) form a ring, and also the ones in $(\ker p, d_{D_*})$, as well as the homomorphisms from (D_*, d_{D_*}) into $(\ker p, d_{D_*})$ form an abelian group; the automation of proofs in such a complicated environment would be hard; some ideas will be proposed in Section 7.

Another possible solution is explained in this section. It involves considering just one simple algebraic structure where all the homomorphisms can be embedded, and then develop tools allowing to convert a homomorphism from this structure into one of the underlying ones. Taking into account that $\ker p$ is a substructure of D_*, we will consider as our only structure the ring of endomorphisms $R = \hom(D_*, d_{D_*})(D_*, d_{D_*})$. Later we will introduce the tools allowing to convert homomorphisms from R into homomorphisms, for instance, of $\hom(\ker p, d_{D_*})(\ker p, d_{D_*})$, but just to illustrate the benefits of this approach we give a brief example here:

Example 1. Proving the fact "assumes $d \in \hom(D_*, d_{D_*})(D_*, d_{D_*})$ and $h \in \hom(D_*, d_{D_*})(D_*, d_{D_*})$ shows $p = dh + hd \in \hom(D_*, d_{D_*})(D_*, d_{D_*})$", due to partiality matters, requires several reasoning steps. When providing homomorphisms with an algebraic structure this is a trivial proof, since rings are closed under their operations.   □

In order to embed homomorphisms into a ring, it is necessary to choose carefully the elements and also the operators. Firstly, there can be only one representant for each homomorphism, and here partiality appears again, since otherwise both $\lambda x.(\text{one } G)$ and $\lambda x.(\text{if } x \in G \text{ then one } G \text{ else } \textit{arbitrary})$ could be identities in this ring. Secondly, operators must be closed over the structure, and thus they rely strongly on the chosen representation for homomorphisms. Without going in further depth, we decided to consider the carrier of the ring R formed by the completions $\lambda x.(\text{if } x \in G \text{ then } fx \text{ else one } G)$, because then the generic composition operator ∘ can be used in Isabelle (whereas this was not possible with the *extensional functions* based on "arbitrary" in Isabelle). At a second stage, we had to implement the tools allowing to convert an element of $\hom(D_*, d_{D_*})(D_*, d_{D_*})$ into one of, for instance, $\hom(\ker p, d_{D_*})(\ker p, d_{D_*})$ (under precise conditions). This would allow to implement proofs for facts such as $\langle \text{id} -p, D, \ker p \rangle \circ \langle h, D, D \rangle = \langle 0, D, \ker p \rangle$ (property (3) in Def. 2, needed for Lemma 2) in a human readable style such as[4]:

---

[4] The $\cdots$ are just an abbreviation meaning "the previous expression" in sequential calculations.

Example 2. assumes $\langle p, D, D \rangle \circ \langle h, D, D \rangle = \langle h, D, D \rangle$
shows $\langle \mathrm{id} - p, D, \ker p \rangle \circ \langle h, D, D \rangle = \langle 0, D, \ker p \rangle$
proof -
have $\langle \mathrm{id} - p, D, D \rangle \circ \langle h, D, D \rangle = (\langle \mathrm{id}\, D, D \rangle - \langle p, D, D \rangle) \circ \langle h, D, D \rangle$
(…)
also have $\cdots = \langle 0, D, D \rangle$
finally have one: $\langle \mathrm{id} - p, D, D \rangle \circ \langle h, D, D \rangle = \langle 0, D, D \rangle$
from prems have two: $\mathrm{im}(\mathrm{id} - p) \subseteq \ker p$
from one and two
have $(\langle \mathrm{id} - p, D, D \rangle \circ \langle h, D, D \rangle = \langle 0, D, D \rangle) \equiv (\langle \mathrm{id} - p, D, \ker p \rangle \circ \langle h, D, D \rangle = \langle 0, D, \ker p \rangle)$
then show $\langle \mathrm{id} - p, D, \ker p \rangle \circ \langle h, D, D \rangle = \langle 0, D, \ker p \rangle$ **qed**   □

In order to obtain this degree of expressiveness, two major modifications must be made. The first one is related to the implementation of homomorphisms. The information about their domain and codomain must be explicit, otherwise it is not possible to convert them from $\langle f, D, D \rangle$ to $\langle f, D, \ker p \rangle$. Secondly, some lemmas allowing to modify these triples must be introduced (and proved) in Isabelle. These lemmas permit to change the domain and codomain of homomorphism provided that certain conditions are satisfied; when the composition of two homomorphisms is equal to a third one, the domain and codomain of the three homomorphisms can be changed and the equality holds. This collection of lemmas can be summarized in the following one, a generic version of all of them where all the algebraic structures (domains and codomains of the homomorphisms) can be changed:

Lemma 7. Laureano's Lemma- *Let* $\langle g, C, D \rangle$ *and* $\langle f, A, B \rangle$ *be two homomorphisms between chain complexes satisfying* $\langle g, C, D \rangle \circ \langle f, A, B \rangle = \langle h, A, D \rangle$ *and let* A' *be a chain subcomplex from* A, B' *a chain subcomplex from* C', $\mathrm{im}\, f$ *contained on* B', *and* $\mathrm{im}\, h$ *contained on* D'. *Then* $\langle g, C', D' \rangle \circ \langle f, A', B' \rangle = \langle h, A', D' \rangle$.

With this lemma and the new proposed representation for homomorphisms a framework with the following advantages is built. From the point of view of capability, it is possible to implement all the properties needed for Lemma 2. It should be also emphasized that the size of the proofs is manageable and sometimes the proofs require some human guidance in order to finish. Moreover, the framework can be transferred to other problems dealing with homomorphisms and particular reasoning about them. Embedding homomorphisms (even between different algebraic structures) in only one algebraic structure can help to easily implement proofs of properties about these homomorphisms (and avoids the implementation of more elaborated algebraic structures); moreover, Lemma 7 can be easily generalized for compositions of n homomorphisms. From the formal point of view, all the computations are carried out in this algebraic structure and all the operations needed can be identified as simplifications inside the ring, or applications of Lemma 7; the abstraction process from the implemented proof to the mathematical proof can be accurately defined.

On the other hand, the amount of concepts that need to be mechanized is rather large. In addition to this, we have observed that the implementation of homomorphisms as records sometimes slows down the interactive proof steps. A likely cause for this loss of performance are inefficiencies in the record package of Isabelle2003[5]. A record encoding a homomorphism is a rather large data structure and contains a lot of redundant information. For instance, in $f\colon D_* \to D_*$, the corresponding record contains four copies of D_*, and D_* itself is already a quite large record. It appears that when several homomorphism are involved in a proof, the redundancies cause a slowed response time of the interactive environment.

6 Instantiating Locales

A final method based on the instantiation of locales (see [1]), a tool recently implemented in Isabelle that could be of great help for this kind of problems, is also considered. Locales, which are light-weight modules, define local scopes for Isabelle; some fixed assumptions are made and theorems can be proved from these assumptions in the context of the locale.

With the instantiation of locales it is not possible to define generic frameworks, but the method has a great expressiveness for concrete problems. The approach presented in this section is based to some extent on the ideas introduced in the first approach, considering homomorphisms embedded in an algebraic structure, but with the clear advantage that now several algebraic structures can be defined at the same time (this will allow us to introduce rings R and R', abelian groups A and A', and so on) and the elements of these structures can be instantiated with their real components. For instance, R could be instantiated with $\mathrm{hom}(D_*, d_{D_*})\,(D_*, d_{D_*})$ as carrier set and the usual composition $\circ$ as product, R' with carrier $\mathrm{hom}(\ker p, d_{D_*})\,(\ker p, d_{D_*})$ and operation $\circ$ as product, A' with carrier $\mathrm{hom}(D_*, d_{D_*})\,(\ker p, d_{D_*})$ and so on. This permits to work with homomorphisms at two different levels. In a first one, they can be considered as elements of the algebraic structures and the computations between the elements of these algebraic structures are identified with the steps of the proofs. In the second level, the operations of the algebraic structures can be instantiated to their definition (for instance, "mult = $\circ$" or "sum = $\lambda f g.\lambda x.fx *_D gx$") as well as the structures (in our case, the differential groups or chain complexes), in order to complete the proof steps requiring this information (for instance, $p|_{\ker p}x = 0$). The structure of the locale needed to implement the proofs of Lemma 2 in Isabelle would now look as follows:

> *locale hom_completion_environment = comm_group G + comm_group K + ring R + ring R0 + comm_group A + comm_group A0 + var p + assumes R = (carrier = hom_complection G G, mult = op $\circ$, one = ...) and R' = ...*
> *defines K = ker p*

[5] These have been removed in the latest release, Isabelle2004, but we have not yet ported our proofs.

First the structures (the variables) needed for the local scope are fixed and then their components are defined. Once all these assumptions have been made, the statement of lemmas will have the following appearance[6]:

> **lemma** *(in hom_complection_environment) reduction_property_one:*
> *assumes $p \in$ carrier R*
> *show (one $R\ominus_3 p)\circ(\lambda x. if x \in$ carrier $\ker p$ then id x else one $\ker p) =$ one R'*

By specifying *(in hom_complection_environment)* we introduce all the facts present in the locale *hom_complection_environment* and then they can be used as theorems in our lemma. The previous statement corresponds to property (2) of Def. 2 specialized for Lemma 2. Using these tools, we get a framework very close to the one in the mathematical lemmas. Moreover, proofs can be mechanized in a readable style quite similar to the mathematical proofs, and only some repetitive steps have to be added to explicitly obtain the value of the fields of the fixed algebraic structures when needed; proofs are easy to read and similar to the ones made "by hand". All these features make this framework quite satisfactory. On the other hand, from the formal point of view there is also a drawback. The operation $\circ$ appearing in the statement of the lemma can not be identified with any of the operations of the framework. It composes elements of different algebraic structures such as $\hom(\ker p, d_{D_*})$ $(\ker p, d_{D_*})$, $\hom(D_*, d_{D_*})$ $(\ker p, d_{D_*})$ but whose relation has not been made explicit at any point inside the locale. The composition $\circ$ is valid in this case since all the elements of the carrier sets are implemented trough functions, and therefore they can be composed. Even in other cases, the operation relating the different structures could be implemented, but there is no mathematical correspondence for this external op $\circ$; this produces a gap in the abstraction function that permits to identify mathematical objects with their representation in the theorem prover.

7   Conclusions and Further Work

A design "from bottom to top" has been used to implement the proof of the BPL in Isabelle. Instead of considering algebraic topology as a whole and formalizing generic concepts of this theory, our approach started from a very concrete (and relevant) lemma (the BPL) in this area that could allow to estimate the real capabilities from Isabelle to work with differential structures and their homomorphisms. We also started from a simple attempt (in Section 3) that gave us helpful ideas for the following approaches about what tools were directly available in Isabelle and those other tools that should be supplied to the system. The Isabelle tools introduced in Section 4 were not enough to produce readable proofs. Then, in Section 5, an original framework where several computations with homomorphisms can be carried out was suggested. The implementation of these tools led us to the problem of what changes must be made into partial

---

[6] In Isabelle syntax, $\ominus_3$ is a reference to the "minus" operation in the third structure of our locale definition, ring R.

functions in order to provide them with an enveloping algebraic structure. Our choice was to introduce completions, identifying the elements outside the domain of the partial function with a distinguished element, the identity of the image structures. Similarly, the *extensional* functions already available in Isabelle could have been used, just with some modifications in the composition of functions. The result is that a generic algebraic structure can not be directly defined from the homomorphisms since there are multiple functions representing each of them. In particular, it was shown that inside this framework implementations could be given for all the properties in Lemma 2, and that from the formal point of view, there was a clear abstraction function between the objects implemented in Isabelle and the mathematical objects required. Finally, in Section 6, the possibilities of a recent Isabelle feature (instantiation of locales) were studied; this new tool fits with several problems and so far would allow to implement all the proofs of the lemmas in Section 2. We have implemented a proof of all the properties in Lemma 2. From this study, we conclude that the approach based on the instantiation of locales is the most promising for further developments in homological algebra.

Some problems remain open, and more work has yet to be done:

1. To complete the proofs of all the lemmas in Section 2 and then give a complete implementation of the proof of the BPL. The proof obtained should satisfy all the requirements needed to extract a program from it. At this point, we could explore the possibilities of extracting code in Isabelle (see, for instance, [2]). In addition to this, a comparison will be needed with other provers where code extraction has been used in non trivial cases (see, for instance, the work on Buchberger's algorithm in Coq [17]).

2. As it has been explained in Section 4, partial functions can be implemented by using equivalence classes. Actually, the solutions proposed here (using completion functions or the *extensional* functions implemented in Isabelle) are just different ways of giving elements to represent these (non-implemented) equivalence classes. The implementation would force to redefine all the concepts regarding homomorphisms (composition, identities,...) but would produce a definitive way to deal with partial functions in the Isabelle/HOL system. Following the instructions given in [13], the definition should be quite feasible.

3. A ringoid (see [10]) is an algebraic structure containing homomorphisms and endomorphisms over different algebraic structures. To give an implementation of it could be useful for several problems in group theory. An attempt to introduce category theory in Isabelle has been already made (see [6]), and some similarities can be found between the implementation of morphisms given there and the representation of homomorphisms that we proposed in Section 5, needed to work in a more effective way with them. Specification of ringoids is not a complicated task in Isabelle. Nevertheless, the benefits and drawbacks of this new approach with respect to our third and fourth approaches should be carefully studied.

4. In order to develop algebraic topology, one of the basic concepts needed (see definitions in Section 2) is infinite complexes (of modules, for instance). Some attempts have been made to produce an implementation of them (see [8]), but definitive results have not been obtained yet. For their implementation, lazy lists or extensible records might be used, but these ideas will be studied in future work.

Acknowledgements

We are grateful to Wolfgang Windsteiger and the anonymous referees for the useful comments on the previous version of this paper.

References

1. C. Ballarin, *Locales and Locale Expressions in Isabelle/Isar*. In Stefano Berardi et al., TYPES 2003, LNCS vol. 3085, pp. 34 - 50.
2. S. Berghofer, *Program Extraction in Simply-Typed Higher Order Logic*, TYPES 2002, LNCS vol. 2646, pp. 21-38
3. R. Brown, *The twisted Eilenberg-Zilber theorem*, Celebrazioni Arch. Secolo XX, Simp. Top. (1967), pp. 34-37.
4. J. Calmet, *Some Grand Mathematical Challenges in Mechanized Mathematics*, In T. Hardin and Renaud Rioboo, editors, Calculemus 2003, pp. 137 - 141.
5. X. Dousson, F. Sergeraert and Y. Siret, *The Kenzo program*, http://www-fourier.ujf-grenoble.fr/~sergerar/Kenzo/
6. J. Glimming, *Logic and Automation for Algebra of Programming*, Master Thesis, Maths Institute, University of Oxford, August 2001, available at http://www.nada.kth.se/~glimming/publications.shtml.
7. V. K. A. M. Gugenheim, *On the chain complex of a fibration*, Illinois Journal of Mathematics 16 (1972), pp. 398-414.
8. H. Kobayashi, H. Suzuki and H. Murao, *Rings and Modules in Isabelle/HOL*, In T. Hardin and Renaud Rioboo, editors, Calculemus 2003, pp. 124 - 129.
9. L. Lambán, V. Pascual and J. Rubio, *An object-oriented interpretation of the EAT system*, Appl. Algebra Eng. Commun. Comput. vol. 14(3) pp. 187-215.
10. S. Mac Lane, *Homology*, Springer, 1994.
11. W. Naraschewski and M. Wenzel, *Object-Oriented Verification based on Record Subtyping in Higher-Order Logic*, In J. Grundy and M. Newey, editors, TPHOLs'98, LNCS vol. 1479, pp. 349-366.
12. T. Nipkow, L. C. Paulson and M. Wenzel, *Isabelle/HOL: A proof assistant for higher order logic*, LNCS vol. 2283, 2002.
13. L. Paulson, *Defining Functions on Equivalence Classes*, Report available at http://www.cl.cam.ac.uk/users/lcp/papers/Reports/equivclasses.pdf
14. J. Rubio and F. Sergeraert, *Constructive Algebraic Topology*, Lecture Notes Summer School in Fundamental Algebraic Topology, Institut Fourier, 1997.
15. J. Rubio, F. Sergeraert and Y. Siret, *EAT: Symbolic Software for Effective Homology Computation*, Institut Fourier, Grenoble, 1997.
16. W. Shih, *Homologie des espaces fibrés*, Publications Mathématiques de l'I.H.E.S. 13, 1962.
17. L. Théry, *Proving and computing: A certified version of the Buchberger's algorithm*, in Proceeding of the 15th International Conference on Automated Deduction, LNAI vol. 1421, 1998.

Algorithm-Supported
Mathematical Theory Exploration:
A Personal View and Strategy

Bruno Buchberger*

Research Institute for Symbolic Computation
Johannes Kepler University Linz
A-4040 Linz, Austria
bruno.buchberger@jku.at

Abstract. We present a personal view and strategy for algorithm-supported mathematical theory exploration and draw some conclusions for the desirable functionality of future mathematical software systems. The main points of emphasis are: The use of schemes for bottom-up mathematical invention, the algorithmic generation of conjectures from failing proofs for top-down mathematical invention, and the possibility to program new reasoners within the logic on which the reasoners work ("meta-programming").

1 A View of Algorithm-Supported
Mathematical Theory Exploration

Mathematical theories are collections of formulae in some formal logic language (e.g. predicate logic). Mathematical theory exploration proceeds by applying, under the guidance of a human user, various algorithmic reasoners for producing new formulae from given ones and aims at building up (large) mathematical knowledge bases in an efficient, reliable, well-structured, re-usable, and flexible way. Algorithm-supported mathematical theory exploration may also be seen as the logical kernel of the recent field of "Mathematical Knowledge Management" (MKM), see [10] and [5]. In the past few decades, an impressive variety of results has been obtained in the area of algorithm-supported reasoning both in terms of logical and mathematical power as wells as in terms of software systems, see for example, ALF [18], AUTOMATH [12], COQ [2], ELF [21], HOL [13], IMPS [1],

---

* Sponsored by FWF (Österreichischer Fonds zur Förderung der Wissenschaftlichen Forschung; Austrian Science Foundation), Project SFB 1302 ("Theorema") of the SFB 13 ("Scientific Computing"), and by RICAM (Radon Institute for Computational and Applied Mathematics, Austrian Academy of Science, Linz). The final version of this paper was written while the author was a visiting professor at Kyoto University, Graduate School of Informatics (Professor Masahiko Sato). I would like to thank my coworkers A. Craciun, L. Kovacs, Dr. T. Kutsia, and F. Piroi for discussions on the contents of this paper, literature work, and help in the preparation of this paper.

B. Buchberger and J.A. Campbell (Eds.): AISC 2004, LNAI 3249, pp. 236–250, 2004.
© Springer-Verlag Berlin Heidelberg 2004

ISABELLE [20], LEGO [17], MIZAR [3], NUPRL [11], OMEGA [4]. We also made an effort in this area, see the THEOREMA [9] system.

However, as a matter of fact, these reasoning systems are not yet widely used by the "working mathematicians" (i.e. those who do math research and/or math teaching). This is in distinct contrast to the current math (computer algebra/numerics) systems like MATHEMATICA, MAPLE, etc. which, in the past couple of years, have finally found their way into the daily practice of mathematicians. In this paper, we want to specify a few features of future systems for algorithm-supported mathematical theory exploration which, in our view, are indispensable for making these systems attractive for the daily routine of working mathematicians. These features are:

- *Integration of the Functionality of Current Mathematical Systems*: Reasoning systems must retain the full power of current numerics and computer algebra systems including the possibility to program one's own algorithms in the system.
- *Attractive Syntax*: Reasoning systems must accept and produce mathematical knowledge in attractive syntax and, in fact, in flexible syntax that can be defined, within certain limitations, by the user.
- *Structured Mathematical Knowledge Bases*: Reasoning systems must provide tools for building up and using (large) mathematical knowledge libraries in a structured way in uniform context with the algorithm libraries and with the possibility of changing structures easily.
- *Reasoners for Invention and Verification*: Reasoning systems must provide reasoners both for inventing (proposing) and for verifying (proving, disproving) mathematical knowledge.
- *Learning from Failed Reasoning*: The results of algorithmic reasoners (in particular, algorithmic provers) must be post-processable. In particular, also the results of failing reasoning attempts must be accessible for further (algorithmic) analysis because failure is the main source of creativity for mathematical invention.
- *Meta-Programming*: The process of (algorithm-supported) mathematical theory exploration is nonlinear: While exploring mathematical theories using known algorithmic reasoners we may obtain ideas for new algorithmic reasoners and we may want to implement them in our system and use them in the next exploration round. Hence, reasoning systems must allow "meta-programming", i.e. the algorithmic reasoners must be programmed basically in the same logic language in which the formulae are expressed on which the reasoners work.

I think it is fair to say that, in spite of the big progress made in the past couple of years, none of the current reasoning systems fulfills all the above requirements. In fact, some of the requirements are quite challenging and will need a lot more of both fundamental research and software technology. It is not the goal of this paper, to compare the various current systems (see the above references) w.r.t. to these six requirements. Rather, we will first summarize how we tried to fulfil the first three requirements in our THEOREMA system (see the web site http://www.theorema.org/ and the papers cited the) and then, in the main part of the paper, we will sketch a few ideas (which are partly implemented and partly being implemented in THEOREMA) that may contribute to the other three requirements.

2   Integration of the Functionality of Current Mathematical Systems

Let us start from the fact that predicate logic is not only rich enough to express any mathematical proposition but it includes also, as a sublanguage, a universal programming language and, in fact, a practical and elegant programming language. For example, in THEOREMA syntax, the following formula

$$\forall_F (\textit{is-Gröbner-basis}[F] \Leftrightarrow \forall_{f,g \in F} \ \textit{reduced}[\textit{S-polynomial}[f,g],F]=0) \qquad (1)$$

can be read as a proposition (which can be proved, by the inference rules of predicate logic, from a rich enough knowledge base K) but it can also be read as an algorithm which can be applied to concrete input polynomial sets F, like $\{x^2y - x - 2, xy^2 - xy + 1\}$. Application to inputs proceeds by using a certain simple subset of the inference rules of predicate logic, which transform

$$\textit{is-Gröbner-basis}[F]$$

into a truth value using a knowledge base that contains elementary Gröbner bases theory and the above formula (1). (Note that THEOREMA uses brackets for function and predicate application.)

Here is another example of a predicate logic formula (in THEOREMA syntax), which may be read as a (recursive) algorithm:

$$\textit{is-sorted}[\langle\rangle]$$
$$\forall_x \ \textit{is-sorted}[\langle x\rangle]$$
$$\forall_{x,y,\overline{z}} \ \textit{is-sorted}[\langle x,y,\overline{z}\rangle] \Leftrightarrow \begin{array}{l} x \le y \\ \textit{is-sorted}[\langle y,\overline{z}\rangle]. \end{array}$$

(Formulae placed above each other should be read as conjunctions.)

In this algorithmic formula, we use "sequence variables", which in THEOREMA are written as overbarred identifiers: The substitution operator for sequence variables allows the substitution of none, one, or finitely many terms whereas the ordinary substitution operator for the ordinary variables of predicate logic (like x and y in our example) allows only the substitution of exactly one term for a variable. Thus, for example,

$$\textit{is-sorted}[\langle 1,2,2,5,7\rangle],$$

by using the generalized substitution operator for the sequence variable $\overline{z}$, may be rewritten into

$$1 \leq 2$$
$$\wedge$$
$$\textit{is-sorted}[\langle 2, 2, 5, 7 \rangle],$$

etc.

(The extension of predicate logic by sequence variable is practically useful because it allows the elegant formulation of pattern matching programs. For example, the formula $p[\overline{v}, w, \overline{x}, w, \overline{y}] = 1$ says that p yields 1 if two arguments are identical. The meta-mathematical implications of introducing sequence variables in predicate logic are treated in [16].)

Hence, THEOREMA has the possibility of formulating and executing any mathematical algorithm within the same logic frame in which the correctness of these algorithms and any other mathematical theorem can be proved. In addition, THEOREMA has a possibility to invoke any algorithm from the underlying MATHEMATICA algorithm library as a black box so that, if one wants, the entire functionality of MATHEMATICA can be taken over into the logic frame of THEOREMA.

3   Attractive Syntax

Of course, the internal syntax of mathematical formulae on which algorithmic reasoners may be based, theoretically, is not a big issue. Correspondingly, in logic books, syntax is kept minimal. As a means for human thinking and expressing ideas, however, syntax plays an enormous role. Improving syntax for the formulae appearing in the input and the output of algorithmic reasoners and in mathematical knowledge bases is an important practical means for making future math systems more attractive for working mathematicians. Most of the current reasoning systems, in the past few years, started to add functionalities that allow to use richer syntax.

The THEOREMA system put a particular emphasis on syntax right from the beginning, see the papers on THEOREMA on the home page of the project http://www.theorema.org/. We allow two-dimensional syntax and user-programming of syntax, nested presentation of proofs, automated generation of natural language explanatory text in automated proofs, hyperlinks in proofs etc. In recent experiments, we even provided tools for graphical syntax, called "logico-graphic" syntax, see [19]. The implementation of these feature was made comparatively easy by the tools available in the front-end of MATHEMATICA which is the programming environment for THEOREMA. Of course, whatever syntax is programmed by the user, the formulae in the external syntax is then translated into the internal standard form, which is a nested MATHEMATICA expression, used as the input format of all the THEOREMA reasoners. For example,

$$\forall_{x,y,\overline{z}} \left(\textit{is-sorted}[\langle x, y, \overline{z} \rangle] \Leftrightarrow \begin{array}{c} x \leq y \\ \textit{is-sorted}[\langle y, \overline{z} \rangle] \end{array} \right)$$

internally is the nested prefix expression:

$$^{\mathrm{TM}}\mathrm{ForAll}[$$
$$\bullet\mathrm{range}[\bullet\mathrm{simpleRange}[\bullet\mathrm{var}[x]],$$
$$\bullet\mathrm{simpleRange}[\bullet\mathrm{var}[y]],$$
$$\bullet\mathrm{simpleRange}[\bullet\mathrm{var}[\bullet\mathrm{seq}[z]]]],$$
$$\mathrm{True},$$
$$^{\mathrm{TM}}\mathrm{Iff}[\mathrm{is\text{-}sorted}[^{\mathrm{TM}}\mathrm{Tuple}[\bullet\mathrm{var}[x], \bullet\mathrm{var}[y], \bullet\mathrm{var}[\bullet\mathrm{seq}[z]]]],$$
$$^{\mathrm{TM}}\mathrm{And}[^{\mathrm{TM}}\mathrm{LessEqual}[\bullet\mathrm{var}[x], \bullet\mathrm{var}[y]],$$
$$\mathrm{is\text{-}sorted}[^{\mathrm{TM}}\mathrm{Tuple}[\bullet\mathrm{var}[y], \bullet\mathrm{var}[\bullet\mathrm{seq}[z]]]]]]]]$$

Translators exist for translating formulae in the syntax of other systems to the THEOREMA syntax and vice versa. A translator to the recent OMDOC [15] standard is under development.

4   Structured Mathematical Knowledge Bases

We think that future math systems need "external" and "internal" tools for structuring mathematical knowledge bases.

External tools are tools that partition collections of formulae into sections, subsections, etc. and maybe, in addition, allow to give key words like 'Theorem', 'Definition' etc. and labels to individual formulae so that one can easily reference and re-arrange individual formulae and blocks of formulae in large collections of formulae. Such tools, which we call "label management tools", are implemented in the latest version of THEOREMA, see [23].

In contrast, internal structuring tools consider the structure of mathematical knowledge bases itself as a mathematical relation, which essentially can be described by "functors" (or, more generally, "relators"). The essence of this functorial approach to structuring mathematical knowledge can be seen in a formula as simple as

$$\forall_{x,y} \left(x \sim y \Leftrightarrow \begin{matrix} x \leq y \\ y \leq x \end{matrix} \right).$$

In this formula, the predicate $\sim$ is defined in terms of the predicate $\leq$. We may want to express this relation between $\sim$ explicitly by defining the higher-order binary predicate AR:

$$\forall_{\sim, \leq} \left(AR[\sim, \leq] \Leftrightarrow \forall_{x,y} \left(x \sim y \Leftrightarrow \begin{matrix} x \leq y \\ y \leq x \end{matrix} \right) \right).$$

We may turn this "relator" into a "functor" by defining, implicitly,

$$\forall_{\leq} \left(\forall_{x,y} \left(AR[\leq][x,y] \Leftrightarrow \begin{matrix} x \leq y \\ y \leq x \end{matrix} \right) \right).$$

(In this paper, we do not distinguish the different types of the different variables occurring in formulae because we do not want to overload the presentation with technicalities.)

Functors have a computational and a reasoning aspect. The *computational* aspect of functors already received strong attention in the design of programming languages, see for example [14]: If we know how to compute with $\leq$ then, given the above definition of the functor AF, we also know how to compute with the predicate $AF[\leq]$.

However, in addition, functors have also a *reasoning* aspect which so far has received little attention in reasoning systems: For example, one can easily prove the following "conservation theorem":

$$\forall_{\sim,\leq} \left(\begin{array}{l} AR[\sim,\leq] \\ is\text{-}transitive[\leq] \end{array} \Rightarrow is\text{-}transitive[\sim] \right),$$

where

$$\forall_{\leq} \left(is\text{-}transitive[\leq] \Leftrightarrow \forall_{x,y,z} \left(\begin{array}{l} x \leq y \\ y \leq z \end{array} \Rightarrow x \leq z \right) \right).$$

In other words, if we know that $\leq$ is in the "category" of transitive predicates and $\sim$ is related to $\leq$ by the relator AR (or the corresponding functor AF) then $\sim$ also is in the category of transitive predicates. Of course, studying a couple of other conservation theorems for the relator AR, one soon arrives at the following conservation theorem

$$\forall_{\sim,\leq} \left(\begin{array}{l} AR[\sim,\leq] \\ is\text{-}quasi\text{-}ordering[\leq] \end{array} \Rightarrow is\text{-}equivalence[\sim] \right),$$

which is the theorem which motivates the consideration of the relator AR.

After some analysis of the propositions proved when building up mathematical theories, it should be clear that, in various disguises, conservation theorems make up a considerable portion of the propositions proved in mathematics.

Functors for computation, in an attractive syntax, are available in THEOREMA, see for example the case study [6]. Some tools for organizing proofs of conservation theorems in THEOREMA where implemented in the PhD thesis [24] but are not integrated into the current version of THEOREMA. An expanded version of these tools will be available in the next version of THEOREMA.

5   Schemes for Invention

Given a (structured) knowledge base K (i.e. a structured collection of formulae on a couple of notions expressed by predicates and functions), in one step of the theory exploration process, progress can be made in one of the following directions:

- invention of *notions* (i.e. axioms or definitions for new functions or predicates),
- invention and verification of *propositions* about notions,
- invention of *problems* involving notions,
- invention and verification of *methods* (algorithms) for solving problems.

The results of these steps are then used for expanding the current knowledge base by new knowledge. For verifying (proving propositions and proving the correctness of methods), the current reasoning systems provide a big arsenal of algorithmic provers. In the THEOREMA system, by now, we implemented two provers for general predicate logic provers (one based on natural deduction, one based on resolution) and special provers for analysis, for induction over the natural numbers and tuples, for geometric theorems (based on Gröbner bases and other algebraic methods), and for combinatorial identities. We do not go into any more detail on algorithmic proving since this topic is heavily treated in the literature, see the above references on reasoning systems. Rather, in this paper, our emphasis is on algorithmic invention. For this, in this section, we propose the systematic use of formulae schemes whereas, in the next section, we will discuss the use of conjecture generation from failing proofs. In a natural way, these two methods go together in an alternating bottom-up/top-down process.

We think of schemes as formulae that express the accumulated experience of mathematicians for inventing mathematical axioms (in particular definitions), propositions, problems, and methods (in particular algorithms). Schemes should be stored in a (structured) schemes library L. This library could be viewed as part of the current knowledge. However, it is conceptually better to keep the library L of schemes, as a general source of ideas for invention, apart from the current knowledge base K that contains the knowledge that is available on the specific notions (operations, i.e. predicates and functions) of the specific theory to be explored at the given stage.

The essential idea of formulae schemes can, again, be seen already in the simple example of the previous section on functors: Consider the formula

$$\forall_{x,y} \left(x \sim y \Leftrightarrow \begin{matrix} x \leq y \\ y \leq x \end{matrix} \right)$$

that expresses a relation between the two predicates $\leq$ and $\sim$. We can make this relation explicit by the definition

$$\forall_{\sim,\leq} \left(AR[\sim,\leq] \Leftrightarrow \forall_{x,y} \left(x \sim y \Leftrightarrow \begin{matrix} x \leq y \\ y \leq x \end{matrix} \right) \right).$$

This scheme (which we may also conceive as a "functor") can now be used as a condensed proposal for "inventing" some piece of mathematics depending on how we look at the current exploration situation:

Invention of a new notion (definition, axiom):

If we assume that we are working w.r.t. a knowledge base K in which a binary predicate constant P occurs, then we may apply the above scheme by introducing a new binary predicate constant Q and asserting

$$AR[Q, P].$$

In other words, application of the above scheme "invents the new notion Q together with the explicit definition"

$$\forall_{x,y}\left(Q[x,y] \Leftrightarrow \frac{P[x,y]}{P[y,x]}\right).$$

Invention of a new proposition:

If we assume that we are working w.r.t. a knowledge base K in which two binary predicate constants P and Q occur, then we may apply the above scheme by conjecturing

$$AR[Q,P],$$

i.e.

$$\left(Q[x,y] \Leftrightarrow \frac{P[x,y]}{P[y,x]}\right).$$

We may now use a suitable (automated) prover from our prover library and try to prove or disprove this formula. In case the formula can be proved, we may say that the application of the above scheme "invented a new proposition on the known notions P and Q".

Invention of a problem:

Given a binary predicate constant P in the knowledge base K, we may ask to "find" a Q such that

$$AR[Q,P].$$

In this case, application of the scheme AR "invents a problem". The nature of the problem specified by AR depends on what we allow as "solution" Q. If we allow any binary predicate constant occurring in K then the problem is basically a "method retrieval and verification" problem in K: We could consider all or some of the binary predicate constants Q in K as candidates and try to prove/disprove $AR[Q,P]$. However, typically, we will allow the introduction of a new constant Q and ask for the invention of formulae D that "define" Q so that, using K and D, $AR[Q,P]$ can be proved. Depending on which class of formulae we allow for "defining" Q (and possible auxiliary operations), the difficulty of "solving the problem" (i.e. finding Q) will vary drastically. In the simple example above, if we allow to use the given P and explicit definitions, then the problem is trivially solved by the formula $AR[Q,P]$ itself, which can be considered as an "algorithm" w.r.t. the auxiliary operation P. If, however, we allow only the use of certain elementary algorithmic functions in K and only the use of recursive definitions then this problem may become arbitrarily difficult.

Invention of a method (algorithm):

This case is formally identical to the case of invention of an axiom or definition. However methods are normally seen in the context of problems. For example if we have a problem

$$PR[Q,R]$$

of finding an operation Q satisfying PR in relation to certain given operations R then we may try the proposal

$$AR[Q,P]$$

as a method. If we restrict the schemes allowed in the definition of Q (and auxiliary operations) to being recursive equalities then we arrive at the notion of algorithmic methods.

Case Studies:

The creative potential of using schemes, together with failing proof analysis, can only be seen in major case studies. At the moment, within the THEOREMA project, three such case studies are under way: One for Gröbner bases theory, [7], one for teaching elementary analysis, and one for the theory of tuples, [8]. The results are quite promising. Notably, for Gröbner bases theory, we managed to show how the author's algorithm for the construction of Gröbner bases can be automatically synthesized from a problem specification in predicate logic. Since the Gröbner bases algorithm is deemed to be nontrivial, automated synthesis may be considered as a good benchmark for the power of mathematical invention methods.

Not every formula scheme is equally well suited for inventing definitions, propositions, problems, or methods. Here are some examples of typical simple schemes in each of the four areas:

A Typical Definition Scheme:

$$\forall_{P,Q} \left(alternating\text{-}quantification[Q, P] \Leftrightarrow \forall_f \left(Q[f] \Leftrightarrow \forall_x \exists_y \forall_z \, P[f, x, y, z] \right) \right).$$

Many of the notions in elementary analysis (e.g. "limit") are generated by this (and similar) schemes.

A Typical Proposition Scheme:

$$\forall_{f,g,h} \left(is\text{-}homomorphic[f, g, h] \Leftrightarrow \forall_{x,y}(h[f[x, y]] = g[h[x], h[y]]) \right).$$

Of course, all possible "algebraic" interactions (describable by equalities between various compositions) of functions are candidates for proposition schemes.

A Typical Problem Scheme:

$$\forall_{A,P,Q} \left(explicit\text{-}problem[A, P, Q] \Leftrightarrow \forall_x \, \begin{array}{c} P[A[x]] \\ Q[x, A[x]] \end{array} \right).$$

This seems to be one of the most popular problem schemes: Find a method A that produces, for any x, a "standardized" form $A[x]$ (that satisfies $P[A[x]]$) such that $A[x]$ is still in some relation (e.g. equivalence) Q with x. (Examples: sorting problem, problem of constructing Gröbner bases, etc.)

Two Typical Algorithm Schemes:

$$\forall_{F,c,s,g,h_1,h_2} \left(\begin{array}{l} divide\text{-}and\text{-}conquer[F, c, s, g, h_1, h_2] \Leftrightarrow \\ \left(F[x] = \begin{cases} s[x] & \Leftarrow c[x] \\ g[F[h_1[x]], F[h_2[x]]] & \Leftarrow \text{otherwise} \end{cases} \right) \end{array} \right).$$

This is of course the scheme by which, for example, the merge-sort algorithm can be composed from a merge function g, splitting functions h_1, h_2, and operations c, s that handle the base case.

$$
\begin{aligned}
&\forall_{G,lc,df} \\
&\left(\begin{array}{l}
\textit{critical--pair--completion}[G, lc, df] \;\Leftrightarrow \\
\left(\begin{array}{l}
\forall_F \; (G[F] = G[F, \text{pairs}[F]]) \\
\forall_F \; (G[F, \langle\rangle] = F) \\
\forall_{F,g_1,g_2,\overline{p}} \\
\left(\begin{array}{l}
G[F, \langle\langle g_1, g_2\rangle, \overline{p}\rangle] = \\
\text{where} \left[f = lc[g_1, g_2], \; h_1 = trd[rd[f, g_1], F], \; h_2 = trd[rd[f, g_2], F], \right. \\
\left. \left\{\begin{array}{ll}
G[F, \langle\overline{p}\rangle] & \Leftarrow h_1 = h_2 \\
G[F \frown df[h_1, h_2], \langle\overline{p}\rangle \asymp \left\langle\langle F_k, df[h_1, h_2]\rangle \big|_{k=1,\ldots,|F|}\right\rangle] & \Leftarrow \text{otherwise}
\end{array}\right. \right]
\end{array}\right)
\end{array}\right)
\end{array}\right).
\end{aligned}
$$

This is the scheme by which, for example, the author's Gröbner bases algorithm can be composed from the function lc ("least common reducible", i.e. the least common multiple of the leading power products of two polynomials) and the function df ("difference", i.e. the polynomial difference in the polynomial context). The algorithm scheme can be tried in all domains in which we have a reduction function rd (whose iteration is called trd, "total reduction") that reduces objects f w.r.t. to finite sets F of objects. The algorithm (scheme) starts with producing all pairs of objects in F and then, for each pair $\langle g_1, g_2\rangle$, checks whether the total reduction of $lc[g_1, g_2]$ w.r.t. g_1 and g_2 yields identical results h_1 and h_2, respectively. If this is not the case, $df[h_1, h_2]$ is added to F.

6   Learning from Failed Reasoning

Learning from failed reasoning can be applied both in the case of proving propositions and in the case of synthesizing methods (algorithms). In this paper, we will sketch the method only for the case of method (algorithm) synthesis.

Let us assume that we are working with some knowledge base K and we are given some problem, e.g.

$$\textit{explicit-problem}[A, P, Q],$$

where the predicates P and Q are "known", i.e. they occur in the knowledge base K. For example, P and Q could be the unary predicate $\textit{is-finite-Gröbner-basis}$ and the binary predicate $\textit{generate-the-same-ideal}$, respectively. (The exact definitions of these predicates are assumed to be part of K. For details see [7]). Then we can try out various algorithm schemes in our algorithm schemes library L. In the example, let us try out the general scheme $\textit{critical-pair-completion}$, i.e. let us assume

$$\textit{critical-pair-completion}[A, lc, df].$$

(It is an interesting, not yet undertaken, research subject to try out for this problem systematically all algorithm schemes that are applicable in the context of polynomial domains and study which ones will work. Interestingly, in the literature, so far all methods for constructing Gröbner bases rely on the *critical-pair-completion* scheme in one way or the other and no drastically different method has been found!)

The scheme for the unknown algorithm A involves the two unknown auxiliary functions lc and df. We now start an (automated) proof of the correctness theorem for A, i.e. the theorem

$$\forall_F \left(\frac{is\text{--}finite\text{--}Gr\ddot{o}bner\text{--}basis\,[A[F]]}{generate\text{--}the\text{--}same\text{--}ideal\,[F, A[F]]} \right).$$

Of course, this proof will fail because nothing is yet known about the auxiliary functions lc and df. We now analyze carefully the situation in which the proof fails and try to guess requirements lc and df should satisfy in order to be able to continue with the correctness proof. It turns out that a relatively simple requirements generation techniques suffices for generating, completely automatically, the following requirement for lc

$$\forall_{g1,g2} \left(\begin{array}{l} lp[g1] \mid lc[g1,g2] \\ lp[g2] \mid lc[g1,g2] \\ \forall_p \left(\begin{array}{l} lp[g1] \mid p \\ lp[g2] \mid p \end{array} \Rightarrow (lc[g1,g2] \mid p) \right) \end{array} \right),$$

where $lp[f]$ denotes the leading power product of polynomial f. This is now again an explicit problem specification, this time for the unknown function lc. We could again run another round of our algorithm synthesis procedure using schemes and failing proof analysis for synthesizing lc. However, this time, the problem is easy enough to see that the specification is (only) met by

$$lc[g_1, g_2] = lcm[lp[g_1], lp[g_2]],$$

where $lcm[p, q]$ denotes the least common multiple of power products p and q. In fact, the specification is nothing else then an implicit definition of lcm and we can expect that an algorithm satisfying this specification is part of the knowledge base K.

Heureka! We managed to get the main idea for the construction of Gröbner bases completely automatically by applying algorithm schemes, automated theorem proving, and automated failing proof analysis. Similarly, in a second attempt to complete the proof of the correctness theorem, one is able to derive that, as a possible df, we can take just polynomial difference.

The current requirements generation algorithms from failing proof, roughly, has one rule. Given the failing proof situation, collect all temporary assumptions $T[x_0, \ldots, A[\ldots], \ldots]$ and temporary goals $G[x_0, \ldots m[\ldots, A[\ldots], \ldots]]$, where m is (one of) the auxiliary operations in the algorithm scheme for the unknown algorithm A and x_0, etc. are the current "arbitrary but fixed" constants, and produce the following requirement for m:

$$\forall_{x,\ldots,y,\ldots} (T[x, \ldots, y, \ldots] \Rightarrow G[x, \ldots, m[\ldots, y, \ldots]]).$$

This rule is amazingly powerful as demonstrated by the above nontrivial algorithm synthesis. The invention of failing proof analysis and requirements generation techniques is an important future research topic.

7   Meta-programming

Meta-programming is a subtle and not widely available language feature. However, we think that it will be of utmost practical importance for the future acceptance of algorithm-supported mathematical theory exploration systems. In meta-programming, one wants to use the logic language both as an object and a meta-language. In fact, in one exploration session, the language level may change several times and we need to provide means for governing this change in future systems. For example, in an exploration session, we may first want to define (by applying the divide–and–conquer scheme) a sorting algorithm

$$\forall_x \left(sort[x] = \begin{cases} x & \Leftarrow is\text{-}short\text{-}tuple[x] \\ merge[x, sort[left[x]], sort[right[x]]] & \Leftarrow otherwise \end{cases} \right).$$

Then we may want to apply one of the provers in the system to prove the algorithm correct, i.e. we want to prove

$$\forall_x \left(\begin{array}{l} is\text{-}sorted[sort[x]] \\ contain\text{-}the\text{-}same\text{-}elements[x, sort[x]] \end{array} \right)$$

by calling, say, a prover *tuple-induction*

$$\forall_x\ tuple\text{-}induction \left[\left(\begin{array}{l} is\text{-}sorted[sort[x]] \\ contain\text{-}the\text{-}same\text{-}elements[x, sort[x]] \end{array} \right), K_0 \right],$$

and checking whether the result is the constant 'proved', where K_0 is the knowledge base containing the definition of *sort* and *is-sorted* and definitions and propositions on the auxiliary operations.

Now, *tuple-induction* itself is an algorithm which the user may want to define himself or, at least, he might want to inspect and modify the algorithm available in the prover library. This algorithm will have the following structure

$$tuple\text{-}induction[\forall_x F, K] =$$
$$and[tuple\text{-}induction[F_{x \leftarrow \langle \rangle}, K],$$
$$tuple\text{-}induction[F_{x \leftarrow \langle x0, \overline{y0} \rangle}, append[K, F_{x \leftarrow \langle \overline{y0} \rangle}]],$$

where $\leftarrow$ is substitution and x_0 and $\overline{y_0}$ are "arbitrary but fixed" constants (that must be generated from x, F, and K).

Also, *tuple-induction* itself needs a proof that could proceed, roughly, by calling a general predicate logic prover in the following way

general-prover
$$\left[\forall_{x,F,K} \left(\left(tuple\text{-}induction[\forall_x F, K] = \text{``proved''} \right) \Rightarrow \left(append[ind, K] \models \forall_x F \right) \right) \right],$$

where *ind* are the induction axioms for tuples and $\models$ denotes logic consequence.

Of course, the way it is sketched above, things cannot work: We are mixing language levels here. For example, the '∀' in the previous formula occurs twice but on different language layers and we must not use the same symbol for these two occurrences. Similarly, the '∀' in the definition of tuple-induction has to be different from the '∀' in the definition of *sort*. In fact, in the above sketch of an exploration session, we are migrating through three different language layers.

We could now use separate name spaces for the various language layers. However, this may become awkward. Alternatively, one has to introduce a "quoting mechanism". The problem has been, of course, addressed in logic, see [22] for a recent discussion but we think that still a lot has to be done to make the mechanism practical and attractive for the intended users of future reasoning systems. In THEOREMA, at present, we are able to define algorithms and formulate theorems, apply algorithms and theorems to concrete inputs ("compute") and prove the correctness of algorithms and theorems in the same session and using the same language, namely the THEOREMA version of predicate logic. For this, the name spaces are automatically kept separate without any action needed from the side of the user. However, we are not yet at the stage where we could also formulate provers and prove their correctness within the same language and within the same session. This is a major design and implementation goal for the next version of THEOREMA. An attractive solution may be possible along the lines of [25] and [26].

8   Conclusion

We described mathematical theory exploration as a process that proceeds in a spiral. In each cycle of the spiral, new axioms (in particular definitions), propositions, problems, and methods (in particular algorithms) are introduced and studied. Both invention of axioms, propositions, problems, and methods as well as verification of proposition and methods can be supported by algorithms ("algorithmic reasoners"). For this, at any stage of an exploration, we have to be able to act both on the object level of the formulae (axioms, propositions, problems, methods) and the meta-level of reasoners. We sketched a few requirements future mathematical exploration systems should fulfil in order to become attractive as routine tools for the exploration activity of working mathematicians.

A particular emphasis was put on the interaction between the use of schemes (axiom schemes, proposition schemes, problem schemes, and algorithm schemes) and the algorithmic generation of conjectures from failing proofs as a general heuristic, computer-supportable strategy for mathematical invention. The potential of this strategy was illustrated by the automated synthesis of the author's Gröbner bases algorithm. The ideas presented in this paper will serve as work plan for the next steps in the development of the THEOREMA system.

References

1. An Interactive Mathematical Proof System. http://imps.mcmaster.ca/.
2. The COQ Proof Assistant. http://coq.inria.fr/.
3. The MIZAR Project. http://www.mizar.org/.

4. The OMEGA System. http://www.ags.uni-sb.de/~omega/.
5. A. Asperti, B. Buchberger, and J. H. Davenport, editors. *Mathematical Knowledge Management: Second International Conference, MKM 2003 Bertinoro, Italy, February 16–18, 2003*, volume 2594 of *Lecture Notes in Computer Science.* Springer-Verlag, 2003.
6. B. Buchberger. Groebner Rings in THEOREMA: A Case Study in Functors and Categories. Technical Report 2003-46, SFB (Special Research Area) Scientific Computing, Johannes Kepler University, Linz, Austria, 2003.
7. B. Buchberger. Towards the automated synthesis of a Gröbner bases algorithm. *RACSAM (Review of the Spanish Royal Academy of Sciences)*, 2004. To appear.
8. B. Buchberger and A. Craciun. Algorithm synthesis by lazy thinking: Examples and implementation in THEOREMA. In *Proc. of the Mathematical Knowledge Management Symposium*, volume 93 of *ENTCS*, pages 24–59, Edunburgh, UK, 25–29 November 2003. Elsevier Science.
9. B. Buchberger, C. Dupré, T. Jebelean, F. Kriftner, K. Nakagawa, D. Vasaru, and W. Windsteiger. The THEOREMA project: A progress report. In M. Kerber and M. Kohlhase, editors, *Proc. of Calculemus'2000 Conference*, pages 98–113, St. Andrews, UK, 6–7 August 2000.
10. B. Buchberger, G. Gonnet, and M. Hazewinkel, editors. *Mathematical Knowledge Management: Special Issue of Annals of Mathematics and Artificial Intelligence*, volume 38. Kluwer Academic Publisher, 2003.
11. R. Constable. *Implementing Mathematics Using the NUPRL Proof Development System.* Prentice-Hall, 1986.
12. N. G. de Bruijn. The mathematical language AUTOMATH, its usage, and some of its extensions. In M. Laudet, D. Lacombe, L. Nolin, and M. Schützenberger, editors, *Proc. of Symposium on Automatic Demonstration, Versailles, France*, volume 125 of *LN in Mathematics*, pages 29–61. Springer Verlag, Berlin, 1970.
13. M. Gordon and T. Melham. *Introduction to* HOL: *A Theorem Proving Environment for Higher-Order Logic.* Cambridge University Press, 1993.
14. R. Harper. Programming in Standard ML. Carnegie Mellon University. www-2.cs.cmu.edu/~rwh/smlbook/online.pdf, 2001.
15. M. Kohlhase. OMDOC: Towards an internet standard for the administration, distribution and teaching of mathematical knowledge. In J.A. Campbell and E. Roanes-Lozano, editors, *Artificial Intelligence and Symbolic Computation: Proc. of the International Conference AISC'2000*, volume 1930 of *Lecture Notes in Computer Science*, pages 32–52, Madrid, Spain, 2001. Springer Verlag.
16. T. Kutsia and B. Buchberger. Predicate logic with sequence variables and sequence function symbols. In A. Asperti, G. Bancerek, and A. Trybulec, editors, *Proc. of the 3rd Int. Conference on Mathematical Knowledge Management, MKM'04*, volume 3119 of *Lecture Notes in Computer Science*, Bialowieza, Poland, 19–21 September 2004. Springer Verlag. To appear.
17. Zh. Luo and R. Pollack. LEGO proof development system: User's manual. Technical Report ECS-LFCS-92-211, University of Edinburgh, 1992.
18. L. Magnusson and B. Nordström. The ALF proof editor and its proof engine. In H. Barendregt and T. Nipkow, editors, *Types for Proofs and Programs*, volume 806 of *LNCS*, pages 213–237. Springer Verlag, 1994.
19. K. Nakagawa. Supporting user-friendliness in the mathematical software system THEOREMA. Technical Report 02-01, PhD Thesis. RISC, Johannes Kepler University, Linz, Austria, 2002.
20. L. Paulson. ISABELLE: the next 700 theorem provers. In P. Odifreddi, editor, *Logic and Computer Science*, pages 361–386. Academic Press, 1990.

21. F. Pfenning. ELF: A meta-language for deductive systems. In A. Bundy, editor, *Proc. of the 12th International Conference on Automated Deduction, CADE'94*, volume 814 of *LNAI*, pages 811–815, Nancy, France, 1995. Springer Verlag.

22. F. Pfenning. Logical frameworks. In A. Robinson and A. Voronkov, editors, *Handbook of Automated Reasoning*, chapter 17, pages 1063–1147. Elsevier Science and MIT Press, 2001.

23. F. Piroi and B. Buchberger. An environment for building mathematical knowledge libraries. In A. Asperti, G. Bancerek, and A. Trybulec, editors, *Proc. of the 3rd Int. Conference on Mathematical Knowledge Management, MKM'04*, volume 3119 of *Lecture Notes in Computer Science*, Bialowieza, Poland, 19–21 September 2004. Springer Verlag. To appear.

24. E. Tomuta. An architecture for combining provers and its applications in the THEOREMA system. Technical Report 98-14, PhD Thesis. RISC, Johannes Kepler University, Linz, Austria, 1998.

25. M. Sato. Theory of judgments and derivations. In Arikawa, S. and Shinohara, A. eds., *Progress in Discovery Science*, Lecture Notes in Artificial Intelligence 2281, pp. 78 – 122, Springer, 2002.

26. M. Sato, T. Sakurai, Y. Kameyama and A. Igarashi. Calculi of meta-variables. In Baaz M. and Makowsky, J.A. eds., *Computer Science Logic*, Lecture Notes in Computer Science 2803, pp. 484 – 497, Springer, 2003.

An Expert System on Detection, Evaluation and Treatment of Hypertension

E. Roanes-Lozano[1], E. López-Vidriero Jr.[2], L.M. Laita[3],
E. López-Vidriero[4,5], V. Maojo[3], and E. Roanes-Macías[1]

[1] Universidad Complutense de Madrid, Algebra Dept.
c/ Rector Royo Villanova s/n, 28040-Madrid, Spain
{eroanes,roanes}@mat.ucm.es
[2] Hospital Universitario "Nuestra Señora de Valme"
Autovía Sevilla-Cádiz s/n, 41004-Sevilla, Spain
lopez-vidriero@telefonica.es
[3] Universidad Politécnica de Madrid, Artificial Intelligence Dept.
Campus de Montegancedo, Boadilla del Monte, 28660-Madrid, Spain
{laita,maojo}@fi.upm.es
[4] Hospital "Gregorio Marañón", Hypertension Unit
Doctor Castelo 49, 28009-Madrid, Spain
[5] Universidad Complutense de Madrid, Internal Medicine Dept.
Facultad de Medicina, Ciudad Universitaria, 28040-Madrid, Spain

Abstract. This article presents the development of an expert system on detection, evaluation and treatment of hypertension (high blood pressure), together with an outline of the associated computational processes. It has been implemented on the computer algebra system *Maple*.

The starting point is the knowledge about detection of major cardiovascular disease (CVD) risk and about detection, evaluation and treatment of hypertension, provided in table and algorithm format by experts from some well known medical societies and committees. As the drug choices for treating hypertension depends on whether the patient suffers, among other diseases, from high CVD risk or not, the expert system consists of two consecutive subsystems. The first one determines the CVD risk level, meanwhile the second one uses the output of the first one to detect, evaluate and, if necessary, suggest a treatment of hypertension.

The knowledge expressed by the experts was revised and reorganized by the authors. Curiously, some errata were found in some of the classifications provided in table format.

The computational processes involved are simple because the cases are already separated (disjoint), so they can be translated into IF...THEN... rules that can be applied using classifications in simple procedures. Therefore, verification is restricted to considering a few cases (the conjunction of all diseases together with different CDV risks). Nevertheless, we think this is a very useful piece of software for practitioners, because, to reach, for instance, the treatment of hypertension, several messy concatenated steps have to be taken.

Keywords: Expert Systems, Hypertension, Computer Algebra Systems, Medical Informatics.

B. Buchberger and J.A. Campbell (Eds.): AISC 2004, LNAI 3249, pp. 251–264, 2004.
© Springer-Verlag Berlin Heidelberg 2004

1   Introduction

A considerable amount of work, including [1] (probably the best known, although not the only example) has been carried out in the field of medical rule-based expert systems. For instance, a formal method for creating appropriateness criteria [2] has been developed at the RAND Health Service (Santa Monica, California): it is based on combining literature review and ratings from expert panels. One problem with the current RAND methodology is to ensure logical consistency: RAND researchers manually develop algorithms from ratings and re-meet with experts to settle inconsistencies [3]. Alternative approaches to this problem have been proposed e.g. in [4–6].

Many health organizations are now developing and using evidence-based policies for medical care. They are created using systematic methodologies. In the field of major cardiovascular disease (CVD) risk and hypertension (high blood pressure), we can underline: the *International Society of Hypertension* of the *World Health Organization* (ISH-WHO), the *Joint National Committee on Prevention, Detection, Evaluation and Treatment of High Blood Pressure* (JNC) of the *National Heart, Lung, and Blood Institute* (*U.S. Dept. of Health and Human Services*)... [7–11]. They are not designed for retrospective evaluation of medical practice, but as a decision-making aid. As far as we know, expert systems directly based on these reports in the field of hypertension haven't been spread. A new proposal of expert system on the topic is detailed below.

2   General Description

This article extracts the information and knowledge from different sources [7–11] and merges it with the knowledge of the three coauthors that are MDs. Let us underline that one of the authors is the chief of the Hypertension Unit of one of the main Spanish hospitals and other of the authors is finishing his PhD in Medicine on this topic. The accurate knowledge representation needed to implement these topics has required a refinement in the details of the tables and algorithms provided. Finally the logic and computational processes are simple but sound.

2.1   Data Acquisition

First, different data from the patient have to be acquired. They are:

- Blood Pressure (BP): Systolic Blood Pressure (SBP) and Diastolic Blood Pressure (DBP) figures.
- Major cardiovascular disease (CVD) risk (Boolean) factors, apart from hypertension (*): obesity, dyslipidemia, diabetes melitus (DM) (†), cigarette smoking, physical inactivity, microalbuminuria (estimated filtration rate < 60mL/min)(‡), age (>55 for men, >65 for women), family history of premature CVD (men age <55, women age <65). Note that hypertension is another CVD risk factor, but it is not a datum, as it is deduced from the SBD and DBP.

- Other cardiovascular disease (CVD) risk-related (Boolean) facts (**): target organ damage (TOD), associated clinical alterations (ACA), diabetes (1).
- Other pathologies suffered by the patient (Boolean) (***): heart failure, post myocardial infarction, high CVD risk (2), diabetes (1), chronic kidney disease (3), recurrent stroke prevention.

2.2   First Subsystem: CDV Risk Evaluation

The patient's SBP and DBP are used to allocate him in a (refined) category of BP (Normal/ Prehypertension/ Stage 1 Hypertension/ Stage 2a Hypertension/ Stage 2b Hypertension). This classification is intermediate between those of [9] and of [10, 11].

The computational processes involved are simple because the cases are already separated (disjoint), so they can be translated into IF...THEN... rules, that can be applied using classifications in simple procedures.

The assessment of the factors in (*) is provided by history, physical examination, laboratory tests, electrocardiograms...

From the refined BP category obtained and the data in (*), a grade of CVD risk is calculated using tables [7, 8], that we have also translated into IF...THEN... rules. The computational processes are simple too (because the cases are again disjoint). The same will happen with the second subsystem.

2.3   Second Subsystem: Hypertension Detection, Evaluation and Treatment

Now the items in (**) and (***) are assessed: (1) has already been asked in (†); (2) has just been evaluated from the data in (*) by the first subsystem; (3) is deduced from (‡) and the information about the rest of the items can be found in the patient's clinical history.

The patient's SBP and DBP are used to allocate him in a category of BP (Normal - Prehypertension - Stage 1 Hypertension - Stage 2 Hypertension). This classification is the one suggested in [10, 11].

From this information, and following a certain treatment algorithm, the authors of [10, 11] recommend:

- individuals who are prehypertensive: practice lifestyle modifications in order to reduce the risk of developing hypertension (no drug therapy)
- individuals with hypertension (stages 1 and 2) be treated with an appropriate drug therapy if a first trial of lifestyle modification fails:
  - individuals with hypertension of stage 1 (not suffering other pathologies) that need drug therapy be treated with thiazide diuretic (THIAZ) or a combination including this therapy
  - individuals with hypertension of stage 2 (not suffering other pathologies) that need drug therapy be treated with a 2-drug combination
  - individuals with hypertension of any stage (also suffering from other pathologies) that need drug therapy should have it adjusted according to those pathologies.

According to [10,11], the goal is to return both SBP and DBP into the Normal or Prehypertensive regions. In case of patients with diabetes or chronic kidney disease, the goal is stricter: SBP<130 mmHg and DBP<80 mmHg.

2.4 Algorithm

The whole process is summarized in the algorithm of Figure 1. The different steps of the process will be analyzed in detail in the next sections.

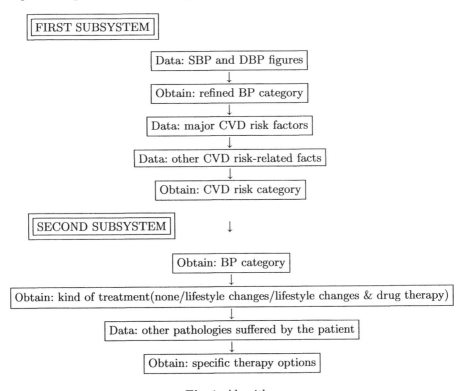

Fig. 1. Algorithm

We plan to inform JNC of the inaccuracies detected in their tables, as well as the existence of this expert system (it is going to be extensively clinically tested prior to spreading).

3 BP Classification

JNC 7 [10,11] proposes the classification of BP for adults aged 18 and more included in Table 1. It will be the one used by our second subsystem. A refined one (Table 2), inspired by that of JNC6 [9] and [8], where Stage 2 is divided into stages 2a and 2b, but suppressing the "optimal" category, is used in the first subsystem.

Table 1. Classification of BP

Category	SBP mmHg		DBP mmHg
Normal	<120	and	<80
Prehypertension	120-139	or	80-89
Hypertension, Stage 1	140-159	or	90-99
Hypertension, Stage 2	≥160	or	≥100

Table 2. Refined classification of BP

Category	SBP mmHg		DBP mmHg
Normal	<120	and	<80
Prehypertension	120-139	or	80-89
Hypertension, Stage 1	140-159	or	90-99
Hypertension, Stage 2a	160-179	or	100-109
Hypertension, Stage 2b	≥180	or	≥110

Table 3. Table 1, once corrected

Category	SBP mmHg		DBP mmHg
Normal	<120	and	<80
Prehypertension	120-139	and	<90
		or	
	<140	and	80-89
Hypertension, Stage 1	140-159	and	<100
		or	
	<160	and	90-99
Hypertension, Stage 2	≥160	or	≥100

It seems clear that the four regions that Table 1 pretends to define are those in Figure 2.

But only the "Normal" and "Stage 2 Hypertension" regions are correctly defined. For instance, if the pseudo-classification of Table 1 was followed strictly, the diagnosis for a patient with $SBP = 130mmHg \wedge DBP = 100mmHg$ would be a double "Prehypertension" and "Stage 2 Hypertension"! The reason is that the regions "Prehypertension" and "Stage 1 Hypertension" defined by those tables don't have the shape shown in Figure 2. For instance, the Prehypertension region really defined by Table 1 is shown in Figure 3. Therefore, the regions are not disjoint, so Table 1 doesn't provide a real classification. Admitting that the intention of those who wrote Table 1 was to describe the regions in Figure 3, a correct description would appear in Table 3.

The same problem arises in Table 2. It is similarly corrected (Table 4). Curiously, these inaccuracies were found when implementing the corresponding IF... THEN... rules, so there has been a certain feedback from the implementation.

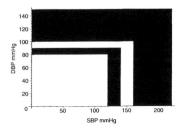

Fig. 2. Regions presumably corresponding to Table 1

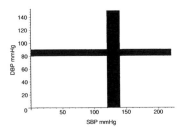

Fig. 3. Prehypertension region, as defined in Table 1

Nevertheless, although a computer language needs correct specifications, these inaccuracies are overcome by practitioners (due to their implicit knowledge).

4   CVD Risk Classification

In [8], when hypertension is present, and partially following [7, 9], Table 5 is proposed for classifying the CVD risk (CVDRF stands for CVD Risk Factors, apart from hypertension; and TOD, DM and ACA were introduced in Section 2.1).

Again, this is a pseudo-classification, as, for instance, "Stage 2a Hypertension" ∧ "0 CVDRF" ∧ "ACA" would also lead to a double: "Medium CDV risk" and "Very High CDV risk".

In our opinion the three first rows should include that there are no ACA and the two first rows should also include that the patient doesn't suffer from DM and there is no TOD. Again, these inaccuracies are overcome by practitioners (due to their implicit knowledge), but we have precised these details in our adapted table (Table 6). We shall represent by nCVDRF the number of CVDRF (hypertension is excluded).

Moreover, as this is going to be a first step of a deduction process, two columns (corresponding to Normal BP and Prehypertension) have been added in the left hand side of the new Table 6, so that we can forward fire all cases. From the medical point of view, these two new columns have no interest, as Table 5 was a classification of CVD Risk when hypertension was present. Specifically,

Table 4. Table 2, once corrected

Category	SBP mmHg		DBP mmHg
Normal	<120	and	<80
Prehypertension	120-139	and	<90
		or	
	<140	and	80-89
Hypertension, Stage 1	140-159	and	<100
		or	
	<160	and	90-99
Hypertension, Stage 2a	160-179	and	<110
		or	
	<180	and	100-109
Hypertension, Stage 2b	≥180	or	≥110

Table 5. Classification of CVD risk when hypertension is present

CVD Risk Factors and other CVD facts	St. 1 Hyp.	St. 2a Hyp.	St. 2b Hyp.
0 CVDRF	Low	Medium	High
1 or 2 CVDRF	Medium	Medium	Very high
≥ 3 CVDRF ∨ TOD ∨ DM	High	High	Very high
ACA	Very high	Very high	Very high

the last row of those first two columns has no sense, as without hypertension there can be no ACA associated to hypertension.

5   Basic Notions on Rule Based Expert Systems

A Rule Based Expert System (RBES) contains *rules*, *facts* and *integrity constraints* that translate the information provided by the experts. Let us describe these items using an elementary set of rules written in bivalued logic (α, β, γ, δ, ε, η, θ, ς represent propositional variables). In addition to these items, the RBES is provided an *inference engine* and, possibly, a *user interface*.

Rule 1. $\alpha \wedge \neg\beta \to \gamma$
Rule 2. $\gamma \to \delta \vee \varepsilon$
Rule 3. $\eta \to \neg\varepsilon$
Rule 4. $\delta \to \theta$
Rule 5. $\varsigma \to \neg\theta$

Letters like α or letters preceded by $\neg$, like $\neg\beta$, are called *literals*.

Rules are implications between a literal or conjunction of literals (*antecedent*), and a literal or disjunction of literals (*consequent*).

A *potential fact* is any literal that appears in at least one antecedent but in no consequent of the rules. In our example, $\alpha, \neg\beta$, η and ς are potential facts.

Table 6. Classification of CVD risk (adapted)

CVD Risk Factors and other CVD facts	Normal	Prehyp.	St.1 Hyp.	St.2a Hyp.	St.2b Hyp.
nCVDRF=0 $\land$ ¬TOD $\land$ ¬DM $\land$ ¬ACA	Low	Low	Low	Medium	High
1≤ nCVDRF ≤ 2 $\land$ ¬TOD $\land$ ¬DM $\land$ ¬ACA	Low	Low	Medium	Medium	Very high
(nCVDRF ≥ 3 $\lor$ TOD $\lor$ DM) $\land$ ¬ACA	Low	Low	High	High	Very high
ACA	Low	Low	Very high	Very high	Very high

A *fact* is any potential fact that the user states to hold. For instance, the user may state that the potential fact ζ is given as a fact.

An *Integrity Constraint* (IC) is any well formed combination of literals and connectives that an expert has asserted to never hold. For instance, suppose that an expert assesses that γ and ¬θ never hold together. We then have new information to be added to the RBES: the negation of the integrity constraint (NIC): ¬$(\gamma \land \neg\theta)$.

It is said that a rule can be *forward fired* if all the literals in the antecedent are facts (or *derived facts*, see below). Forward firing corresponds to the formal logic rule of "modus ponens". The inference engine of the RBES fires the rules.

In some cases, the consequent of a rule may be part of the antecedent of another rule, like γ in Rules 1 and 2. If α, ¬β and η are facts, then forward firing Rule 1 outputs γ, being γ an example of what is called a *derived fact*. Now, forward firing Rule 2 and Rule 3 output $\delta \lor \varepsilon$ and ¬ε, so δ and ¬ε are also derived facts (because δ is a tautological consequence of $\delta \lor \varepsilon$ and ¬ε).

There are two main types of logical inconsistencies.

If we are given the RBES composed of rules 4 and 5 and facts δ and ζ, forward firing outputs the logical contradiction $\theta \land \neg\theta$.

Suppose that we are given the facts α, ¬β and ζ in a RBES composed only by Rules 1 and 5, to which the NIC ¬$(\gamma \land \neg\theta)$ is added. In this case, forward firing leads to the IC $\gamma \land \neg\theta$, so the logical contradiction IC $\land$ NIC is obtained.

In the RBES composed of rules R1, R2, R3, R4 and R5, if α, ¬β, η and ζ are given as facts, both types of contradictions are reached.

Observe that if a logical contradiction is inferred (by forward firing) from the rules and ICs in a RBES and a *consistent set of facts* (that is, a set of facts that doesn't include together a literal and its negation), we say that the RBES is *inconsistent*. In such case all formulae written in the language of the RBES do follow from the rules and ICs in the RBES and that consistent set of facts.

6   Detailing the Expert System

A RBES can use a Boolean logic, a multi-valued modal logic, a fuzzy logic... Boolean logic should not be used when a level of certainty is to be assigned to the data (what is usually the case when dealing with medical information). This is not the case here, as the input data have a True/False format and data acquisition is precise (all input data can be clearly determined from the history, physical examination, laboratory tests, electrocardiograms...). Moreover, proceeding with only part of the input data could be very dangerous (e.g. the suggested medication could be incompatible).

As said in the introduction, this work is mainly based on the conclusions of the committees of the ISH-WHO [7], JNC 6 [9] and JNC 7 [10, 11]. These committees release conclusions based on large-scale clinical trials and sound statistical studies, and are considered as the main reference in the field. This widely accepted knowledge is what our program automatices. Our inference engine doesn't add any new rule, it just incorporates the conclusions of these committees, so we don't have to deal with confidence thresholds of new added knowledge.

6.1 Translation of the Corrected BP Classification Tables to a Set of Production Rules

Now that the semialgebraic regions defined by Table 3 are disjoint, the information contained in it 3 can be translated as the following set of rules:

IF SBP<120 AND DBP<80 THEN Normal

IF 120≤SBP<140 AND DBP<90 THEN Prehypertension

IF SBP<140 AND 80≤DBP<90 THEN Prehypertension

IF 140≤SBP<160 AND DBP<100 THEN Stage 1 Hypertension

IF SBP<160 AND 90≤DBP<100 THEN Stage 1 Hypertension

IF SBP≥160 THEN Stage 2 Hypertension

IF DBP≥100 THEN Stage 2 Hypertension

The information in Table 4 can be translated similarly.

In previous occasions we have implemented expert systems devoted to other illnesses where the rules really had to be fired once and again and again, as new derived facts were obtained each time. Then we used a Gröbner bases-based inference engine [5, 6, 12, 13] based on a application to RBES of these authors [14], itself based on previous works for Boolean logic [15, 16] and modal multi-valued logics [17]. Such an approach would be infraused here, as rules have to be fired only one time. Therefore we have decided to implement simple procedures in *Maple* using IF...THEN...ELSE... nested conditionals, like the following procedure, that determines the particular BP case.

```
> class_BP:=proc()
>    global BP;
>    if SBP>=160 or DBP>=100
>       then BP:=Hypertension2
>       elif SBP>=140 or DBP>=90
```

```
>                then BP:=Hypertension1
>                elif SBP>=120 or DBP>=80
>                    then BP:=Prehypertension
>                    else BP:=Normal
>        fi;
>      NULL;
>    end:
```

Observe that, comparing with the tables, the nesting has been done from bottom to top, so that shorter conditions (almost identical to those in the original table) could be used.

6.2   Translation of the Adapted CVD Classification Table to a Set of Production Rules

The information in Table 6 can also be translated as a set of rules. We list afterwards some condensed ones (we have include OR connectives in the antecedents in order to shorten the number of rules, although this doesn't agree the definition in Section 5):

IF BP=Normal OR BP=Prehypertension THEN Low

IF ACA AND (BP = St.1 Hyp. OR BP = St.2a Hyp. OR BP = St.2b Hyp.) THEN Very High

IF $\neg ACA$ AND (nCVDRF > 0 OR TOD OR DM) THEN Very High.

IF $\neg$ACA AND nCVDRF = 0 AND $\neg$TOD AND $\neg$DM THEN High.

...

For the sake of brevity, the corresponding *Maple* code is not included (it has also been implemented using nested conditionals).

6.3   Translating the Kind of Treatment

As mentioned in section 2.3, the kind of treatment is based in the BP of the patient, but a distinction in the borders between regions is made if the patient suffers from diabetes or chronic kidney disease by the authors of [10, 11]. The different recommendations were detailed in Section 2.3. The computational approach is similar to those described in sections 6.1 and 6.2.

6.4   Specific Therapy Options

For instance, in [10, 11] the kind of treatment is based on the possible occurrence of other pathologies suffered by the patient. Of the different initial therapy options: THIAZ (thiazide diuretic), ACEI (angiotensin converting enzime inhibitor), ARB (angiotensin receptor blocker), BB (beta blocker) CCB (calcium channel blocker), ALDO ANT (aldosterone antagonist), only some are adequate

Table 7. Specific therapy options

Pathology	Therapy options
Heart Failure	THIAZ,BB,ACEI,ARB,ALDO ANT
Post miocardial Infarction	BB,ACEI,ALDO ANT
CVDR=High or CVDR=Very High	THIAZ,BB,ACEI,CCB
Diabetes	THIAZ,BB,ACEI,ARB,CCB
Chronic kidney disease	ACEI,ARB
Recurrent stroke prevention	THIAZ,ACEI

for each of the pathologies that the patient could suffer that interact with the hypertension's drug therapy. We have followed the recommendations of [10, 11], summarized in Table 7.

What the corresponding procedure we have implemented does is to consider sets of drug therapies for each pathology, and to intersect them (using *Maple*'s intersect command), according to the individual's sufferings.

7 Verification

Logic verification of the first subsystem (CDV risk evaluation) is not necessary, as disjoint cases are described. Almost the same can be said about the second subsystem (detection, evaluation and treatment of hypertension), where it is enough to check some details, e.g. the compatibility of medications when different pathologies appear simultaneously.

We have checked the conjunction of all pathologies, together with the different BP categories, and there were no anomalies found. For instance, if an individual with hypertension of stage 1 would answer "yes" to all questions, then we would have (input lines are preceded by ">", meanwhile output lines are centered):

```
> treat_HBP();
           Lifestyle modifications suggested.
                  Therapy Options:
                       {ACEI}
```

8 User Interface

A friendly interface, similar to the one implemented for the *CoCoA*-based Anorexia detection expert system [12], that simplifies data introduction and hides the computations carried out, has been implemented using *Maple*'s *Maplets* (Figure 4). In order to spread this work, we plan to translate the code into a compilable computer language and produce a stand-alone application.

9 Conclusions

We should stress on that interaction with the experts is necessary in order to search for the best representation of knowledge from the medical viewpoint.

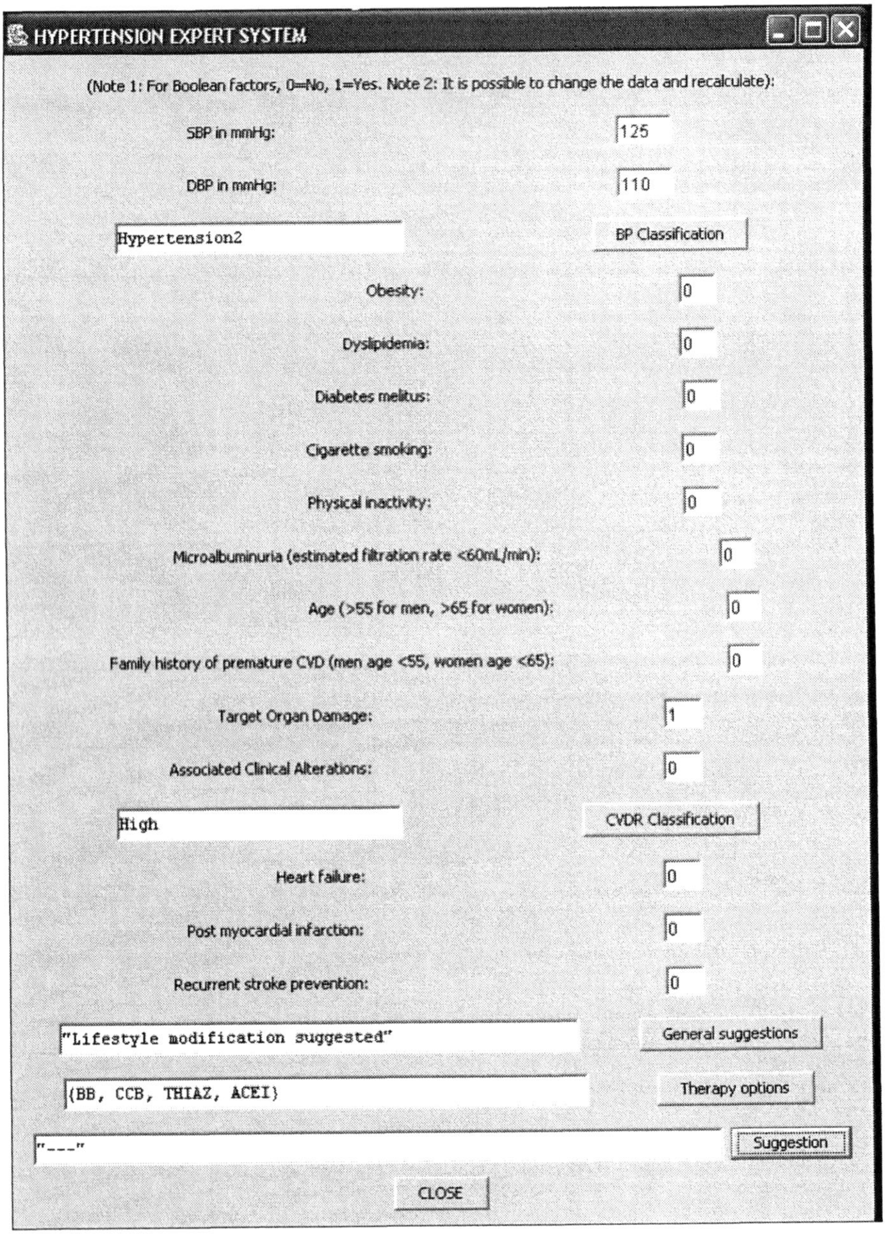

Fig. 4. User interface

What was unexpected was the feedback from the computer scientists that minor errata appeared in the tables provided by the doctors.

Using a Computer Algebra System has made possible to shorten the development period, thanks to its wide range of possibilities.

Finally, we think this can be a very useful piece of software for practitioners, because, to reach, for instance, the treatment of hypertension, several messy concatenated steps have to be taken. Practitioners can use the results in this work to improve their clinical practice (although the patient should be forwarded to a hypertension specialist). Moreover, specialists in the particular field of hypertension can use the results generated by this expert system to compare them with their personal opinion.

Acknowledgments

This work was partially supported by project PR3/04-12410 (Universidad Complutense de Madrid, Spain).

References

1. B. Buchanan and E. H. Shortliffe, eds., *Rule Based Expert Systems: the MYCIN Experiments of the Stanford Heuristic Programming Project* (Addison Wesley, New York, 1984).
2. M. Field and M., K. Lohr, eds., *Guidelines for Clinical Practice. From Development to Use* (National Academy Press. Washington, D.C. 1992).
3. S. Bernstein and J. Kahan, *Personal communication* (RAND Health Services, 1993).
4. R.N. Shiffman and R.A. Greenes, Improving Clinical Guidelines with Logic and Decision-table Techniques: Application to Hepatitis Immunization Recommendations. *Med. Decision Making* **14(3)** (1994) 245-254.
5. L. M. Laita, E. Roanes Lozano and V. Maojo, Inference and Verification in Medical Appropriateness Criteria, in J. Calmet and J. Plaza, eds., *Artificial Intelligence and Symbolic Computation, Procs. of AISC'98* (Springer-Verlag LNAI-1476, Berlin 1998) 183-194.
6. L. M. Laita, E. Roanes-Lozano, V. Maojo, E. Roanes-Macías, L. de Ledesma and L. Laita, An Expert System for Managing Medical Appropriateness Criteria Based on Computer Algebra Techniques, *Comps. & Maths. with Appls.* **42(12)** (2001) 1505-1522
7. Guidelines Subcommittee. 1999 International Society of Hypertension - World Health Organization (ISH-WHO), Guidelines for the management of hypertension. *Hypertension* **17** (1999) 151-183.
8. Anonymous, *Guía razonada de algoritmos médicos. Hipertensión arterial.* (Pfizer, Madrid, 2002).
9. JNC 6. National High Blood Pressure Education Program. The Sixth Report of the Joint National Committee on Prevention, Detection, Evaluation and Treatment of High Blood Pressure. *Arch. Intern. Med.* **157** (1997) 2413-2446.
10. Joint National Committee on Prevention, Detection, Evaluation and Treatment of High Blood Pressure (JNC), *JNC 7 Express. The Seventh Report of the Joint National Committee on Prevention, Detection, Evaluation and Treatment of High Blood Pressure.* (National Heart, Lung, and Blood Institute, U.S. Dept. of Health and Human Services, 2003).

11. A. V. Chobanian et al., JNC 7 Complete Version. Seventh Report of the Joint National Committee on Prevention, Detection, Evaluation and Treatment of High Blood Pressure. *Hypertension* **42** (2003) 1206-1252.

12. C. Pérez. L.M. Laita, E. Roanes-Lozano, L. Lázaro, J. González, L. Laita, A Logic and Computer Algebra-Based Expert System for Diagnosis of Anorexia, *Math. Comp. Simul.* 58 (2002) 183-202.

13. V. Maojo, L. M. Laita, E. Roanes Lozano, J. Crespo and J. Rodríguez Pedrosa, A New Computerized Method to Verify and Disseminate Medical Appropriatenes Criteria, in R. W. Brause and E. Hanisch, eds., *Medical Data Analysis, Procs. of ISMDA '2000* (Springer-Verlag LNAI-1933, Berlin 2000) 1212-217.

14. E. Roanes Lozano, L. M. Laita and E. Roanes Macías, A Polynomial Model for Multivalued Logics with a Touch of Algebraic Geometry and Computer Algebra. *Math. Comp. Simul.* **45(1/2)** (1998) 83-99.

15. J. Hsiang, Refutational Theorem Proving using Term-rewriting Systems, *Artificial Intelligence*, **25** (1985), 255-300.

16. D. Kapur and P. Narendran, An Equational Approach to Theorem Proving in First-Order Predicate Calculus, 84CRD296, *General Electric Corporate Research and Development Report* (Schenectady, NY, March 1984, rev Dec 1984). Also in *Proceedings of IJCAI-85*, (1985) 1446-1156.

17. J. Chazarain, A. Riscos, J. A. Alonso and E. Briales, Multivalued Logic and Gröbner Bases with Applications to Modal Logic. *J. Symb. Comput.*, **11** (1991) 181-194.

Two Revision Methods Based on Constraints: Application to a Flooding Problem

Mahat Khelfallah and Belaïd Benhamou*

Université de Provence, Laboratoire LSIS
39 Rue Joliot-Curie 13453 Marseille, France
{mahat,benhamou}@cmi.univ-mrs.fr

Abstract. In this paper, we are interested in geographic information revision in the framework of a flooding problem. We show how to express and how to revise this problem by using simple linear constraints. We present two revision strategies based on linear constraints resolution: the partial revision and the global revision methods. We apply these approaches on both a real-world flooding problem and random flooding instances.

Keywords: Revision, Linear constraints, Geographic information.

1 Introduction

Many research works have been done in the field of knowledge revision (see [1, 5] for overviews). Revision is the restoration of the knowledge base consistency by considering more reliable information. It identifies the inconsistencies, then corrects them by keeping a maximum of the initial information unchanged.

In this paper, we are interested in geographic knowledge revision based on linear constraints resolution in the framework of a flooding problem. We show how the flooding problem is expressed by linear constraints and propose two revision methods: the *partial revision* and the *global revision* methods.

The rest of this paper is organized as follows. We describe in section 2 the flooding problem and show how it can be represented by linear constraints. In section 3, the revision steps. We propose in section 4 two revision methods. We experiment, in section 5, both revision methods on a real flooding application and on random flooding instances. Section 6 concludes the work.

This paper is a condensed version of [4].

2 Description and Representation of the Flooding Problem

During a flooding in the Hérault valley (in the south of France), a part of this area was studied in order to get correct estimates of the water heights (above the

---

* This work has been supported by the REVIGIS project, IST-1999-14189.

B. Buchberger and J.A. Campbell (Eds.): AISC 2004, LNAI 3249, pp. 265–270, 2004.

sea level) in the flooded parcels. Two kinds of information are available: *(i)* The estimates of the water height in each parcel. *(ii)* The observations on hydraulic relations between some adjacent parcels.

The flooding problem is represented by the linear constraint network $N = (X, C)$ where X is a set of continuous variables, X_i corresponding to the water height in parcel i. C is the set of constraints representing both estimates on water heights and the hydraulic relations. The constraints are defined as follows:

Estimates on water heights : Each parcel i is associated with the constraints $l_i \leq X_i - X_0$ and $X_i - X_0 \leq u_i$, (X_0 represents the sea height, $X_0 = 0$). The scalar l_i and u_i are respectively the lower and the upper bounds of the interval delimiting the water height in the parcel i.

Hydraulic relations : An observed flow from the parcel i to the parcel j is expressed by the constraint $X_j - X_i \leq 0$. A hydraulic balance between parcels i and j is represented by the constraints: $X_j - X_i \leq 0$ and $X_i - X_j \leq 0$.

We associate to the linear constraint network $N = (X, C)$ a directed edge-weighted graph, $G_d = (X, E_d)$, called the *distance graph*. X is the set of vertices corresponding to the variables of the network N, and E_d is the set of arcs representing the set of constraints C. Each constraint $X_j - X_i \leq a_{ij}$ of C is represented by the arc $i \rightarrow j$, which is weighted by a_{ij}. In the sequel, n is the number of the vertices of the distance graph G_d and e is the number of its arcs.

Each of the two considered sources of information is consistent separately, but conflicts appear when both sources are merged. A conflict is detected when a flow is observed from the parcel i to the parcel j while the estimated water height in i is strictly less than the estimated water height in j. Since the observations on hydraulic relations are considered more reliable than the estimates on water heights, these estimates are *revised* to restore the consistency.

3   Revision

The revision of a linear constraint network consists in the detection of all the conflicts in the network, then the identification of a subset of constraints whose the correction restores the consistency, and finally the correction of such constraints.

3.1   Detection of Conflicts

We present a method which detects the conflicts of a linear constraint network. This method is based on the following theorem.

Theorem 1. *A linear constraint network is consistent if and only if its corresponding distance graph does not contain elementary negative circuits.*

Theorem 1 states that the removal of all the elementary negative circuits in the distance graph, restores the consistency of its corresponding linear constraint network. To detect the (elementary) negative circuits in the distance graph of a flooding problem, we use the result of the following proposition.

Proposition 2. *An elementary circuit in the distance graph of a flooding problem is negative if and only if it includes a negative path $\{i, 0, j\}$.*

To find all the conflicts, we have to enumerate all the pairs (i, j) of vertices involved in a negative path $\{i, 0, j\}$ and check existence of an elementary path from j to i which does not include the vertex 0. The complexity of the conflict detection procedure is $O(n^2 e)$ in the worst case.

3.2 Representation of the Conflicts

We recall that we are interested in revising estimates on water heights in the parcels. Each conflict between the constraints $l_i \leq X_i - X_0$ and $X_j - X_0 \leq u_j$ is represented by the tuple $(i, j, u_j - l_i)$. The set of conflicts is represented by the graph $G_c = (V, E)$. V is the set of vertices $V = \{Low_i, Upp_i : 1 \leq i \leq n\}$, where Low_i (respectively Upp_i) corresponds to the constraint $l_i \leq X_i - X_0$ (respectively to the constraint $X_i - X_0 \leq u_i$). Each conflict $(i, j, u_j - l_i)$ is represented by the edge (Low_i, Upp_j). G_c is called the *graph of conflicts*.

3.3 Computing a Subset of Constraints to Correct

To remove all the detected conflicts, some constraints involved in them have to be revised. We have to look for a subset of constraints whose the revision is sufficient to restore the consistency. A minimal revision of the problem needs to find a minimal subset of constraints whose the correction restores the consistency. This amounts to find a minimal vertex cover of the conflict graph.

Looking for a vertex cover of a fixed size is an NP-Complete problem [2], and looking for minimal vertex covers is NP-Hard.

To compute a minimal vertex cover of the graph of conflicts, we use the *Minimal-Cover* algorithm (see [4] for details) whose complexity is $O(n_c 2^{n_c})$ in the worst case, where n_c is the number of vertices of the conflict graph.

Another alternative is to consider a "good" vertex cover rather than a minimal one. Such cover is computed by the *Good-Cover* algorithm (see [4] for details) which consider only the vertices of highest degree in the graph of conflicts. The complexity of the *Good-cover* algorithm is $O(n_c^2)$ in the worst case.

3.4 Revision of the Conflicting Constraints

We shall see now, how to perform the corrections. Let $(i, j, u_j - l_i)$ be a conflict between the constraint $l_i \leq X_i - X_0$ and the constraint $X_j - X_0 \leq u_j$. The elimination of this conflict needs the revision of one of these constraints. The following proposition states how this operation is done.

Proposition 3. *Let $c = (i, j, u_j - l_i)$ be a conflict between the constraints $X_j - X_0 \leq u_j$ and $l_i \leq X_i - X_0$. Replacing the constraint $X_j - X_0 \leq u_j$ (respectively, the constraint $l_i \leq X_i - X_0$) by the constraint $X_j - X_0 \leq l_i$ (respectively, the constraint $u_j \leq X_i - X_0$) corrects the conflict c.*

The revision of a constraint cannot generate new conflicts and we have the following theorem.

Theorem 4. *If the constraints corresponding to a vertex cover of the conflict graph of a flooding problem are corrected, then the consistency of the problem is restored.*

4   Revision Methods

We propose in the following, two revision mothods : the *partial revision* and the *global revision* which offer a good compromise between minimality and efficiency of revision.

4.1   Partial Revision

The *partial revision* method consists first in identifying the list L of all the conflicts (i, j, d) and in sorting it according to the ascending order of the distances d. In the second phase, it takes a bundle L' formed of the n first conflicts of the list L, then computes a minimal vertex cover C_m of the conflict graph corresponding to L'. The constraints of C_m are revised and all the corresponding conflicts are removed from L. The second phase is repeated until all the conflicts of L are removed. The complexity of the *partial revision* algorithm is $O(n^2 2^{2n})$ in the worst case (see [4] for details).

4.2   The Global Revision

The *global revision* method detects all the conflicts, then computes a "good" subset of constraints whose the correction restores the consistency. This constraint subset corresponds to a "good" vertex cover of the conflict graph which is computed by the *Good-Cover* procedure. The complexity of the *global revision* algorithm is $O(n^2 e)$ ([4]).

5   Experimental Results

The revision algorithms presented in this paper are implemented in C and run on a pentium 4, 2.4 MHz with 512 MB of RAM. They are both tested on the real-world flooding problem in the Hérault valley and on random flooding instances.

The real flooding problem contains 180 parcels and 630 constraints. Both methods solve efficiently the problem (in less than 1 second) and their performances are comparable. We then experiment them on random instances of the flooding problem.

Generation of random flooding problems is based on two parameters: the number of variables n, and the constraint density d which is defined by $d = \frac{number\ of\ constraints}{n(n-1)}$. The thightness t of the constraints is represented by the

Table 1. Results on the global and partial revision of random flooding instances

Revision		Global revision			Partial revision		
		Density			Density		
		0.2	0.5	0.8	0.2	0.5	0.8
$n{=}500$	# conflicts	61844	62110	62049	61543	62034	61851
	# corrections	305	306	308	314	318	317
	Time (s)	3	1	1	3	2	2
$n{=}1000$	# conflicts	248277	249962	247737	248640	247275	247718
	# corrections	619	622	619	635	633	632
	Time (s)	25	14	9	33	21	16
$n{=}2000$	# conflicts	994792	993225	996207	998079	998293	993822
	# corrections	1247	1244	1247	1264	1257	1269
	Time (s)	202	109	75	306	228	187

interval $[100,300]$ where the upper and lower bounds of the water height estimates are generated. A sample of 50 problems is generated for each tuple (n, d, t) and the mesures are taken in average.

We can see in table 1 that the density does not affect significantly the number of detected conflicts. However, when the density grows, the revision becomes faster for both methods. This is due to the fact that when the density grows, the conflicts share more constraints, and their elimination is faster. We can see also that the number of corrected constraints in the *global revision* is always less than the one in the *partial revision*. The *global revision* is also faster than the *partial revision*. This is due to the minimal cover computing complexity.

6   Discussion

In a recent work [3], we have proposed different revision methods in the framework of the flooding problem. The first one is the *all conflicts method* which performs a minimal revision, but is applicable only to small instances of random flooding problems. The *global revision* method proposed in this paper is more efficient although it is not minimal. The *partial revision* outperforms the *hybrid revision* method proposed in [3]. The heuristic which selects first the conflicts having the smallest distances in each iteration, improves significantly the performances of the method. In future, we will try to extend this work to revise any problem expressed as a linear constraint network.

References

1. C. Alchourròn, P. Gärdenfors, D. Makinson, On the logic of theory change : Partial meet functions for contraction and revision, Journal of Symbolic Logic 50 (1985) 510–530.
2. M. Garey, D. Johnson, Computers and Intractability: A Guide to the Theory of NP-Completeness, W. H. Freeman and co., San Francisco, 1979.

3. M. Khelfallah, B. Benhamou, Geographic information revision based on constraints, accepted in the 16th European Conference on Articifial Intelligence (ECAI'04), Valencia, Spain, Aug. 24-27, 2004.

4. M. Khelfallah, B. Benhamou, Two revision methods based on constraints: Application to a flooding problem, Unpublished report available on request from the authors, 2004.

5. B. Nebel, How hard it is to revise a belief base? In D. Gabbay and Ph. Smets, eds, Handbook on Defeasible Reasoning and Uncertainty Management Systems, volume III: Belief Change, pp 77-145. Kluwer Academic, 1998.

Abstraction-Driven Verification
of Array Programs

David Déharbe[1], Abdessamad Imine[2], and Silvio Ranise[2]

[1] UFRN/DIMAp, Natal, Brazil
david@dimap.ufrn.br
[2] LORIA & INRIA-Lorraine, Nancy, France
{imine,ranise}@loria.fr

Abstract. We describe a refutation-based theorem proving algorithm capable of checking the satisfiability of non-ground formulae modulo (a combination of) theories. The key idea is the use of abstraction to drive the application of (i) ground satisfiability checking modulo theories axiomatized by equational clauses, (ii) Presburger arithmetic, and (iii) quantifier instantiation. A prototype implementation is used to discharge the proof obligations necessary to show the correctness of some typical programs manipulating arrays. On these benchmarks, the prototype automatically discharge more proof obligations than *Simplify* – the prover of reference for program checking – thereby confirming the viability of our approach.

1 Context and Motivation

Satisfiability procedures for equality and theories of standard data-types, such as arrays, lists, and arithmetic are at the core of most state-of-the-art verification tools (e.g., DPLL(T) [6] and *Simplify* [3]). These are required for a wide range of verification tasks and are fundamental for efficiency. Satisfiability problems have the form $T \wedge \phi$, where ϕ is a Boolean combination of ground literals, T is a *background theory*, and the goal is to prove that $T \wedge \phi$ is unsatisfiable. A *satisfiability procedure* for a theory T is an algorithm capable of checking whether $T \wedge \phi$ is satisfiable or not, for any ground formula ϕ.

The task of designing, proving correct, and implementing satisfiability procedures for decidable theories of practical interest is quite difficult. First, most problems involve more than one theory, so that one needs to combine satisfiability procedures (see e.g. [7]). Second, every satisfiability procedure needs to be proved correct: a key step is to show that whenever the algorithm reports "satisfiable," its final state represents a model of $T \wedge \phi$. Unfortunately, model-construction arguments can be quite complex (see e.g., [10]).

Although designing and combining decision procedures are necessary and very important activities to build practically useful reasoning tools, they are not sufficient. In fact, many proof obligations encountered in routine verification problems require a degree of flexibility which is not provided by actual state-of-the-art tools. The main problem is that only a tiny portion of such proof obligations falls exactly into the domain the procedures are designed to solve. As an

B. Buchberger and J.A. Campbell (Eds.): AISC 2004, LNAI 3249, pp. 271–275, 2004.
© Springer-Verlag Berlin Heidelberg 2004

example, consider a satisfiability procedure for the union of (the quantifier-free fragment of) the theory $\mathcal{E}$ of equality, the quantifier-free fragment of Presburger arithmetic $\mathcal{PA}$, and the formula

$$a < b \wedge \max(a, b) = a \wedge \forall X, Y.(X < Y \Rightarrow \max(X, Y) = Y). \qquad (1)$$

The available procedure will fail to detect the unsatisfiability of (1) since it does not know how to instantiate the quantified variables. Here, we propose a mechanism to augment decision procedures to cope with quantifiers.

2   Abstraction-Driven Refutation Theorem Proving

For lack of space, we assume the basic notions of first-order logic [4], the superposition calculus [8], and the Nelson-Oppen combination schema [7].

Handling Ground Formulae. Recently, there has been a lot of interest around theorem proving algorithms to discharge the proof obligations arising in various verification problems, which are large ground formulae with a complex Boolean structure to be checked satisfiable modulo a background theory $\mathcal{T}$. An integration of propositional solving (SAT, for short) and satisfiability checking modulo $\mathcal{T}$ has been advocated to efficiently discharge these formulae. The *abstract-check-refine* algorithm underlying such integrations is depicted in Figure 1, where $gfol2prop(\phi_g)$ returns the propositional abstraction of the ground formula ϕ_g (e.g. the abstraction of $(a = b \wedge b = c) \Rightarrow f(a) = f(c)$ is $(p \wedge q) \Rightarrow r$, p is the abstraction of $a = b$, q of $b = c$, and r of $f(a) = f(c)$), *prop2gfol* is its inverse, and *check_assign* is such that $check\_assign(\mathcal{T}, \beta) = unsat$ iff β is unsatisfiable modulo $\mathcal{T}$; otherwise, $check\_assign(\mathcal{T}, \beta) = sat$. Many refinements are necessary to make this schema efficient (see e.g. [6] for details).

Handling Non-ground Formulae. Although the algorithm of Figure 1 has proved to be very effective for hardware verification problems (see again [6] for experimental evidence of this), its applicability in program verification is limited

---

function check_ground (ϕ_g: *ground formula*)
1    $\phi^p \longleftarrow gfol2prop(\phi_g)$
2   **while** $\phi^p \neq false$ **do**
3   **begin**
4          $\beta^p \longleftarrow$ pick a propositional assignment of ϕ^p
5          $\rho \longleftarrow check\_assign(\mathcal{T}, prop2gfol(\beta^p))$
6   **if** $\rho = sat$ **then return** *sat*
7          $\phi^p \longleftarrow \phi^p \wedge \neg gfol2prop(\beta^p)$
8   **end**
9   **return** *unsat*
end

Fig. 1. A refutation-based algorithm for **ground** formulae

function check_fol (ϕ_0: *first-order formula*)
1    $\phi_0' \longleftarrow innermost\_univ(top\_ex(\phi_0))$
2    $(\phi_g, \Delta) \longleftarrow abs\_univ(\phi_0')$
3   **return** check_ground$'(\phi_g)$
end
function check_comb ($EqCls$: *set of clauses*, $\mathcal{PA}$: *first-order theory*,
$\qquad\qquad\qquad\qquad$ β: *conjunction of ground literals*)
1    $(\beta_1, \beta_2) \longleftarrow purify(\beta)$
2   **repeat**
3        $\alpha \longleftarrow$ pick an arrangement for β_1 and β_2
4        $(\rho_e, \xi_e) \longleftarrow check\_assign(EqCls, \beta_1 \cup \alpha)$
5        $\rho_a \longleftarrow check\_assign(\mathcal{PA}, \beta_2 \cup \alpha \cup \xi_e)$
6   **if** $(\rho_e = sat) \wedge (\rho_a = sat)$ **then return** sat
7   **until** all possible α's have been considered
8   **return** $unsat$
end

Fig. 2. A refutation-based algorithm for **non-ground** formulae

since the proof obligations arising in such a context frequently contain quantified variables, whose instantiation requires some ingenuity. In order to overcome this difficulty, we propose the algorithm check_fol in Figure 2, which augments check_ground with the capability of handling quantifiers. Let ϕ_0 be a non-ground formula to be checked for unsatisfiability modulo $\mathcal{ET} \cup \mathcal{PA}$, where $\mathcal{ET}$ is a finite set of equational clauses. First, we move the existential quantifiers not in the scope of a universal quantification to the top-most position in the formula, by using obvious rules such as $\exists x.P(x) \vee \exists x.Q(x)$ rewrites to $\exists x, y.(P(x) \vee Q(y))$ (cf. *top_ex*, line 1). Afterwards, we minimize the scope of the remaining quantifiers by using the rules for antiprenexing of [9] (cf. *innermost_univ*, line 1). For example, the formula $\forall x.\exists y.(P(x) \wedge Q(y))$ is transformed to $(\forall x.P(x)) \wedge Q(c_y)$, where c_y is a Skolem constant. Afterwards, we transform the formula ϕ_0' into a ground formula ϕ_g by replacing the quantified sub-formulae of ϕ_0' with fresh propositional letters and recording the association between propositional letters and the quantified sub-formulae in a mapping δ (cf. *abs_univ*, line 2) s.t. $\Delta(\phi_g) = \phi_0'$, where Δ denotes the homomorphic extension of δ to ϕ_g over the Boolean connectives. Then, we invoke the function check_ground$'$, which is a modified version of check_ground where the call to *check_assign* (at line 5 in Figure 1) is replaced by the following invocation of check_comb (cf. Figure 2):

$$5' \quad \rho \longleftarrow \text{check\_comb}(\mathcal{ET} \cup \hat{\delta}(\beta^p), \mathcal{PA}, prop2gfol(\beta^p)),$$

where the function $\hat{\delta}$ is s.t. $\hat{\delta}(\beta) = \{\delta(\lambda) \mid \lambda \in \beta$ and is in the domain of $\delta\}$, for any set β of ground literals. We assume that *check_assign* for $\mathcal{ET}$ is implemented by using a superposition prover as described in [1] and it is also capable of returning the ground facts which are derived by the prover (cf. ξ_e at line 4 of check_comb, Figure 2). The main difference between check_comb and the Nelson-Oppen combination schema is that the two component procedures communicate *also* ground facts on the signature of $\mathcal{PA}$ and *not only* the equalities over shared constants (cf. line 5 of check_comb). The function $purify$ is such that $purify(\beta) = (\beta_1, \beta_2)$, β is satisfiable iff $\beta_1 \wedge \beta_2$ is, β_1 (β_2, resp.) is a conjunction of literals not containing terms of $\mathcal{PA}$ ($\mathcal{ET}$, resp.), and α is an arrangement for β_1 and β_2 (see [11] for a definition).

Notice that the superposition prover plays two rôles. First, it implements the satisfiability procedure for $\mathcal{ET}$. Second, it finds ground instances of the quantified sub-formulae in ϕ'_0, which are then sent to the decision procedure for $\mathcal{PA}$. Notice that if the calls to the superposition prover returns, then check_comb terminates since there are only finitely many arrangements α's (see again [11] for details) and only finitely many clauses are generated by superposition (given its termination). So, check_fol is *terminating* whenever the call to check_comb returns. Finally, the algorithm is obviously sound but it is *incomplete* for an arbitrary $\mathcal{ET}$.

Table 1. Experimental Results on Array Programs

	$Simplify$	haRVey
Find (20)	14	20
Selection (14)	12	14
Heap (22)	13	17

3   Experiments: Verification of Array Programs

We have built a prototype version of check_fol on top of haRVey[1] [2] by adding an implementation of check_comb and a procedure for the quantifier-free fragment of $\mathcal{PA}$ based on the Fourier-Motzkin method (see e.g. [12]). We have used the Why[2] tool to generate the proof obligations encoding the (total) correctness of the following programs: Hoare's Find, Selection sort, and Heap sort [5]. Why is capable of generating proof obligations in the formats of several tools, among which haRVey and the state-of-the-art theorem prover for program verification, $Simplify$ [3]. In Table 1, the first column reports the identifier of the algorithm and the total number of proof obligations encoding its correctness; the second (third) column gives the number of proof obligations proved by $Simplify$ (our algorithm, resp.). For Find and Selection sort, $Simplify$ fails to prove about 25% of the proof obligations whereas our system proves them all. For Heap sort, $Simplify$ proves 13 over 22 formulae whereas our system proves 17. If we add three lemmas about an inductive predicate recognizing heaps (see [5] for details), then $Simplify$ proves 16 formulae whereas our system goes to 22. For Find, 19 of the 20 proof obligations are ground after the invocation of top_ex at line 1 in Figure 2. On the sole non-ground proof obligation, we have terminated $Simplify$ after half an hour of work; our system, instead, terminates with the correct answer in about 10 seconds. For Selection sort and Heap sort, no proof obligation becomes

---

[1] http://www.loria.fr/~ranise/haRVey
[2] http://why.lri.fr

ground after the invocation of *top_ex*. Both our system and *Simplify* are quite successful in finding suitable instances to the quantified variables but our system automatically discharges more proof obligations than *Simplify*.

We believe that the main reason for the success (measured in number of proof obligations automatically discharged) of our algorithm over *Simplify* is that we use superposition as the instantiation mechanism rather than the heuristic matching algorithm described in [3]. In fact, superposition performs inferences also on the quantified formulae: this may generate ground instances of facts which are otherwise impossible to derive by requiring that at most one formula participating to an inference is non-ground as it is the case for *Simplify*.

References

1. A. Armando, S. Ranise, and M. Rusinowitch. A Rewriting Approach to Satisfiability Procedures. *Info. and Comp.*, 183(2):140–164, June 2003.
2. D. Déharbe and S. Ranise. Light-Weight Theorem Proving for Debugging and Verifying Units of Code. In *Proc. of the 1st Int. Conf. on Software Engineering and Formal Methods (SEFM'03)*. IEEE Computer Society Press, 2003.
3. D. Detlefs, G. Nelson, and J. B. Saxe. Simplify: A Theorem Prover for Program Checking. Technical Report HPL-2003-148, HP Lab., 2003.
4. H. B. Enderton. *A Mathematical Introduction to Logic*. Academic Pr., 1972.
5. J.-C. Filliâtre and N. Magaud. Certification of Sorting Algorithms in the System Coq. In *Theorem Proving in Higher Order Logics: Emerging Trends*. 1999.
6. H. Ganzinger, G. Hagen, R. Nieuwenhuis, A. Oliveras, and C. Tinelli. DPLL(T): Fast Decision Procedures. In *Proc. of the 16th Int. Conf. on Computer Aided Verification (CAV'04)*, LNCS. Springer, 2004.
7. G. Nelson and D. C. Oppen. Simplification by cooperating decision procedures. *ACM TOPLAS*, 1(2):245–257, 1979.
8. R. Nieuwenhuis and A. Rubio. Paramodulation-based theorem proving. In A. Robinson and A. Voronkov, editors, *Hand. of Automated Reasoning*. 2001.
9. A. Nonnengart and C. Weidenbach. *Handbook of Automated Reasoning*, chapter Computing Small Clause Normal Forms. Elsevier Science, 2001.
10. N. Suzuki and D. R. Jefferson. Verification Decidability of Presburger Array Programs. *J. of the ACM*, 27(1):191–205, 1980.
11. C. Tinelli and M. T. Harandi. A new correctness proof of the Nelson–Oppen combination procedure. In *Front. of Combining Systems: Proc. of the 1st Int. Workshop*, pages 103–120. Kluwer Academic Publishers, 1996.
12. P. van Hentenryck and T. Graf. Standard Forms for Rational Linear Arithmetics in Constraint Logic Programming. *Ann. of Math. and Art. Intell.*, 5:303–319, 1992.

Singularities in Qualitative Reasoning

Björn Gottfried

Centre for Computing Technologies
University of Bremen, Germany
bg@tzi.de

Abstract. Qualitative Reasoning is characterised by making knowledge explicit in order to arrive at efficient reasoning techniques. It contrasts with often intractable quantitative models. Whereas quantitative models require computations on continuous spaces, qualitative models work on discrete spaces. A problem arises in discrete spaces concerning transitions between neighbouring qualitative concepts. A given arrangement of objects may comprise relations which correspond to such transitions, e.g. an object may be neither left of nor right of another object but precisely aligned with it. Such singularities are sometimes undesirable and influence underlying reasoning mechanisms. We shall show how to deal with singular relations in a way that is more closely related to commonsense reasoning than treating singularities as basic qualitative concepts.

1 Introduction

In this essay we shall discuss problems arising by describing arrangements of objects qualitatively. We are concerned with the relations depicted in Fig. 1, which have been introduced in [3]. We refer to the set of these relations as $\mathcal{BA}$. The relations in $\mathcal{BA}$ describe arrangements of intervals in the two-dimensional plane qualitatively; they can be considered as the two-dimensional analogue of Allen's one-dimensional interval relations [1]. $\mathcal{BA}$ is distinguished from other qualitative representations (cf. [2]) in that it comprises only *disconnection relations*. Relations between disconnected objects are of interest in a number of areas, mainly when spatiotemporal interactions between objects are to be described. It could be argued that connection relations are equally important. But there are no connections, for example, between road-users in traffic, pedestrians walking in a market square, sportsmen playing on a pitch, or generally between objects forming patterns of spatiotemporal interactions. Sometimes the distances between objects become very small, but they still remain detached from one another and can generally change their orientation and position independently of other objects. We are simply interested in possible relations between objects that are not connected.

It is less a question of motivating the necessity of different disconnection relations, than of restricting the relations to those in general positions. In $\mathcal{BA}$ the endpoints of all intervals are in general positions. The examples on the right of Fig. 1 show relations in singular positions. These correspond to special cases

B. Buchberger and J.A. Campbell (Eds.): AISC 2004, LNAI 3249, pp. 276–280, 2004.
© Springer-Verlag Berlin Heidelberg 2004

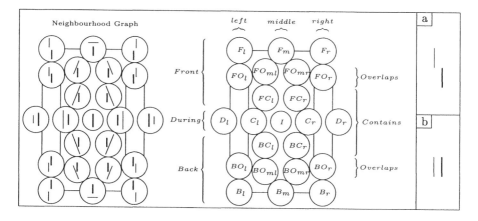

Fig. 1. Left: Interval relations embedded in two dimensions; the vertical reference interval is displayed bold. Middle: A mnemonic description; Right: Singular relations

in which intervals are precisely aligned with each other. But in $\mathcal{BA}$ there are no such singular relations explicitly defined. For instance, there is no relation between F_l and FO_l which would correspond to an interval in which one endpoint is located exactly level with an endpoint of the other interval (see Fig. 1.(a)). The question arises as to how we deal with such singular arrangements, in which one endpoint lies precisely at a location which marks the transition between qualitative concepts, as between F_l and FO_l. This is important, since no possible arrangement of intervals should remain undefined. This is the issue we are interested in.

2   Singular Relations

First of all we show why singular relations exist at all. A qualitative representation is the result of an abstraction process, which can be regarded as the partitioning of a continuous space into a number of equivalence classes. As a side-effect of this, singularities emerge as transitions between neighbouring classes. For example, the continuous space of interval arrangements in $\mathbb{R}^2$ can be conceived as consisting of all metrically distinct interval arrangements. A special abstraction of this continuous space distinguishes the relations of $\mathcal{BA}$, in which each equivalence class is a binary relation between two intervals. In the neighbourhood graph in Fig. 1, the transition between two neighbouring classes marks a singularity, for instance, between F_l and FO_l. We refer to the interval relations which fall into these transitions as singular relations. Fig. 1.(a) serves as an example. An arbitrary small change in the position of one interval which forms a singular relation with another interval transforms it in most cases into a general relation. Such a small change applied to a general relation would not normally change this relation.

We would like to argue that singular relations should not have the status of basic relations in a qualitative representation. By contrast to $\mathcal{BA}$, which comprises only 23 relations, there exist 226 relations when we additionally consider singular relations [3] – a significant difference since all these relations need to be distinguished when analysing and interpreting situations. More importantly, singular relations are somewhat misplaced in the context of qualitative reasoning. We are not at all interested in whether objects are precisely aligned. We focus on coarse relations between objects, which are simple to obtain and which allow efficient commonsense reasoning. For instance, we want to know whether one object is to the left of another one, whether it is moving in the same direction, and the like. What distinguishes qualitative relations from metrical relations is that they can be recognised easily by perception. However, this does not apply to singular relations, which require precise measurements. We conclude that singular relations are not compatible with commonsense reasoning, although there are exceptions in those fields where singular relations are as easy to obtain as general relations. For example, in the case of events we often know whether one event follows another one directly: after the performance a reception is held in the foyer; there is no time to go shopping between the performance and the reception – these events *meet* in time.

2.1   Representing Singular Relations

Having said that singular relations are incompatible with the idea of commonsense reasoning – more so in two dimensions than in one – we must show how to deal appropriately with singular relations. We cannot simply exclude them, since we need to represent every conceivable arrangement of intervals. One way of dealing with them consists in assigning singular relations to similar general relations. The singular relation in the first example could be assigned to F_l, since there is only one point that is not actually in relation F_l; this may be an appropriate solution in applications in which coarse reasoning is performed. But when such a precise distinction matters we are outside the scope of qualitative reasoning.

The second example, Fig. 1.(b), is more difficult to handle. If we regard this arrangement as D_l then we are heading for a problem. What about the converse relation? If it is regarded as D_r, then it holds for both intervals that each is contained in the other one – a quite awkward situation. For this reason we have to proceed as we do whenever we encounter indeterminate information in any qualitative representation: by sets of possible relations. In Fig. 1.(b) we would represent the singular relation by $\{D_l, C_l, FO_l, BO_l\}$ and its converse by $\{D_r, C_r, FO_r, BO_r\}$. In this way, we can deal with parallel intervals which are equal in length. The representation does not seem to be very precise, but precision is exactly what we want to avoid in a qualitative representation. When can we be sure whether parallel lines really are equal in length? Only when we have precise measuring tools. Isn't there always a little uncertainty left when working without such tools? At most we know that two lines in a given arrangement are *likely* to be equal in length, but at the same time we also know that they may be something else – something similar. Similar relations form a neighbourhood in the $\mathcal{BA}$-graph, and such neighbourhoods circumscribe the singular relations.

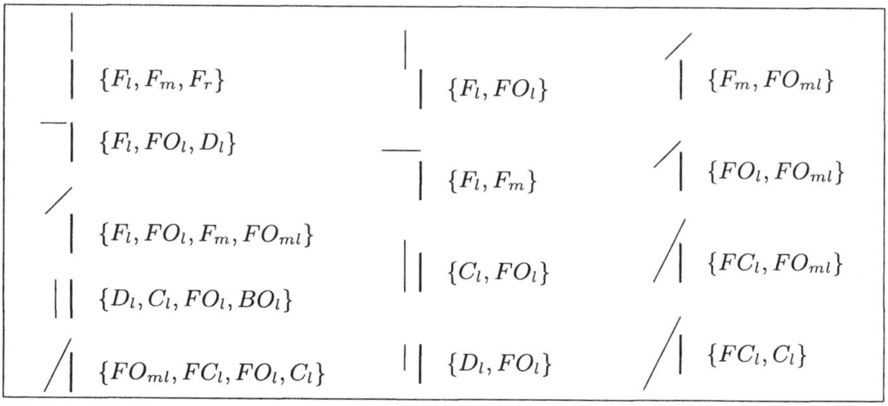

$\{F_l, F_m, F_r\}$

$\{F_l, FO_l, D_l\}$

$\{F_l, FO_l, F_m, FO_{ml}\}$

$\{D_l, C_l, FO_l, BO_l\}$

$\{FO_{ml}, FC_l, FO_l, C_l\}$

$\{F_l, FO_l\}$

$\{F_l, F_m\}$

$\{C_l, FO_l\}$

$\{D_l, FO_l\}$

$\{F_m, FO_{ml}\}$

$\{FO_l, FO_{ml}\}$

$\{FC_l, FO_{ml}\}$

$\{FC_l, C_l\}$

Fig. 2. Singular relations defined by sets of general relations

Accordingly $\{D_l, C_l, FO_l, BO_l\}$ would seem to be quite an appropriate description of what we really know about two parallel lines which are *probably* equal in length.

Fig. 2 shows how singular relations are represented by sets of general relations. Only a quarter of all relations are depicted, since the other relations are symmetrical to those in Fig. 2. As with the disconnection relations of $\mathcal{BA}$, only disconnected singularities are considered. *Apparently connected singularities* are treated as *apparently connected general relations*, i.e. they are conceived as disconnected relations in which distances become arbitrarily short. Our knowledge gets more uncertain near singular relations – this uncertainty is represented by sets comprising a number of possible relations rather than only one relation. In particular, if two endpoints are in singular positions then these sets consist of three or four general relations, depending on whether the endpoints lie on the same singularity, e.g. $\{F_l, F_m, F_r\}$ in Fig. 2, or on different singularities, e.g. $\{D_l, C_l, FO_l, BO_l\}$. By contrast, if there is only one endpoint in singular position the sets consist of only two general relations. We observe that all singularities are uniquely identified by this technique.

2.2   Reasoning with Singular Relations

How does this representation of singular relations affects reasoning processes? Let us consider the example in Fig. 3. We assume that we know the relations between x and y as well as those between y and z. For the position of y with respect to x we write x_y, and accordingly we write y_z for the position of z with respect to y. Our goal is to infer the relationship between z and x, i.e. x_z. We do this by the composition operation which was defined in [3]: for each pair of general relations the transitivity relation is given. The left hand side of Fig. 3 shows x_y in singular relation; the composition result is indeterminate. In comparison, the right hand side of Fig. 3 shows x_y in general relation; here the

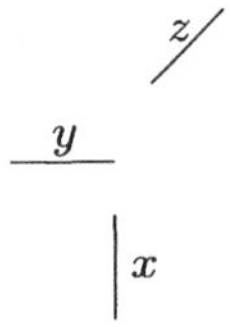

$$
\begin{array}{l}
x_y = \{F_l, F_m\} \\
y_z = F_l \\
x_z = x_y \circ y_z \\
\quad = \{F_l, F_m\} \circ F_l \\
\quad = F_l \circ F_l \cup F_m \circ F_l \\
\quad = \{F_l, F_m, F_r\} \cup \{F_r\} \\
\quad = \{F_l, F_m, F_r\}
\end{array}
\qquad
\begin{array}{l}
x_y = F_m \\
y_z = F_l \\
x_z = x_y \circ y_z \\
\quad = F_m \circ F_l \\
\quad = F_r
\end{array}
$$

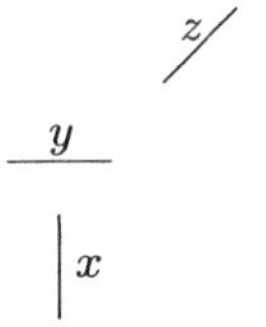

Fig. 3. Transitivity with a singular relation (left), and without any singularity (right)

composition result is less indeterminate. Note that we assume that x_z cannot be perceived directly, as is actually the case in this figure.

3   Discussion

Hitherto qualitative representations have treated singular relations as being on a par with general relations. This is useful in some areas, for example, in order to distinguish whether an event happens *before* another event, or whether it immediately follows (*meets*) another one [1]. We have argued that singular relations are not as important as general relations in some applications, and that they form a different sort of relation since they do not accord with common-sense reasoning. Characterising singularities on the basis of neighbourhoods, we have treated them as relations of second order rather than basic relations. As a consequence, the endpoints of basic relations always lie in general positions. Indeed $\mathcal{BA}$ forms a set of relations which covers all possible situations when circumscribing singular relations by neighbourhoods of general relations – $\mathcal{BA}$ leaves nothing undefined. This also holds for other qualitative representations.

To summarise, we have identified singularities as artefacts in qualitative representations. They are problematic in some areas, in that they require precise measurements whereas precision is normally avoided in qualitative reasoning. We have outlined how to deal with singularities by means of sets of possible relations, i.e. by defining singularities as sets of general relations.

References

1. J. F. Allen. Maintaining knowledge about temporal intervals. *Communications of the ACM*, 26(11):832–843, 1983.
2. A. G. Cohn and S. M. Hazarika. Qualitative spatial representation and reasoning: An overview. *Fundamenta Informaticae*, 43:2–32, 2001.
3. B. Gottfried. Reasoning about intervals in two dimensions. In *IEEE Int. Conference on Systems, Man and Cybernetics*, The Netherlands, 2004.

From a Computer Algebra Library to a System with an Equational Prover

Serge Mechveliani

152020, Program Systems Institute, Pereslavl-Zalessky, Russia
mechvel@botik.ru

Abstract. We consider joining to our computer algebra library a certain prover based on the TRW machine with an order-sorted algebraic specification language for input. A resource distribution approach helps to automate the proof tactics. Inductive reasoning is organized through adding of equalities. The unfailing Knuth-Bendix completion combined with special completion for the Boolean terms enables proofs in predicate logic. The question is how to develop further a language for equational reasoning/computing, the prover library and special provers to make possible the solution of more substantial problems than the ones mentioned in the examples.

Keywords: computer algebra, equational prover, term rewriting.

For several years we have been developing a computer algebra (CA) library called DoCon (www.botik.ru/~mechvel/papers.html). It is written in the Haskell language and implements a good-sized piece of commutative algebra. Now, in order to extend the ability to operate with explicit knowledge about domains, we aim at the following two goals. **(1)** To develop and implement an adequate "object" language (**OL**) based on order-sorted equational specifications (OSTRW), a prover related to OL, and to incorporate the existing CA library and prover into one system. **(2)** To enrich the existing tactics, the prover library, etc., in order to make the system more effective in solving problems.

The project also has relevance to the area of partial evaluation, since such a prover also allows automated reasoning about functional programs.

Our program is called Dumatel (a joke Russian word from the novel "Skazka o troike" by brothers Strugatsky, it stands for "thinker").

As an introduction, some ideas related to the project can be found in [5], [3], [4], [6], [1], and the manual for the Maude system. Among the projects with a similar direction, we can mention Theorema (www.theorema.org). There is also the Maude system (maude.cs.uiuc.edu) remarkable for its treating of OSTRW, *reflexion*, **AC** (associative and commutative) operators. In this four-page paper we cannot describe the design principles in much detail. More information will be available in

Reference: the initial open source Dumatel system version, together with some expanded documentation, is expected to appear during 2004. The relevant URL is www.botik.ru/~mechvel (/papers.html).

B. Buchberger and J.A. Campbell (Eds.): AISC 2004, LNAI 3249, pp. 281–284, 2004.

On Languages: Haskell is chosen as the *implementation language* (**IL**); the prover is written in IL, and it reasons about specifications expressed in OL. The prover strategies are formulated in the *strategy language* (**SL**), for which we have again found Haskell to be adequate. When OL develops a sufficient richness, it has the potential to become an SL.

Why do not we choose a prover tool from a great number of existing ones? After one year of investigation we failed to find a tool satisfying our requirements (and personal taste): (1) being open source, (2) having sufficient OSTRW to be an OL, (3) providing a good high-level functional language for implementation, and suitable universality for a strategy language, (4) having a rich CA library and TRW library (Maude was candidate No. 1). And our taste also does matter: an encouraging six-year experience with Haskell, with its "laziness", clarity and a balance between its being high-level and the efficiency of the Glasgow Haskell tool, with implementing in it a large CA library (to link to the prover) – all impel us to continue in this direction. As a more distant perspective, we intend the future OL to become a really adequate language for expressing strategic-level knowledge in programs.

Illustrating the System with an Example.

The following simple problem is taken from Example 3.7 in [2]. Let G be a group with operations e, *, i, and P a non-empty subset in G such that if X and Y belong to P then X*(i Y) belongs to P. Prove that it follows from these axioms that **(1)** e ∈ P, **(2)** for all X (X ∈ P ==> i X ∈ P)

To solve this problem, denote x ∈ P as (P x) and specify a *theory* as a many-sorted specification in OL:

```
groupTask = theoryUnion (bool lpo) groupWithP
  where
  groupWithP =
    Theory {sorts     = [G],
            operators = [e  : -> G,       i_: G -> G,       -- inversion
                         _*_: G G -> G,   P_: G -> Bool],   -- for subset
            variables = [[X, Y, Z] : G],
            equations = [(X*Y)*Z = X*(Y*Z),                 -- Group
                         e * X   = X,       (i X)*X = e],   -- laws
            opPrecedence = preced,  greaterTerm = (lpo preced)
            btLaws       = [],      btOrdering  = ...        }
  preced = [P, i, *, e, false],    -- P > i > ...
```

This specification is expressed as an IL data. The TRW interpreter treats such specifications as direct term reduction programs. Other parts of the prover, like *completion*, treat it as a theory from which to derive the consequences. The term ordering (greaterTerm) has the operator precedence table as a parameter. theoryUnion is a function joining theories in a sensible way; here it joins the theory for Boolean algebra. This example is processed by the prover, with the result printed to a string, by the following IL program:

```
shows              (ukbbGoalRem res) . ("\n\n"++) .
showsUKBBHistory (ukbbHistory res) ""                    :: String
where
res    = proveByUKBBRefutation Infinity skolemArgs groupTask formula
formula = parse (" (exist [X] (P X)) &
                  (forall [X,Y] (P X & P Y ==> P (X*(i Y))))
                  ==> P e  &  (forall [X] (P X ==> P (i X)))  ")
```

Here Infinity means the infinite resource given to the goal, and skolemArgs specify the way to add the operators appearing after skolemization. Our predicate calculus formula is presented as a *term* built with the *operators* "exist", ..., "==>", "P", "*", "i". The proof starts by reducing the formula term. The function proveByUKBBRefutation forms the negation, skolemizes it (extending the theory to groupTask'), converts it to a list of what we call btDisjuncts, each disjunct represented as a certain Boolean term (**BT**) [2]. A Theory also keeps a list of BT. The formulas join the btLaws part of the theory in the form of such a list. BT-s form an associative and commutative algebra, with the idempotence law, with respect to the operations xor, &. For more efficiency, the prover applies to BT a special completion method (the function ukbb). Our example yields the four disjuncts in a BT form:

```
btDisjuncts = [(P XSk) xor 1, (P (X*(i Y)))&(P Y)&(P X) xor (P Y)&(P X),
               (P XSk0)&(P e) xor (P e),    (P i XSk0)&(P e)          ]
```

XSk, XSk0 are the constants returned by skolemization. btDisjuncts join the theory, making it groupTask3, and then there applies the completion: ukbb _ groupTask3 ([],[1])   – "complete the equations and BT of the given theory aiming at the BT goal [1]". Deriving a BT which reduces 1 to 0 would mean a successful proof by refutation. ukbb combines the usual *unfailing completion* [3], [6] with the superpositions of kind equation+BT, BT+BT [2], and with a special reduction on BT. It also accumulates a proof *history*. In our example, the result prints as

```
ukbbGoalRem =  ([],[])       -- [true] -> [false] -> []   done
History     =
[[2] (P XSk) xor 1         [4] (P XSk0)&(P e) xor (P e)
 [5] (P i XSk0)&(P e)      [3] (P Y)&(P X)&(P (X*(i Y))) xor (P Y)&(P X)
 [1] e*X  -> X
 [6] (P Y)&(P i Y)&(P e) xor (P Y)&(P e)  from [1],  [3]   ...
 [7] P ((i (Y0* i XSk))*Y0)               from [..],  [2]   ...
 [8] 1                                    from [1],  [7] ]
```

The labels of type [Integer] serve as the references on print-out.

The ukbb method implements mainly the idea of RN+ strategy given in [2]. But Dumatel applies a stronger method for the BT reduction. This is based on the monomial ordering by the *monomial scheme*: power product made of the corresponding top predicate symbols. Its relation to graded algebras and Gröbner bases is a new computer algebra subject in our design.

Sketching Some of the Remaining Features

The resource distribution approach means that most important functions in the prover are given a resource argument: a bound for the number of "steps"; they return the "current state" and the resource remainder. Resource exhaustion is one of the flags for such a function to stop. The resource is distributed between the goals according to the strategy heuristics. All this helps to automate the tactics of proof, breaking off any unsuccessful proof branch when it grows too long (recall that a generic "solving" task is algorithmically undecidable).

The inductive reasoning is organized through adding of equalities (see, for example, the materials on Inductive Prover (M.Clavel) at the Maude system site).

The first approach strategy (FAS) for the *formula proof by the given theory* (the function `prove`) consists of transforming the list of *goals*, trying in a certain sequence the inference **rules** (attempts) of Simplification (`ukbb`), Constants Lemma, Implication Elimination, and Induction by expression value (by **construction** of such value). The resource is distributed between the goals in the simplest way (far from being optimal).

Examples of What It Can and Cannot Do: the FAS strategy proves automatically, with small resource cost, such tasks as, for example, (1) `forall [N,M]` (N+M = M+N) for the unary natural number arithmetics specified by recursive definitions, (2) `reverse reverse xs = xs` for the list reverse defined via concatenation, and the latter defined via `CONS`. But (2) needs a certain simple lemma to be introduced as a *hint*. For although FAS proves the lemma and all the rest, it fails to guess to put this lemma as a subgoal. We intend to improve FAS to the extent that it will become capable of tasks like proving the equivalence of various sorting algorithms.

Early Future Plans: to link the CA library to the prover (they are implemented in the same language), move OL from many-sorted TRW to order-sorted, add high-order operators, functoriality, develop AC completion, and develop further strategies (e.g. with setting an ordering on the subgoal formulas, heuristics for choosing the induction parameter) and special prover libraries (e.g. for the polynomial algebra).

References

1. Goguen, J., Meseguer, J.: Order-Sorted Algebra I: Equational Deduction for Multiple Inheritance, Polymorphism, and Partial Operations. (1988?).
   http://citeseer.nj.nec.com/goguen92ordersorted.html
2. Hsiang, J.: Refutational theorem proving using term-rewriting systems. Artificial Intelligence, 1985, v. **25** (255–300).
3. Hsiang, J., Rusinowitch, M.: On word problems in equational theories. In Proc. of 14-th Conference on ALP, Karlsruhe, LNCS **267**, (54–71), 1987.
4. Huet, G., Oppen, D.: Equations and rewrite rules. A Survey. In "Formal languages: perspectives an open problems" (349–405). New York, Pergamon Press, 1980
5. Knuth, D., Bendix, P.: Simple word problems in universal algebras. In "Computational Problems in Abstract Algebra", (263–297), Pergamos Press, 1970.
6. Löchner, B., Hillenbrand, T.: A Phytography of Waldmeister. AI Communications, IOS Press, (**15**) (2,3) (2002) (127–133).

Author Index

Lecture Notes in Artificial Intelligence (LNAI)

Vol. 2972: R. Monroy, G. Arroyo-Figueroa, L.E. Sucar, H. Sossa (Eds.), MICAI 2004: Advances in Artificial Intelligence. XVII, 923 pages. 2004.

Vol. 2969: M. Nickles, M. Rovatsos, G. Weiss (Eds.), Agents and Computational Autonomy. X, 275 pages. 2004.

Vol. 2961: P. Eklund (Ed.), Concept Lattices. IX, 411 pages. 2004.

Vol. 2953: K. Konrad, Model Generation for Natural Language Interpretation and Analysis. XIII, 166 pages. 2004.

Vol. 2934: G. Lindemann, D. Moldt, M. Paolucci (Eds.), Regulated Agent-Based Social Systems. X, 301 pages. 2004.

Vol. 2930: F. Winkler (Ed.), Automated Deduction in Geometry. VII, 231 pages. 2004.

Vol. 2926: L. van Elst, V. Dignum, A. Abecker (Eds.), Agent-Mediated Knowledge Management. XI, 428 pages. 2004.

Vol. 2923: V. Lifschitz, I. Niemelä (Eds.), Logic Programming and Nonmonotonic Reasoning. IX, 365 pages. 2004.

Vol. 2915: A. Camurri, G. Volpe (Eds.), Gesture-Based Communication in Human-Computer Interaction. XIII, 558 pages. 2004.

Vol. 2913: T.M. Pinkston, V.K. Prasanna (Eds.), High Performance Computing - HiPC 2003. XX, 512 pages. 2003.

Vol. 2903: T.D. Gedeon, L.C.C. Fung (Eds.), AI 2003: Advances in Artificial Intelligence. XVI, 1075 pages. 2003.

Vol. 2902: F.M. Pires, S.P. Abreu (Eds.), Progress in Artificial Intelligence. XV, 504 pages. 2003.

Vol. 2892: F. Dau, The Logic System of Concept Graphs with Negation. XI, 213 pages. 2003.

Vol. 2891: J. Lee, M. Barley (Eds.), Intelligent Agents and Multi-Agent Systems. X, 215 pages. 2003.

Vol. 2882: D. Veit, Matchmaking in Electronic Markets. XV, 180 pages. 2003.

Vol. 2871: N. Zhong, Z.W. Raś, S. Tsumoto, E. Suzuki (Eds.), Foundations of Intelligent Systems. XV, 697 pages. 2003.

Vol. 2854: J. Hoffmann, Utilizing Problem Structure in Planing. XIII, 251 pages. 2003.

Vol. 2843: G. Grieser, Y. Tanaka, A. Yamamoto (Eds.), Discovery Science. XII, 504 pages. 2003.

Vol. 2842: R. Gavaldá, K.P. Jantke, E. Takimoto (Eds.), Algorithmic Learning Theory. XI, 313 pages. 2003.

Vol. 2838: N. Lavrač, D. Gamberger, L. Todorovski, H. Blockeel (Eds.), Knowledge Discovery in Databases: PKDD 2003. XVI, 508 pages. 2003.

Vol. 2837: N. Lavrač, D. Gamberger, L. Todorovski, H. Blockeel (Eds.), Machine Learning: ECML 2003. XVI, 504 pages. 2003.

Vol. 2835: T. Horváth, A. Yamamoto (Eds.), Inductive Logic Programming. X, 401 pages. 2003.

Vol. 2821: A. Günter, R. Kruse, B. Neumann (Eds.), KI 2003: Advances in Artificial Intelligence. XII, 662 pages. 2003.

Vol. 2807: V. Matoušek, P. Mautner (Eds.), Text, Speech and Dialogue. XIII, 426 pages. 2003.

Vol. 2801: W. Banzhaf, J. Ziegler, T. Christaller, P. Dittrich, J.T. Kim (Eds.), Advances in Artificial Life. XVI, 905 pages. 2003.

Vol. 2797: O.R. Zaïane, S.J. Simoff, C. Djeraba (Eds.), Mining Multimedia and Complex Data. XII, 281 pages. 2003.

Vol. 2792: T. Rist, R.S. Aylett, D. Ballin, J. Rickel (Eds.), Intelligent Virtual Agents. XV, 364 pages. 2003.

Vol. 2782: M. Klusch, A. Omicini, S. Ossowski, H. Laamanen (Eds.), Cooperative Information Agents VII. XI, 345 pages. 2003.

Vol. 2780: M. Dojat, E. Keravnou, P. Barahona (Eds.), Artificial Intelligence in Medicine. XIII, 388 pages. 2003.

Vol. 2777: B. Schölkopf, M.K. Warmuth (Eds.), Learning Theory and Kernel Machines. XIV, 746 pages. 2003.

Vol. 2752: G.A. Kaminka, P.U. Lima, R. Rojas (Eds.), RoboCup 2002: Robot Soccer World Cup VI. XVI, 498 pages. 2003.

Vol. 2741: F. Baader (Ed.), Automated Deduction – CADE-19. XII, 503 pages. 2003.

Vol. 2705: S. Renals, G. Grefenstette (Eds.), Text- and Speech-Triggered Information Access. VII, 197 pages. 2003.

Vol. 2703: O.R. Zaïane, J. Srivastava, M. Spiliopoulou, B. Masand (Eds.), WEBKDD 2002 - MiningWeb Data for Discovering Usage Patterns and Profiles. IX, 181 pages. 2003.

Vol. 2700: M.T. Pazienza (Ed.), Extraction in the Web Era. XIII, 163 pages. 2003.

Vol. 2699: M.G. Hinchey, J.L. Rash, W.F. Truszkowski, C.A. Rouff, D.F. Gordon-Spears (Eds.), Formal Approaches to Agent-Based Systems. IX, 297 pages. 2002.

Vol. 2691: V. Mařík, J.P. Müller, M. Pechoucek (Eds.), Multi-Agent Systems and Applications III. XIV, 660 pages. 2003.

Vol. 2684: M.V. Butz, O. Sigaud, P. Gérard (Eds.), Anticipatory Behavior in Adaptive Learning Systems. X, 303 pages. 2003.

Vol. 2682: R. Meo, P.L. Lanzi, M. Klemettinen (Eds.), Database Support for Data Mining Applications. XII, 325 pages. 2004.

Vol. 2671: Y. Xiang, B. Chaib-draa (Eds.), Advances in Artificial Intelligence. XIV, 642 pages. 2003.

Vol. 2663: E. Menasalvas, J. Segovia, P.S. Szczepaniak (Eds.), Advances in Web Intelligence. XII, 350 pages. 2003.

Vol. 2661: P.L. Lanzi, W. Stolzmann, S.W. Wilson (Eds.), Learning Classifier Systems. VII, 231 pages. 2003.

Vol. 2654: U. Schmid, Inductive Synthesis of Functional Programs. XXII, 398 pages. 2003.

Vol. 2650: M.-P. Huget (Ed.), Communications in Multiagent Systems. VIII, 323 pages. 2003.

Vol. 2645: M.A. Wimmer (Ed.), Knowledge Management in Electronic Government. XI, 320 pages. 2003.

Vol. 2639: G. Wang, Q. Liu, Y. Yao, A. Skowron (Eds.), Rough Sets, Fuzzy Sets, Data Mining, and Granular Computing. XVII, 741 pages. 2003.